T0351075

Digital Satellite Navigation and Geophysics

Bridge the gap between theoretical education and practical work experience with this hands-on guide to GNSS. A clear, practical presentation of GNSS theory is provided, with emphasis on GPS, GLONASS, and QZSS, together with the key applications in navigation and geophysics.

Whether you are a practicing engineer, a researcher, or a student, you'll gain a wealth of insights from the authors' 25 years of experience in the GNSS field. You'll also get hands-on user experience with a bundled real-time software receiver and signal simulator, enabling you to create your own GNSS lab for research or study.

Numerous practical examples and case studies are provided, which you can explore using the real signal data provided or generated by you using the signal simulator. Also covered are issues related to GNSS signal propagation and its use in geophysics, including ionosphere mapping, atmosphere monitoring, scintillation measurements, earthquake prediction, and more.

Ivan G. Petrovski works on GNSS applications development at iP-Solutions, Japan. He has been involved in the GNSS field for more than 25 years. Prior to working at iP-Solutions, he worked as Associate Professor with Moscow Aviation Institute (MAI), as Japan Science and Technology Agency (STA) Fellow with Japan National Aerospace Laboratory (NAL), directed the Institute of Advanced Satellite Positioning in Tokyo University of Marine Science and Technology (TUMST), and led GNSS-related R&D for DX Antenna and GNSS Technologies Inc. He received his Ph.D. in aerospace navigation from MAI in 1993.

Toshiaki Tsujii is the Head of Navigation Technology Section, Aviation Program Group, at the Japan Aerospace Exploration Agency (JAXA), where he has been investigating aspects of satellite navigation and positioning for more than 20 years. He was at the Satellite Navigation and Positioning (SNAP) Group, University of New South Wales, Australia, as a visiting research fellow from 2000 to 2002. He received his Dr. Eng. in applied mathematics and physics from Kyoto University in 1998.

This book provides an excellent introduction to satellite navigation and the technologies that make it possible. The authors set forth both fundamentals and practical aspects in a manner that allows even newcomers to GNSS to utilize it effectively. The topics covered in this book address the most important elements of current-day GNSS applications.
Sam Pullen, Stanford University

Digital Satellite Navigation and Geophysics

A Practical Guide with GNSS Signal Simulator and Receiver Laboratory

IVAN G. PETROVSKI
iP-Solutions, Japan

TOSHIAKI TSUJII
Japan Aerospace Exploration Agency (JAXA)

CAMBRIDGE
UNIVERSITY PRESS

University Printing House, Cambridge CB2 8BS, United Kingdom

One Liberty Plaza, 20th Floor, New York, NY 10006, USA

477 Williamstown Road, Port Melbourne, VIC 3207, Australia

314-321, 3rd Floor, Plot 3, Splendor Forum, Jasola District Centre, New Delhi - 110025, India

79 Anson Road, #06-04/06, Singapore 079906

Cambridge University Press is part of the University of Cambridge.

It furthers the University's mission by disseminating knowledge in the pursuit of education, learning and research at the highest international levels of excellence.

www.cambridge.org
Information on this title: www.cambridge.org/9780521760546

First published 2012

A catalogue record for this publication is available from the British Library

Library of Congress Cataloging in Publication data
Petrovski, Ivan G., 1962–
Digital satellite navigation and geophysics / Ivan G. Petrovski, Toshiaki Tsujii.
p. cm.
Includes bibliographical references and index.
ISBN 978-0-521-76054-6 (hardback)
1. GPS receivers. 2. Radio – Transmitters and transmission. 3. Global Positioning System. 4. Radio wave propagation. 5. Electronics in geophysics. I. Tsujii, Toshiaki. II. Title.
TK6560.P4854 2012
623.89′33–dc23

2011053320

ISBN 978-0-521-76054-6 Hardback

Additional resources for this publication at www. cambridge.org/9780521760546

Contents

Foreword

I built my first crystal radio kit when I was 9 years old. I became hooked on radio technology, even at that tender age, and later went on to build other radios as a teenager, including a shortwave receiver from a kit. I learned how radios worked by building them and tinkering with them. I learned much later on that the famous American physicist, Richard Feynman, also had an interest in radios when he was 11 or 12. He would buy broken radios at rummage sales and try to fix them. That's the way he learned how they worked. As he says in the first chapter of his book *Surely You're Joking Mr. Feynman!–Adventures of a Curious Character*, 'The sets were simple, the circuits were not complicated . . . It wasn't hard for me to fix a radio by understanding what was going on inside, noticing that something wasn't working right, and fixing it.' As we know, Feynman went on to unravel the nature of quantum mechanics amongst other accomplishments. And, all his life, he took great pleasure in finding things out.

My interest in radio and electronics together with a love of physics led me eventually into a teaching and research career in space geodesy and precision navigation. Over the years, I have been involved with a number of mostly radio-based space geodetic techniques. As a Ph.D. student, I worked with very long baseline interferometry (VLBI) and satellite Doppler (the US Navy Navigation Satellite System or Transit) and as a postdoctoral fellow at MIT, I worked with lunar laser ranging data and was introduced to one of the very first civil GPS receivers: the Macrometer™. Upon arriving at the University of New Brunswick's then Department of Surveying Engineering in 1981, it was back to satellite Doppler. But within a year or two, GPS captured my attention and it has been virtually my sole interest ever since.

While most of my writing about GPS has been for journals, the proceedings of technical meetings and *GPS World* magazine, I was fortunate to be a co-author of one of the first, if not *the* first, textbooks on GPS: *Guide to GPS Positioning*, published by Canadian GPS Associates in 1986. It was a technical best-seller, with over 12 000 copies sold worldwide in the English version alone.

There have been many advances in GPS technology since that seminal book was published. In fact, with the almost simultaneous development of the Russian GLONASS satellite navigation system and, more recently, the initial development of the European Galileo system, the Chinese BeiDou/Compass system, the Japanese Quasi-Zenith Satellite System (QZSS), and a number of satellite-based augmentation systems, we now need to use the term 'global navigation satellite systems' or GNSS. And GPS itself is going through a modernization exercise with the introduction of new signals, new satellite designs, and new ground infrastructure.

Much of the research associated with current developments in GNSS is published, in a timely fashion, in technical papers in journals and meeting proceedings. But these papers, typically written by experts for other experts, are often difficult for students to follow – especially those just starting out in the field. There is still the need for good textbooks about GPS and the other GNSS, and there are several currently in print. They do a good job of explaining the theory of how GNSS work but there are very few resources matching the theory to actual receivers so that a student, for example, might learn by doing as I did years ago when learning how shortwave radios worked. In part, this is due to the fact that modern GNSS receivers are extremely complex devices and most are not designed to be teaching tools.

With their new book, Ivan Petrovski and Toshiaki Tsujii have filled this education gap. *Digital Satellite Navigation and Geophysics: A Practical Guide with GNSS Signal Simulator and Receiver Laboratory* is not only a text that clearly describes the generation of GNSS signals and how they are processed by a receiver, but it is also accompanied by both a GNSS software signal simulator and a software GNSS receiver. Now it is possible to learn how GNSS work *by doing*. After reading about an aspect of signal design, transmission, propagation or reception, the student can simulate it and see how a receiver responds to it, immediately seeing 'cause and effect.'

The uniqueness of the book is revealed, in part, by the words in its title. Let's start with 'digital.' The authors carefully and clearly describe the generation, transmission, and reception of GNSS signals, which are inherently digital, although modulated onto radio-frequency (RF) analogue carriers. And, in a modern GNSS receiver, the front end includes an analogue to digital (A/D) converter that transforms the signal to baseband where it can be manipulated in a totally digital way. The authors thoroughly discuss the generation of GNSS signals in the satellites and in signal simulators, as well as the operation of the receiver's RF front end, including A/D conversion, and baseband processing. Not only are the legacy GPS signals fully described, but also those of modernized GPS, GLONASS, QZSS, and Galileo.

Then 'satellite navigation and geophysics.' GNSS users can be very broadly classified into two groups: navigators whose positioning needs can be met by using pseudorange measurements on one frequency to achieve metre-level accuracies and high accuracy users, including surveyors, engineers and scientists who make use of the more precise carrier-phase measurements (in addition to pseudoranges), typically on two frequencies. Positioning accuracy at the centimeter-level or better is often the goal. The authors not only describe the various aspects of GNSS for navigation, but also for the more demanding applications including some of those in geophysics. The use of GNSS for ionospheric scintillation monitoring, for measuring Earth rotation and assessing tectonic plate motion are all discussed.

And finally 'practical guide.' The authors present not only the theory of how GNSS signals are generated and processed under ideal conditions, but also how a receiver responds to various real-world effects such as ionospheric disturbances and multipath. And these and other factors can be studied with a hands-on approach using the software signal simulator and software receiver. The authors provide several case studies and most chapters end with a relevant student project. A useful feature of the book is the schematic or flowchart, repeated at intervals, showing how each chapter is related to the operation of a GNSS receiver or its applications.

Ivan Petrovski and Toshiaki Tsujii are experts in GNSS. Ivan has been in the GNSS field for almost 30 years. He worked as an Associate Professor with the Moscow Aviation Institute (State University of Aerospace Technologies) before being invited in 1997 by the Japan Science and Technology Agency to join the National Aerospace Laboratory as a research fellow. Subsequently, he was Head of R&D at GNSS Technologies Inc. and Director of the Institute of Advanced Satellite Positioning at Tokyo University of Marine Science and Technology. Since the fall of 2007, he has been developing a real-time software GNSS receiver, a signal simulator, and instant positioning technology at iP-Solutions, Tokyo, Japan.

Likewise, Toshiaki has had a long association with GNSS. As Head of the Navigation Technology Section of the Operation and Safety Technology Team in the Japan Aerospace Exploration Agency, he has been investigating aspects of satellite navigation and positioning for more than 20 years. In addition to GNSS aircraft navigation, his group is working on airship- and helicopter-based pseudolite applications.

I had the pleasure of working with both Ivan and Toshiaki when they authored an article for my *GPS World* 'Innovation' column on a significant advance in GNSS technology: instant GPS positioning. That article introduced the iPRx snapshot software receiver for instant positioning to the GNSS community. And, it is the academic version of a new generation of that receiver, along with the companion ReGen signal generator, that is bundled with this book.

Like Richard Feynman, I have always taken pleasure in finding things out – whether it be why the sky is blue or how the atmosphere affects GNSS signals. And, with new signal structures being introduced along with new ways of acquiring and tracking those signals, I'm still learning how a GNSS receiver works. This book will help.

Professor Richard B. Langley
Department of Geodesy and Geomatics Engineering
University of New Brunswick, Fredericton, New Brunswick, Canada

Preface

This book is about ***global navigation satellite systems*** (GNSS), their two main instruments, which are a receiver and a simulator, and their applications. The book is based on an operational off-the-shelf real-time software GNSS receiver and off-the-shelf GNSS signal simulator. The academic versions of these tools are bundled with this book and free for readers to use for study and research.

The GNSS is probably unique in that it combines such a diverse variety of humanity's technological achievements. In order to understand a system better one needs to look at it from a wider perspective. Therefore we always try to present a theory behind each aspect of GNSS in general, which not only allows a better understanding of GNSS, but may also make it useful for specialists engaged in other fields.

The book structure is schematically presented on Figure P.1. Chapter 1 describes general methods of using GNSS. Chapter 2 looks at GNSS satellites and deals with their orbital mechanics. Chapter 3 discusses GNSS signals and how they are generated in satellites, simulators, and pseudolites. Chapters 4,7,10 describe GNSS signal propagation. Chapter 4 looks at where the GNSS signals are in relation to other electro-magnetic signals and how their specifics affect their propagation. Chapter 7 deals with multipath and Chapter 10 is devoted to the very interesting subject of signal scintillation. Chapters 5 and 6 describe in detail a software GNSS receiver front end and a baseband processor. Chapter 8 treats the subject of creating and improving various GNSS observables. Chapters 9,11,12 discuss how these observables are used in navigation and geophysics applications.

Among all GNSS we discuss first of all the US GPS, which is the main tool in satellite navigation and geophysics. We also consider the Russian GLONASS, which after a few years of being a partially filled constellation has now achieved full operational status, providing not only adequate coverage, but also new signals on L2 and L3. Even during the years of partial availability GLONASS was nevertheless regularly used in geodetic applications. Now GLONASS is also entering mass markets of cellular phone applications, car navigation, and so on. We also describe general principles and main features of the European Galileo, those which make it different from other GNSS. The space devoted to Galileo in this book is rather limited because we feel that the system is not yet finalized: the signal structure, the design, and even some concepts, particularly those related to open and restricted access signals, may yet be changed. Nonetheless, we think that the book provides enough information on GNSS in general for a reader to start work with the Galileo system and its applications.

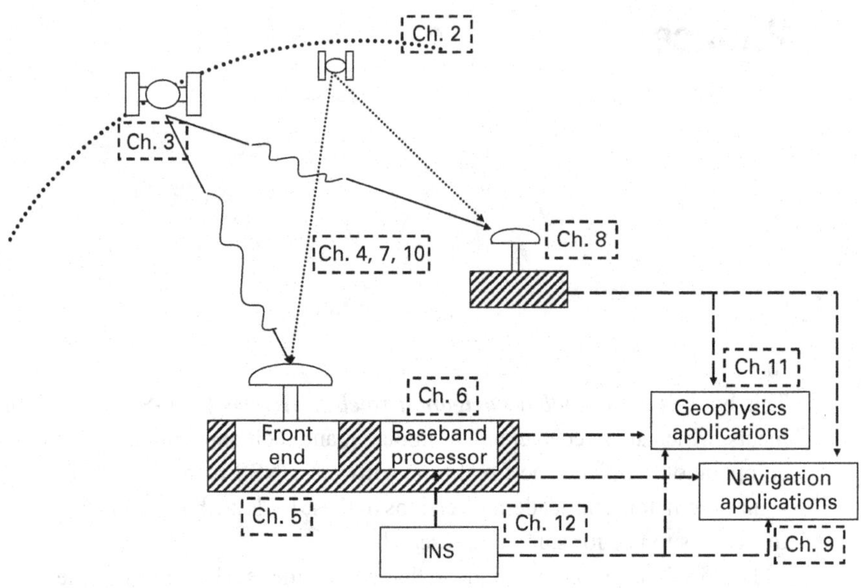

Figure P.1 Structure of the book.

We have tried to make this book ***unique in several respects***:

1. We have tried to implement a practical and systematic approach to GNSS theory and technology. It is based on our experience in creation and usage of software, and in GNSS receivers and signal simulators in particular in airborne applications.
2. The book is accompanied by ***free academic versions*** of a ***real-time software receiver*** and ***signal simulator***.
3. Matching ***hardware*** for the bundled software is described and readily available for interested users. For example, a ***receiver front end*** allows the bundled receiver real-time positioning with live satellites.
4. The book website contains ***prerecorded flight test data***, including ***GPS signal records*** and matching ***INS raw data*** outputs, which readers can process with the receiver while working with the book examples and projects.
5. In this book we describe a GNSS simulator design. An understanding of how a GNSS simulator works gives insights into the operation of real satellite systems. A simulator is a widely used tool nowadays for R&D, testing, and in manufacturing.
6. We have tried to make this book fun to read. There are many interesting facts hidden in the philosophical and physical backgrounds of GNSS, which are rarely or never discussed.
7. We have tried to structure this book to be as useful as possible for both students and experienced engineers alike. The book contains all the material necessary to understand GNSS in depth.
8. Throughout the book we try to give clear physical explanations of various important phenomena, such as why code delay and phase advance of GNSS signals are opposite in sign and equal in value, why ionospheric scintillation amplitude and phase cross-correlation has a negative value, and so on.

In conclusion, we would like to acknowledge some friends and colleagues who have helped us along the way.

First we would like to thank those who helped us with material for this book, especially:

- Mr. Graham Ockleston, Rakon Ltd., for his continuous and highly professional support and for providing illustrative materials,
- Mr. Keisuke Matsunaga, Dr. Susumu Saito, and many colleagues at the Electronic Navigation Research Institute (ENRI) for providing GPS test data under scintillation and for useful discussions,
- Dr. Takuya Tsugawa, National Institute of Information and Communications Technology (NICT), Professor Pornchai Supnithi, King Mongkut's Institute of Technology, Ladkrabang, Thailand, and Professor Chalermchon Satirapod, Chulalongkorn University, Thailand, for their help and collaboration in our ionospheric researches,
- Dr. Horng-Yue Chen, Professor Masataka Ando, Institute of Earth Sciences, Academia Sinica, Taiwan, and Professor Ryoya Ikuta, Shizuoka University, for providing materials for this book,
- Professor Sueo Sugimoto and Professor Yukihiro Kubo, Ritsumeikan University, for providing materials and for fruitful discussions,
- Dr. Jianxin Li, Mr. Andrew Walker, Spirent Communications, and Dr. Kaoru Miyashita, Ibaraki University, for providing illustrative material,
- Mr. Robin Spivey, Bangor University, for making nice home-made antennas out of wire. We use these antennas as an illustration in Chapter 6.

We would like to thank Ms. Natalia Petrovskaia, Cambridge University, for the exceptional illustrations created especially for this book.

The first of us (Ivan Petrovski) would like to thank:

- Dr. Harumasa Hojo, Tokyo University of Marine Science and Technology, for his strong support and continuous help over many years,
- Dr. Takuji Ebinuma, University of Tokyo, for his consultancy on various aspects during GNSS receiver design,
- Mr. Ken Satoh, AmTechs, and Professor Yasuda, Tokyo University of Marine Science and Technology, for their help on many occasions.

I would also like to thank Mr. H.Torimoto, GNSS Technologies Inc, for his support during the years I was working with him.

I especially would like to thank my wife Tanya for many great ideas and continuous support.

The second of us (Toshiaki Tsujii) would like to thank:

- Dr. Takeshi Fujiwara, Mr. Yoshimitsu Suganuma, Mr. Tetsunari Kubota, Mr. Hiroshi Tomita, Dr. Masatoshi Harigae, and many colleagues who have been working together at JAXA, for their consultation and assistance,

– Dr. Masaaki Murata, former Professor of the National Defense Academy of Japan, for his professional guidance and advice when I started research on GNSS,
– Professor Chris Rizos and many colleagues at the Satellite Navigation and Positioning (SNAP) Group, University of New South Wales, Australia, for their professional advice and discussions during and after my stay at the group.

I would also like to express hearty thanks to my wife and children: Kayoko, Honami, Yukino, Gousuke, and Yuto, for their encouragement and support.

Chapters 1–6 were written mostly by Ivan Petrovski, Chapters 8,9,12 mostly by Toshiaki Tsujii, and Chapters 7,10, and 11 by us both.

Although we have endeavored to avoid errors, some may have found their way into this book. The process of continuous refinement is a never-ending task.

Ivan Petrovski, Toshiaki Tsujii

Artist credits

Figures 1.8, 1.10, 1.11, 2.0, and 4.5 are based on artwork created especially for this edition by Natalia Petrovskaia, BA (Cantab.), MPhil (Cantab.). © Natalia Petrovskaia, 2012.

1 Methods of positioning with navigation satellites

"It is impossible to achieve theoretical understanding of the universe without instruction"
Claudius Ptolemy, Almagest.

In this book we consider ***Global Navigation Satellite Systems (GNSS)*** and their applications in navigation and geophysics. First established and up until today the main system is the American GPS, followed a little bit later by the Russian GLONASS. Both were created for navigational purposes. In this chapter we consider the principles and methods of navigation with GNSS satellites. We also give some main definitions here, which are used throughout the book.

1.1 Global and regional satellite navigation systems

Satellite navigation systems can be divided into two main categories: global and regional. There are also local systems, which are based on pseudolites. These systems, though not strictly satellite systems, are also considered in this book, because they may be used in conjunction with GNSS or they may use the same type of signal and the same user equipment as GNSS.

Global Navigation Satellite Systems (GNSS) provide global satellite coverage over the Earth and usually use satellites placed on medium Earth orbits (MEO) with an approximate altitude of 20 000 kilometers above the Earth's surface. The revolution period for MEO satellites is about 12 hours. There are two fully operational global navigation satellite systems today. They are: American Global Positioning System (GPS) [1] and Russian **GLO**bal **NA**vigation **S**atellites **S**ystem (GLONASS) [2] or **GLO**bal'naia **NA**vigacionnaia **S**putnikovaia **S**ystema in Russian. The revolution period is about 11 hours 58 minutes for GPS and about 11 hours 16 minutes for GLONASS. European GNSS Galileo and Chinese Beidou (Compass) systems are under development. The main function of these systems is to provide a ***ranging service***. The ranging service from a satellite gives a user an ability to measure distance to the satellite at each moment of time. The satellites may also transmit correction data for their ranging service. The correction data if transmitted are embedded in the satellite signal which is used for the ranging service. Satellite systems can also have secondary functions, which are not related to positioning. These secondary functions are not considered in this book.

1.1.1 GPS

The GPS development started in the mid-seventies as a U.S. Department of Defense (DoD) multi-service program. The GPS was completed in the early nineties and in 1994 had been officially accepted by the US Federal Aviation Administration (FAA) for use in aviation. GPS employs Code Division Multiple Access (CDMA), the same principle which is used today by cellular phone systems. All satellites broadcast their signals on the same frequency but with a different spreading code. The spreading code allows user equipment to acquire the signal, which is below the noise floor, and to distinguish between various satellites. The system is undergoing modernization to implement new signals for the benefit of civilian users and for aviation applications. Some of these new signals are already available.

Since the mid-nineties GPS has enjoyed tremendous success, becoming probably the second most widely applicable technology after the Internet. GPS has helped to create a huge number of jobs worldwide, has revolutionized navigation and geodesy, and has had a major impact on many other areas of science and technology including geophysics, geography, fleet management, travel, and so on. Much of the success of the system is due to the public accessibility of the signals and information about the system, allied to the absence of any royalties and of any hidden fees for a user. In year 2000, removal of accuracy limitations for civilian users also played a huge role in the overall GPS success. All this made GPS an essential and probably main system for any multi-GNSS equipment. This most likely will be the case for many years to come, even when more GNSSs become available.

1.1.2 GLONASS

GLONASS development was started soon after GPS by the former USSR. It was declared operational in 1996. GLONASS employs Frequency Division Multiple Access (FDMA) on top of CDMA. The GLONASS spreading code can be the same for all satellites, because the satellites are distinguished by slightly different frequencies. Though the frequencies for each satellite may be different, the signal structure is similar to GPS, because the spreading code is necessary in order to detect a very low power satellite signal.

FDMA had been implemented in GLONASS probably because the system mostly was envisioned as a military response to GPS by the Soviet Union. FDMA improves system anti-jamming capabilities, because it would require much more power to jam signals on a broader frequency band. However, FDMA also leads to more complicated, bulky, and heavy user equipment, which has an adverse impact on consumer applications. Current levels of radio technology, however, would allow to overcome this shortcoming almost completely. Nowadays, user equipment is becoming not only multi-system but also multi-frequency by definition.

On the other hand, the frequency band for GLONASS has been narrowed. The same frequencies are used for satellites located on different sides of the globe. There is a plan to employ CDMA for new GLONASS signals. Today GLONASS has already started to move to CDMA. A new GLONASS-K satellite started to transmit a CDMA signal in

2011. The new GLONASS-K satellite is based on a non-pressurized platform. Consequently GLONASS-K is smaller and considerably lighter than previous models, allowing use of a wider range of launch vehicles and thus making them less costly to put into orbit. The weight of a GLONASS-K satellite falls to 700 kg instead of the 1415 kg of previous satellites. After the complete constellation is deployed, it will require one Soyuz launch per year to maintain the constellation in full [3]. The estimated service life is significantly increased, and the satellites broadcast a CDMA signal on an additional third civilian L-band frequency. At the same time, the existing legacy FDMA signals are guaranteed to be kept "until the last receiver which uses them stops working" [2].

Under the same conditions, GPS currently provides users with more accurate position estimation than GLONASS, in particular because GLONASS broadcast ephemerides are less accurate. The tracking stations for GLONASS are located regionally rather than globally. The regional tracking station distribution affects the overall potential accuracy of orbit parameter estimation. However, as we show in the next chapter, GLONASS orbits are less affected by irregularities of the Earth's gravitational field, which partly compensates for the regional character of the tracking station network. The difference in accuracy is much less significant in differential mode, when corrections to orbits, clocks, and propagation in the atmosphere are estimated using local reference stations.

Another drawback of FDMA is a necessity to process more data, because a wider frequency band requires a higher sampling rate. This can be compensated for by a trade-off with accuracy.

With a full constellation and new signals, GLONASS today is becoming an attractive and essential component of multi-GNSS equipment worldwide.

1.1.3 Galileo

The Galileo development was started recently by the European Union. The Galileo system is also based on CDMA. Many receiver manufacturers are already supporting Galileo signals. With receiver manufacturers tending to move toward software receiver technology it becomes easier to incorporate new signals into existing receiver architecture. In this book we describe software receiver technology based on existing GPS and GLONASS. But the basic principles of GNSS signal generation and reception given here can be easily adapted to the Galileo signals as well as modernized GPS and GLONASS.

We use the ReGen software signal simulator, a free ultra-light version of which is bundled with the book. Figure 1.1 shows satellite distribution over the Earth for GPS and Galileo generated using ReGen software. User position is depicted by a pentagon.

1.1.4 Regional satellite systems

Regional satellite navigation systems provide regional coverage and are usually implemented as geostationary or ***Highly Eccentric Orbit (HEO)*** satellites. The geostationary satellites are located on ***Geo-synchronous Earth Orbit (GEO)*** with altitude of 35 856 kilometers and should keep their position constant relative to the Earth's surface.

The regional systems include American WAAS, Japanese MSAS, Japanese QZSS, European Global Navigation Overlay System (EGNOS), Indian GAGAN (GPS Aided

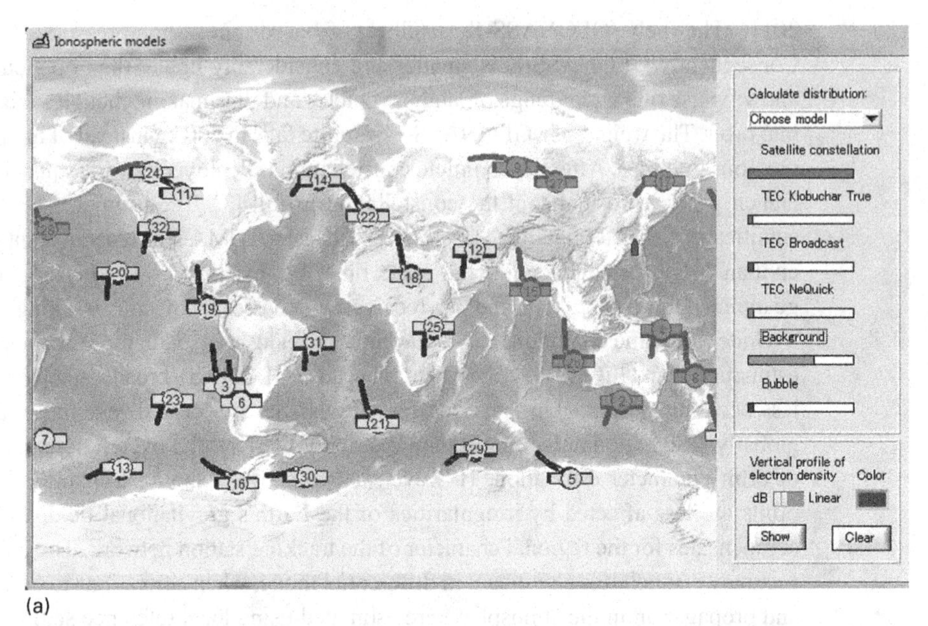

(a)

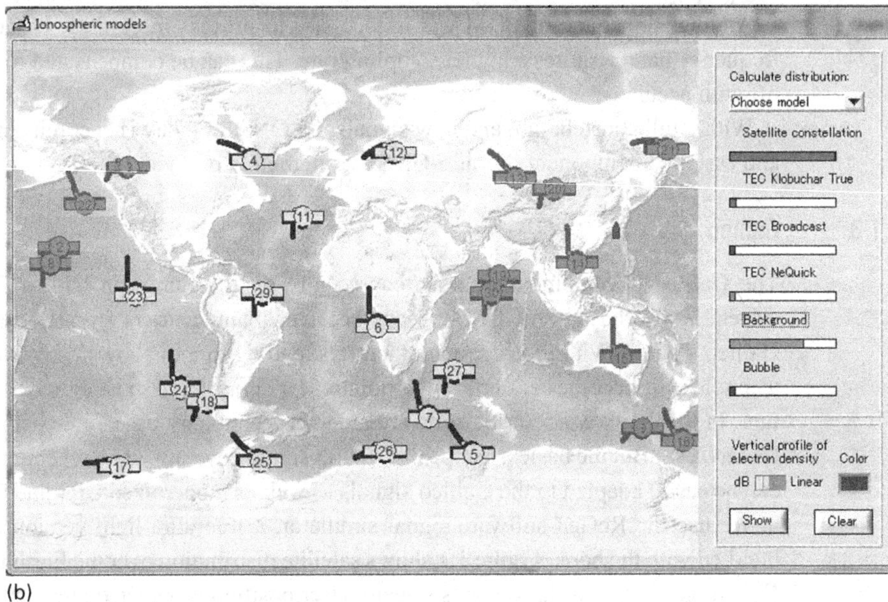

(b)

Figure 1.1 GPS (a) and Galileo (b) constellation ground tracks for a one hour interval. ReGen GUI screenshot.

Geo-Augmented Navigation), Chinese Beidou (Compass in Chinese), and Russian Luch (Beam in Russian). The Chinese Compass is designed to combine components from global and regional systems with some satellites on MEO and some on GEO.

The regional systems have three main functions:

1. To provide corrections for global system users in order to improve the ranging service from global systems.

2. To provide system integrity.
3. To provide an additional ranging service.

To provide ***integrity*** means to guarantee that there is a certain time interval within which a user will be notified if there is any fault in GNSS signals which are used for positioning. This interval is set to six seconds for WAAS.

The Russian regional satellite system Luch is going to be introduced with three satellites in GEO. Their designated location is at orbital slot at 16° west longitude (2011), 95° east longitude (2012), and 167° east longitude (2014). The coverage from GEO satellites excludes northern regions. Therefore it is planned to enhance coverage with HEO satellites, in particular using Molniya orbits. The system will also provide an Internet-based service.

1.1.5 GNSS structure

All these systems can be seen as composed of two parts, a ***space segment***, which is a satellite constellation, and a ***ground segment*** (see Figure 1.2). Sometimes a ***user segment*** is also included as a part of the system. A user segment comprises all service users, including those on the surface of the Earth and in space. The ground segment is an essential component and it consists of:

- a network of GNSS tracking stations,
- master clock,
- control center,
- upload facilities.

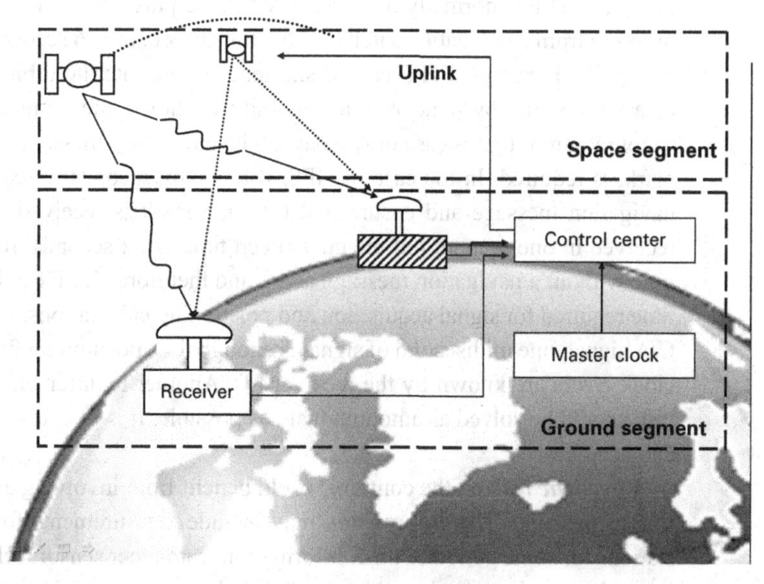

Figure 1.2 GNSS structure.

The ground segment main task as far as positioning is concerned is to define and predict satellite orbits, and upload their parameters to the satellites. The satellites then broadcast their orbit parameters within the satellite navigation message. The ground segment is also a cornerstone of the system control and maintenance. Its essential component is a network of stations, which measure distance to satellites. The distance to a satellite can be measured by using the satellite ranging signal, in which case the station consists of a satellite signal receiver. The distance can also be measured by other means, for example, by measuring a reflected laser signal from a satellite, which is carrying a special mirror for that purpose. This two-way measurement system was first introduced in GLONASS satellites. The satellite orbit measurements can also be augmented by inter-satellite range measurements.

First let us consider what tasks GNSS can solve and what the requirements for these particular tasks are.

1.2 Positioning tasks in navigation and geodesy

Satellite navigation systems solve two primary positioning tasks, which we can roughly describe as being related to areas of navigation and geodesy. These two tasks imply different requirements and therefore different underlying technology.

The ***navigation task*** normally requires sub-meter to meter level accuracy, and instant delivery of the positioning solution, which besides coordinates can also include velocity and attitude information. The solution must be supplied in real time and an initial position fix should be achieved as soon as possible. Specification parameters for navigation solutions usually include accuracy and ***time to first fix (TTFF)***.

TTFF is an important specification for navigation today, especially for hand-held devices. TTFF is normally limited to the time required to receive a complete navigation message from a navigation satellite. This time is equal to 36 seconds in the case of a GPS navigation message. TTFF can be shortened if the satellite orbit parameters and clock data are provided by some other means and reception of the complete navigation message is not required. In this case only a part of the navigation message, which provides a time mark, is required. In the case of GPS, it takes six seconds to receive one frame of the navigation message and ensure that the time mark is received. Of course, it may be received in one second, but the guaranteed time is six seconds. It is possible to resolve time without a navigation message at all, and therefore TTFF can be limited only by the time required for signal acquisition and positioning calculations. In the case of a GPS L1 C/A signal, one millisecond of signal is enough for a positioning fix, if satellite orbits and clock errors are known by the receiver [4]. Another requirement is that the navigation task should be solved as autonomously as possible. It is of course a trade off with TTFF and accuracy.

A ***geodetic task*** on the contrary would benefit from involving as much extra information as possible. This information may include measurements from multiple reference stations for some period of time. Information from other sensors related to the state of the ionosphere and troposphere can also be used in order to mitigate atmospheric effects on signal propagation. The geodetic task is normally solved in post-processing. The requirements for coordinate accuracy level are much higher and generally on the level of millimeters. The geodetic task may also include finding of various geodetic and physical parameters, such as Earth orientation parameters, atmospheric parameters, and so on.

Until recently there were two distinctive streams of applications related to these tasks. These streams were developing relatively independently. The navigation stream probably began together with human history. However, until the seventeenth century navigation was always based on integration with dead reckoning systems. Only in the seventeenth century did it become possible to define coordinates without dead reckoning. Prior to that the coordinates where derived only by dead reckoning and people were defining a position in relation to the starting point. Geodesy abandoned dead reckoning even later. At some point both geodesy and navigation began to use GNSS as a major tool, though the methods, algorithms, and user equipment in applications were quite different. Recently these two streams have come very close and it is no longer possible to learn about one without learning about the other. Geodetic features have become available for real-time applications, and navigation tasks have come closer to geodetic accuracy level [4].

One of the two main subjects of this book is to solve a positioning task with GNSS. We define ***positioning with navigation satellites in a wide sense*** as a task of finding object coordinates, linear and angular, and their derivatives. The coordinates also include time. We can consider an object either as a body or as a point mass. If we consider the object as a body, then coordinates include angular coordinates and their derivatives. If we consider the object as a point mass, then only linear coordinates and their derivatives are included in the positioning task. Three linear coordinates are required to describe an object position as a point mass in a coordinate frame in a three-dimensional world. Six coordinates are required to describe an object as a body in a three-dimensional coordinate frame at a specific instant of time, which is called an ***epoch***. In satellite navigation, whether or not an object is dynamic, it is always required to consider its position in a time-frame as well. Therefore, time variable and coordinate derivatives must always be considered.

It is important that the same object can often be described either as a point mass or a body depending on the specific task. In a navigation task, for example, a navigation satellite is considered as a point mass, whereas in geodesy, the satellite is considered as a body in order for a model to account, for example, for solar pressure.

We define ***positioning with satellite navigation in a narrow sense*** as a process of finding three coordinates of a receiver's antenna phase center. An antenna phase center is a mathematical abstraction and as such is always a point mass. Note that user coordinates are always in fact coordinates of the user receiver antenna phase center. It means that if we have a user with a receiver located in one place and the receiver antenna located at some distance from the receiver, for example at a few kilometers distance connected to the receiver by fiber optical cable, then the user will be measuring the coordinates of the antenna phase center. The coordinates will be calculated without any extra errors caused by the cable, besides those caused by power loss in the cable and a sequentially lower signal to noise ratio. Such systems can be used, for example, for landslide monitoring and in other hazardous environments. In this case only antennas are located in the hazardous environment, whereas all other monitoring equipment, including the receiver, can be located at the safe place [5].

It is also important to note that the antenna phase center depends on the direction from which a satellite signal comes. The antenna phase center may vary as a function of satellite elevation and azimuth. High-end geodetic antennas are calibrated and have elevation-dependent phase center corrections, which should be accounted for when high accuracy is required. Neglecting this effect may account for up to 10 cm error in height [6].

Coordinates describe an object's position in a certain coordinate system, and a coordinate system comprises a reference coordinate system and timeframe.

1.3 Reference coordinate systems

We can define coordinates only in relation to a specific reference coordinate system. The choice of coordinate system affects the accuracy of our coordinates and convenience of their usage. A GNSS user defines position in a coordinate system fixed to the Earth and rotating with it. These ***Earth centered, Earth fixed (ECEF)*** coordinates can be defined by a network of GNSS tracking stations, and therefore can also include the GNSS constellation as part of their definition. That is so far as the navigation task goes. For geodetic tasks, however, we have to consider that these GNSS tracking stations do not comprise a rigid structure on the millimeter level of accuracy. The coordinate system is constantly breathing. An ECEF is then related to fixed stars, which define an inertial coordinate system. However, the inertial system as far as it can be defined is also slowly moving. Main reference coordinate systems are given as described below.

1.3.1 Earth centered inertial (ECI) frame

An ***Earth centered inertial (ECI)*** coordinate frame (see Figure 1.3) is a Cartesian coordinate frame with an origin placed in the Earth's center of mass, and axes fixed relative to the stars. The Z axis coincides with Earth's spin axis and the X axis is defined by the direction from the Earth to the Sun on the first day of spring, when the Sun crosses the Earth's equatorial plane. This point of intersection between the Sun trajectory (ecliptic) and the Earth's equatorial plane is called the ***vernal equinox*** or First Point of Aries. The second name comes from the time of naming, thousands of years ago, when the vernal equinox was in the zodiacal constellation the Ram (Aries). The Aries zodiacal symbol ♈ is still used to mark the vernal equinox.

There are several ECI realizations. The ***International Celestial Reference Frame (ICRF)*** is realized by a catalogue of extragalactic stars based on very long baseline interference (VLBI) observations.

The vernal equinox is moving; therefore it is apparent that the ECI is also moving relative to the stars. This vernal equinox precession has a period of 26 000 years, which results in an ECI precession rate of 0.014° per year. This drift makes it necessary to

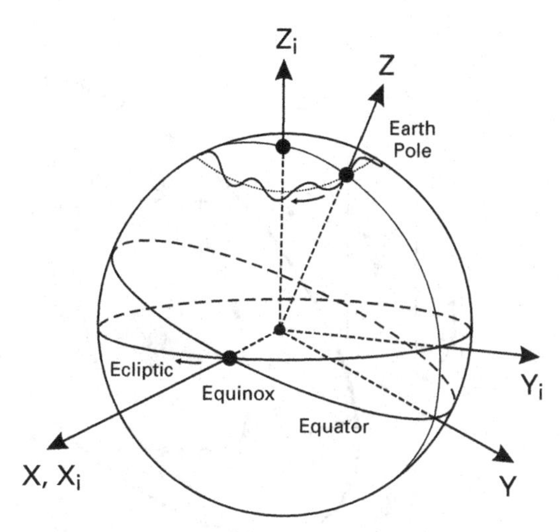

Figure 1.3 Earth centered inertial (ECI) coordinate frame.

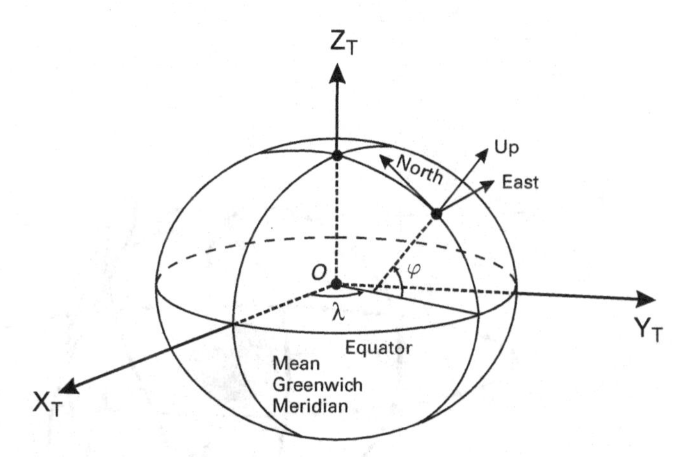

Figure 1.4 Earth centered, Earth fixed (ECEF) coordinate frame and topocentric horizon frame.

reference coordinate frames to a certain date. For example, the European CODE (Centre for Orbit Determination in Europe) – Analysis Center for the International GNSS Service (IGS) – uses the system J2000.0. The vernal equinox coordinates are then adjusted to the epoch of interest through precession and nutation transformations, which are given as sequences of rotations [7].

1.3.2 Earth centered, Earth fixed (ECEF)

The ECEF is a Cartesian coordinate frame with an origin placed in the Earth's center of mass. The Z axis coincides with Earth's spin axis, the X axis goes through the Greenwich meridian. The ECEF is rotating with Earth (see Figure 1.4).

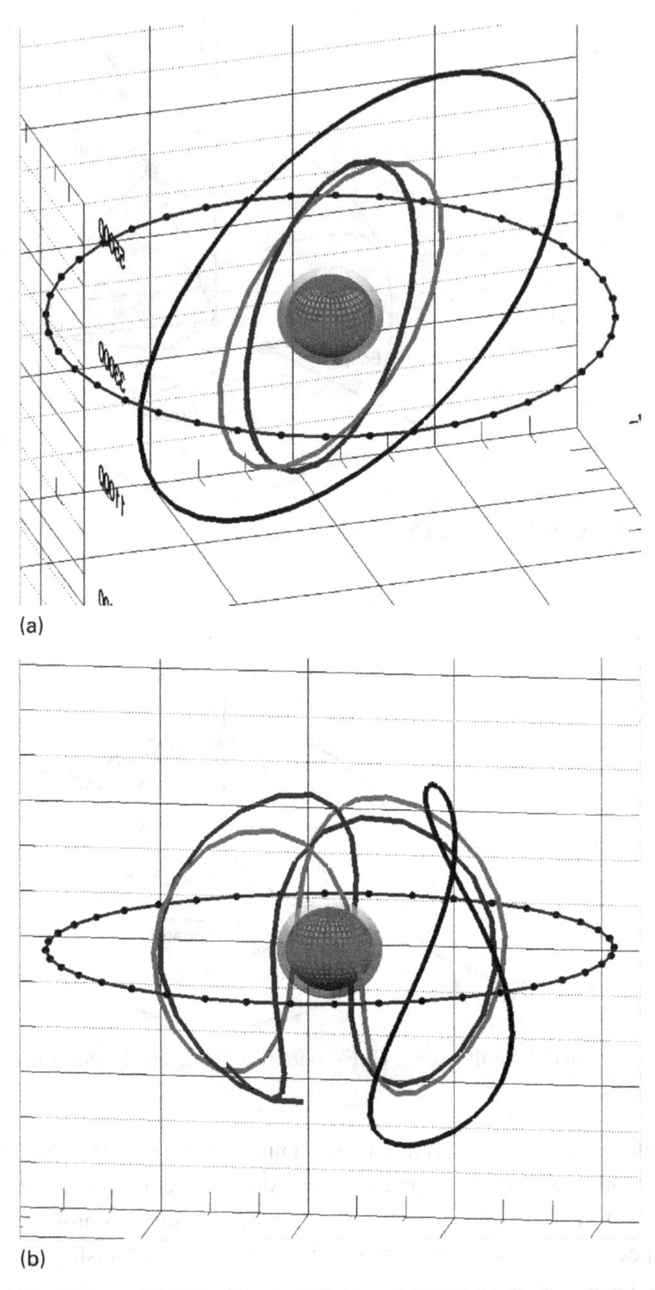

Figure 1.5 GPS, GLONASS, GEO, and QZSS orbits in ECI (a) and ECEF (b). ReGen GUI screenshot.

The satellite orbit parameters are given in one of the ECEF systems. That is convenient because they are defined in the ECEF of the tracking network. The satellite orbit parameters are considered in the inertial frame, because an orbit mathematical presentation in the inertial frame is much simpler, as we can see from Figure 1.5. In order to be useful for a user located on the Earth's surface, satellite coordinates in the inertial frame must be transformed to ECEF. For navigation purposes this transformation is straightforward and can be described only by the Earth's rotation, as follows:

$$\bar{X}_{ECEF} = \begin{bmatrix} x_{ECEF} \\ y_{ECEF} \\ z_{ECEF} \end{bmatrix} = \begin{bmatrix} \cos(\Omega_e \cdot t) & -\sin(\Omega_e \cdot t) & 0 \\ \sin(\Omega_e \cdot t) & \cos(\Omega_e \cdot t) & 0 \\ 0 & 0 & 1 \end{bmatrix} \times \bar{X}_{ECI}, \tag{1.1}$$

where Ω_e is the angular velocity of the Earth's rotation. Positioning tasks with navigation satellites are solved by finding a user position in relation to satellite positions known in an ECEF frame.

The ***international terrestrial reference frame (ITRF)*** is an ECEF frame, which corresponds to ICRF. Although there are various realizations of ITRF, we are in particular interested in the ***International GNSS Service (IGS)*** realization of ITRF, which is based on GNSS observations. The IGS ITRF consists of coordinates and velocities of a set of globally distributed tracking stations, which are in fact GNSS receivers, for specific time instances (epochs). The ITRF frame we can say is fixed to a set of IGS tracking stations on the Earth. As far as a navigation task is concerned, that is all we need to understand about the ITRF frame.

If we consider a geodetic task, which would require much higher accuracy, such as at the millimeter level, we have to consider more subtle effects and the transformation between ICRF and ITRF becomes much more complicated than the one given by (1.1). This transition is defined in [7]. The ITRF and ICRF are connected through ***the Earth orientation parameters***. The Earth orientation parameters are a set of five parameters, including time, which is also provided by IGS.

It is apparent that the motion of satellites is in fact easier to describe in ECI (compare orbits in ECI and ECEF in Figure 1.5). So the satellite coordinates are calculated in an inertial frame and then transferred into ECEF either through the angular velocity of the Earth's rotation (1.1) in the case of a navigation task or through the Earth's orientation parameters in the case of a geodetic task.

As far as GNSS is concerned, we define coordinates in one of the ECEF frames. The choice of ECEF frame depends on which ECEF frame the satellite orbits are given in. GPS satellites broadcast their orbits in WGS-84, GLONASS in PZ-90, and Galileo is going to use ITRF.

The ***World Geodetic System 1984 (WGS-84)*** is the US Department of Defense (DoD) reference system. The WGS-84 is realized through globally distributed GPS tracking stations. GPS ephemerides broadcast by GPS satellites are in WGS-84. WGS-84 has been aligned to the more accurate ITRF and is now expected to be close to ITRF, namely to within about ten centimeters.

Parameters of Earth *(Zemlya in Russian)* ***1990 (PZ-90)*** is the coordinate frame utilized by GLONASS [8]. The latest version of PZ-90 has also been aligned to ITRF.

For navigation purposes it is enough to define coordinates in ITRF, WGS-84, or PZ-90 using only the angular velocity of the Earth's rotation, without involving ICRF and the Earth orientation parameters. One can use ITRF orbits as they are calculated by IGS. The IGS orbits are most precise and available freely through the Internet. The usage of IGS orbit data is free even for commercial use and is not restricted in any way [9]. IGS provides one with precise orbits and satellite clock errors for previous days, which can be used for post-processing, and with predicted orbits and clocks, which can be used for real-time positioning. It is important to note that predicted orbits are still given with very high accuracy, because most of the orbit disturbing forces are well accounted for. However, the same cannot be said about satellite clock errors. Satellite clock errors may degrade comparatively rapidly and less predictably. Currently IGS provides predicted orbits only for a period of one day. It is of course possible to calculate predicted orbits for a longer period using data from the IGS receiver network.

1.3.3 Geodetic spherical coordinates

It is inconvenient to use Cartesian coordinates for most of the applications. To define a position on the globe we use spherical coordinates. Geodetic spherical coordinates are latitude, longitude, and height. When it comes to geodetic coordinates, one must also define a model to represent the Earth, because these coordinates only work on a surface of a three-dimensional object. The Earth model in order of precision can be given as a sphere, a three-dimensional ellipsoid of revolution, and a geoid (see Figure 1.6). ***The geoid*** is defined as a surface at any point normal to the vector of the Earth's gravitational field. It can be easily imagined if we invoke a model in which all the Earth's surface is under water. Then the water surface will give us the geoid, which can be modeled mathematically by a power series. Usually an acceptable geoid model requires a series with at least 18th power. One of the more sophisticated models from the Defense Mapping Agency has power up to 180th order and 32 755 coefficients [10]. The geoid in particular can be measured using special satellites, in which other forces such as solar radiation pressure are minimized (Figure 1.6). An example of such a satellite is the Russian Etalon satellite. Etalon is a spherical satellite with diameter about 1.3 meter and

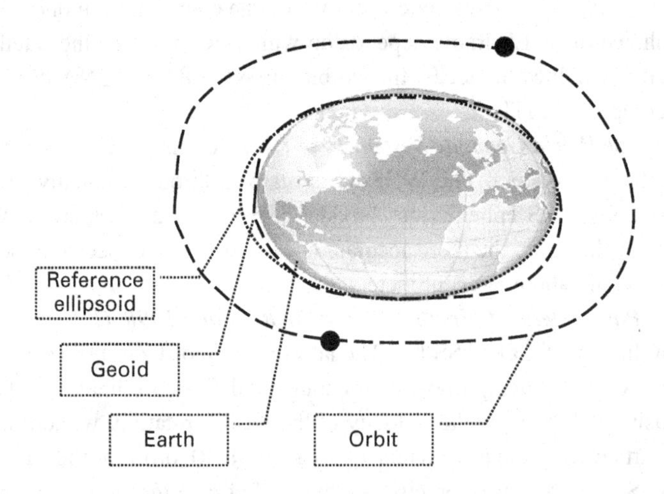

Figure 1.6 Earth model.

its surface is covered by reflectors to facilitate two-way range measurements by lasers and reduce effects of non-gravitational forces. Two Etalon satellites have been placed on GLONASS orbit in 1989 to measure gravitational forces, which affect GLONASS satellites [11]. These satellites are still in use by the international geodetic community.

The Earth models give us different definitions for a height and a vertical. Geodetic coordinates are based on the ellipsoid of revolution model. The reference ellipsoid of revolution may be chosen differently in different countries in order to represent better those parts of the Earth's surface where the country is located. As a result the geodetic coordinates vary in different countries.

1.3.4 Map projections of spherical coordinates

Furthermore, the spherical coordinates must be transferred to a plane surface of a chart. Up to the seventeenth century people related to navigation were rather unhappy with the inadequacies of plane charts, because geometrical projections at that time did not account for the Earth's curvature.

A Mercator projection

In 1569 Gerard Mercator had introduced a chart based on a projection that has solved the above problem. His projection was especially good for marine navigation because a ***loxodrome*** or ***rhumb line***, which is a line of constant compass bearing, was projected onto a chart as a straight line. It did not become generally accepted until after Edward Wright in 1599 had presented the mathematics behind it in his work *Certain Errors in Navigation* [12].

Mercator projection belongs to ***conformal*** or ***orthomorphic*** projections. This type of projection preserves the shape of any small area on the globe, when putting it on a flat map. This type of projection is very good for transforming spherical coordinates to a plane coordinate system of a relatively small area of interest. The angles are preserved during the projection, and the same scale factor is applied to all distance measurements, if the area is not too large.

In Mercator projection the scale factor is changed along the meridians as a function of latitude to keep the scale factor equal in all directions:

$$\kappa = \sec(\phi), \tag{1.2}$$

where ϕ is latitude.

If the mapped area is small, or we use the Mercator projection for illustrative purposes, such as a map in the ReGen simulator, then we can use a simple spherical Earth model. In the case of the spherical Earth the separation between two closely spaced parallels can be expressed as follows [13]:

$$dy = R \cdot d\phi, \tag{1.3}$$

where R is the Earth's radius.

A transformation of this distance through a Mercator projection to the map involves its multiplication by scale factor κ. Then the distance to any particular parallel with latitude ϕ_y from the Equator can be obtained by integration as follows:

$$y = \int_0^{\phi_y} R \cdot \sec(\phi) d\phi = R \cdot \ln\left[\tan(\frac{\pi}{4} + \frac{\phi}{2})\right]. \tag{1.4}$$

The other coordinate is given simply as

$$x = \lambda - \lambda_0, \tag{1.5}$$

where λ_0 is the longitude of the center of the map.

When Mercator projection is used for navigation and large-scale maps are involved, the more precise Earth model should be chosen. In this case, transformations from geodetic spherical coordinates to Mercator coordinates are given in special tables.

B Miller projection

Miller projection was developed by Osborn Miller in 1942, almost 400 years after Mercator projection, as an alternative to it. It was designed to eliminate the major drawback of Mercator projection, which is large area exaggeration toward the Earth's poles (see schematic presentation in Figure 1.7a). Mercator projection in fact is not able to show poles at all.

Miller projection uses different horizontal and vertical scale factors:

$$\begin{cases} \kappa_x = \sec(\phi), \\ \kappa_y = \sec(0.8 \cdot \phi). \end{cases} \tag{1.6}$$

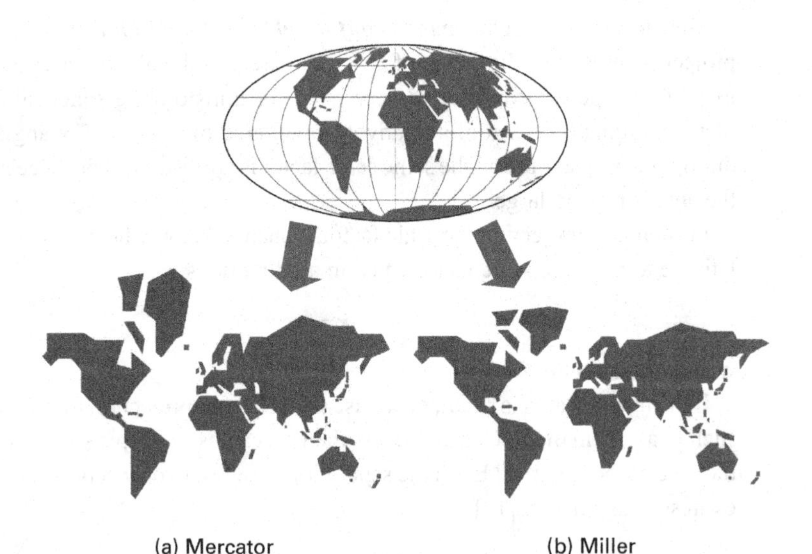

Figure 1.7 Schematic presentation of Mercator (a) and Miller (b) projections, which shows the difference between them.

The y coordinate is calculated as follows:

$$y = R \cdot \frac{\ln\left[\tan\left(\frac{\pi}{4} + 0.4 \cdot \phi\right)\right]}{0.8}. \tag{1.7}$$

Figure 1.7 allows us to see the major difference between Mercator and Miller projections. We can see that the high latitudes on the Mercator projection are very deformed. The map in the bundled simulator software also uses Miller projection (see Figure 1.1).

1.4 Timeframe and timekeeping

Precise timekeeping is a cornerstone of GNSS. A GNSS receiver is in fact measuring the time of signal propagation from a satellite to a user receiver. Satellite onboard clocks must be synchronized and be extremely stable in order for the positioning with satellites to work. For example, a nanosecond error in a satellite onboard clock would lead to 30 centimeters error in range measurements ($\Delta r = C \cdot \Delta t$), which leads, as we will see later in this chapter, to an even larger positioning error.

Precise timekeeping was initially introduced also in order to solve a navigation task, one of finding longitude in the seventeenth century. John Harrison produced the first accurate chronometer, winning the prize for developing a method and instruments for finding longitude. This resulted in the ability to define coordinates for the first time in history without a dead reckoning, by using time reckoning instead.

The GNSS time scale can be defined in various ways. It can be given by one master clock as in the case of GLONASS, or on the contrary, be defined as an average over an ensemble of clocks, as in the case of GPS. An ensemble may also include satellite clocks.

The time scales of individual satellites in the constellation are synchronized not only physically, but also analytically. The physical clocks onboard satellites are normally not corrected within some margin. Instead a satellite broadcasts corrections, which are necessary to apply in order to compensate for any shift between system time and satellite time. These corrections include drift of the satellite clock and its derivatives, predicted for some period of time. It is usually 12 or 24 hours in the case of GPS. These corrections are broadcast as part of a navigation message embedded in the satellite signal.

Today time estimation has become one of the most popular GPS applications. GPS plays a significant role in dissemination of precise time to worldwide users [14]. It has allowed easy and relatively cheap access to the precise time for everybody. Even a simple version of a GNSS receiver allows one in principle to detect time with accuracy on the level of tens of nanoseconds, basically with an error down to receiver positioning accuracy ($\Delta t = \Delta r / C$). The most difficult part is how to keep this accuracy on the time signal's way from the inside of receiver to the user, which is why there are special GPS receivers for time applications.

1.5 Accuracy, precision, and normal distribution

The navigation task is usually incorporated into a GNSS receiver, while the geodetic task is left for post-processing. There are also hybrid field devices that are bringing geodetic tasks to near real time by combining a computer with post-processing facilities with a receiver in a hand-held device. These instruments are very popular in geographical information system (GIS) applications. One of the main parameters in the specification of GNSS devices in general is accuracy. The specification of accuracy in geodesy and navigation is slightly different. We mostly concentrate on accuracy specification as defined for navigation tasks. The reason is that the navigation task is embedded in a GNSS receiver, and therefore GNSS receiver specifications are always given in terms of accuracy for a navigation task. One should distinguish accuracy from precision. Accuracy defines how close our position estimates are to the true position. Precision defines how close our consecutive position estimates are to each other. Specification parameters are all defined as statistical parameters. They are expected to be within certain values only with certain probability. When we measure parameters, we also have to account for their stochastic nature. This stochastic nature is a result of our insufficient knowledge of all models involved.

1.5.1 Significance and pitfalls of normal distribution

Suppose that we need to solve a positioning task in a narrow sense and define estimation of antenna phase center coordinates. A receiver outputs a number of position fixes, which are slightly different from each other. We choose a ***population mean***, or ***average***, or a ***mathematical expectation*** as an estimate of the sought antenna position $(\hat{x}, \hat{y}, \hat{z})$ as follows:

$$\hat{x} = \frac{\sum_{i=0}^{n} x_i}{n}. \tag{1.8}$$

There are similar expressions for $\hat{y}, \hat{z}$ coordinates. There is a paradox related to our estimate. There are many practical cases when a population mean gives a misleading picture. Let us look at populations in Figure 1.8 and an average height and wealth over the shown population. We can see here that the population mean is not what we call an average or typical in everyday life. In those cases the population median will give a more correct representation. ***The population median*** is a value of a sample from the middle of the population re-sorted by value.

The population median coincides with the population mean if the population has a ***normal distribution***. ***Central limit theorem*** explains why we can encounter a normal distribution very frequently. It states that if a number of almost independent values with almost similar distribution increase, then a distribution of their sum approaches a normal distribution. It takes from five (in the case of smooth marginal distributions) to 30 independent values in order for the distribution of a sum to become a normal. Famous French mathematician and physicist Henri Poincaré said that everybody believes that the

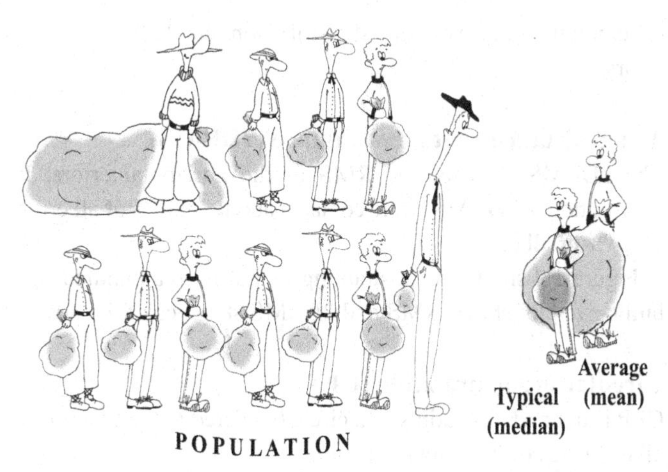

Figure 1.8 A mean value paradox.

normal distribution is universal: physicists believe that mathematicians have proved it theoretically, and mathematicians believe that physicists have proved it by experiments. In 1960 John Tukey published a paper on contaminated or mixed normal distribution, where he has shown that it takes very little to change a normal distribution in a way such that it still appears to be a bell-shaped normal distribution, but its statistics will be very different from the normal distribution statistics. The example in Figure 1.8 and the theory of mixed normal distribution are very closely connected to the way ***outliers*** in GNSS measurements should be handled. Assuming that our position estimates follow normal distribution, a probability of where our position estimate is on the coordinate axis relative to a true position can be defined as the area under the ***power spectrum density (PSD)*** curve.

1.5.2 Measures of accuracy

A Root mean square (RMS)

RMS is a square root of a mean of the squares of the errors:

$$RMS = \sqrt{\frac{\sum_{i=0}^{n} (r_i - r_{\text{true}})^2}{n}}, \tag{1.9}$$

where $r_i = \sqrt{x_i^2 + y_i^2}$.

If, instead of true position, we consider the population mean, than we find a value of a standard deviation σ_x (the similar expression is for σ_y) as follows:

$$\sigma_x = \sqrt{\frac{\sum_{i=0}^{n} (x_i - \hat{x})^2}{n}}. \tag{1.10}$$

We can estimate what percentage of errors will be inside the RMS circle by measuring an area under the PSD curve. We should note that if the distribution is not normal, we can still measure the percentage of errors in the same way, though the number may become different. In the case of normal distribution, the RMS circle contains about 68.27% of all errors.

B Twice distance root mean square (2DRMS)

The 2DRMS is twice the RMS of the horizontal errors. In the case of a normal distribution, a 2DRMS circle contains about 95.45% of all errors. Note that $\pm 3\sigma$ contains 99.73% of all errors.

Figure 1.9 shows the positioning output from a bundled iPRx software receiver. The bull's-eye plot shows RMS and 2DRMS statistics of the position estimates.

C Circular error probable (CEP)

CEP is defined as a radius of a circle (centered at the true position) that encloses 50% of all estimates of horizontal position.

D Spherical error probable (SEP)

SEP is defined as a radius of a sphere (centered at the true position) that encloses 50% of all position estimates.

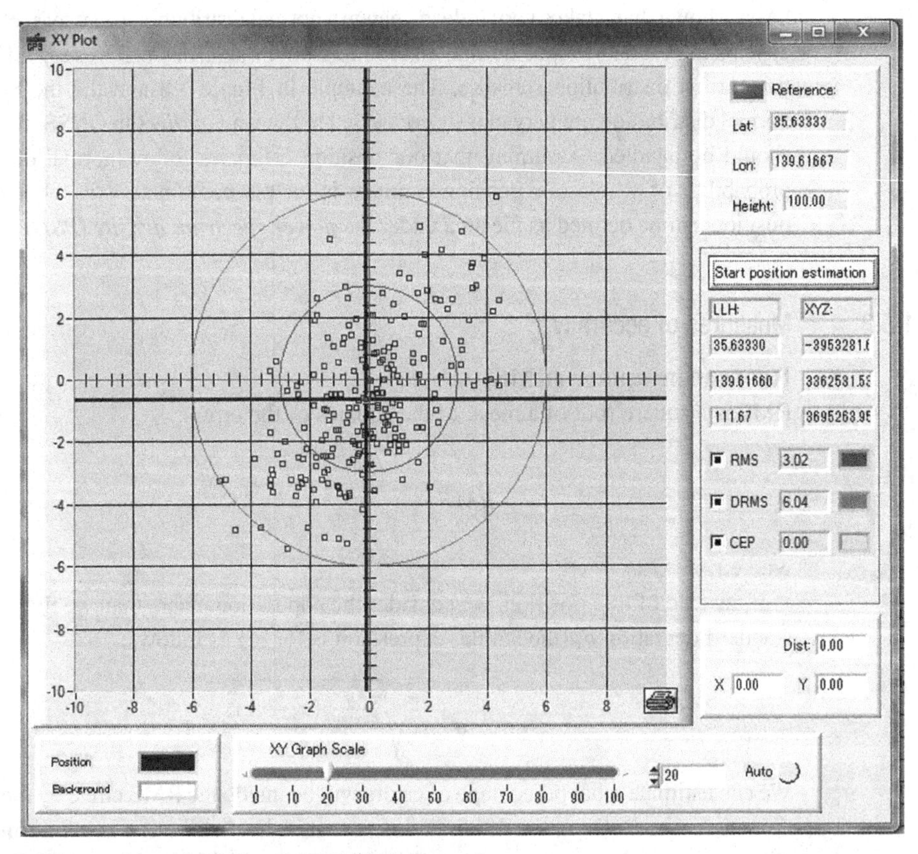

Figure 1.9 Bull's-eye plot with RMS and 2DRMS circles.

Theses probability levels can be directly converted to each other using linear error conversion factors.

1.6 Principle of positioning with satellites and GNSS equations

The principle of operation is based on range measurements from signal sources, of which coordinates are known, to an unknown user position. This method is known as ***trilateration***.

To introduce the principle, let us imagine a two-dimensional being living in a two-dimensional world, as depicted in Figure 1.10. This being is trying to get its location using two satellites in two-dimensional space. If the distances to those satellites are known and coordinates of those satellites are known as well, then the being can find its position on the intersection of two circles with radiuses equal to the distance between it and the satellites. There are two possible positions in the two-dimensional world. In a similar way a user in our three-dimensional world can find his or her location in the intersection of three spheres. The intersection of three spheres again gives two possible positions. The user can choose the correct one easily if the distances to the satellites are large, using common sense.

However, a user makes distance to the satellite measurements by measuring the time of signal propagation from the satellite (Figure 1.11). The synchronized-to-system time, highly stable satellite clocks can be assumed for the moment to have negligible error. However, a less expensive user clock has a much larger error and therefore all distances will be measured with the error (see Figure 1.11a). To find and remove this error, the user must measure the distance to one extra satellite, which allows him/her to solve for this extra unknown (Figure 1.11b). Three satellites will again give two solutions, which in this case also include two solutions for receiver clock errors (Figure 1.11c).

The equations for distances between the user and all satellites can be written as follows:

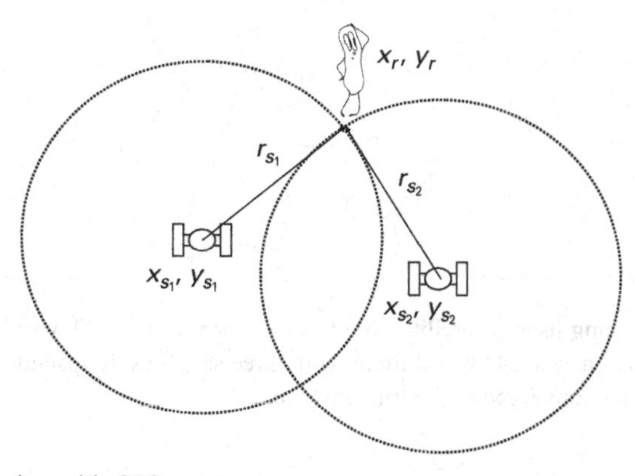

Figure 1.10 Principle of positioning with GPS.

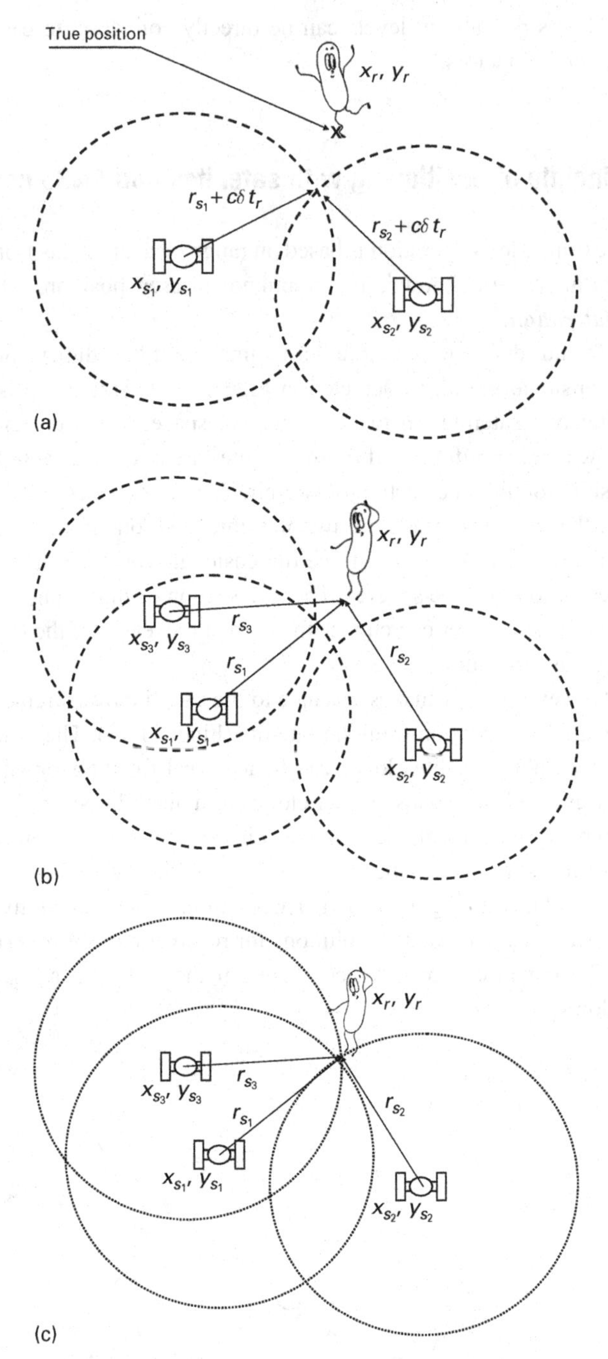

Figure 1.11 Determining user to satellite distance: (a) position in a 2-D world found with two satellites, (b) position in a 2-D world found with three satellites, (c) pseudo ranges corrected for a receiver clock and converged in the true position.

$$r_{S_i} = \sqrt{(x_r(t_r) - x_{S_i}(t_{S_i}))^2 + (y_r(t_r) - y_{S_i}(t_{S_i}))^2 + (z_r(t_r) - z_{S_i}(t_{S_i}))^2}, i = 1, \ldots n, \tag{1.11}$$

where r_{S_i} is the distance between the user and the ith satellite in three-dimensional space, n is the number of observable satellites, x_r, y_r, z_r are coordinates of the user at a time (epoch) of signal receiving t_r, and $x_{S_i}, y_{S_i}, z_{S_i}$ are coordinates of the ith satellite at the epoch of signal transmission t_{S_i}.

The measured distance is that which a satellite signal travels from the satellite position at the time (epoch) of signal transmission to the user position at the epoch of signal receiving. All epochs should be in the same GNSS timeframe. For most applications we can assume that satellite time is precisely synchronized to a system timeframe. However, epoch t_r is in the receiver timeframe, and epochs t_{S_i} are defined from the signal and therefore in the GNSS timeframe. Normally user clocks are not synchronized with satellite time precisely enough for any positioning task. Satellites are equipped with very stable and very expensive atomic clocks. Even those clocks are not stable enough, and a ground segment (see Section 1.1.5) constantly measures their drift and keeps them synchronized with the system time within a specified allowance. Then drift parameters are broadcast to the user in a satellite navigation message. However, these broadcast parameters are still not good enough for precision applications. In which case, the satellite clock errors are estimated along with the unknown receiver antenna coordinates and receiver clock error.

In order to put epoch t_r into a GNSS timeframe, a difference between time of reception in the receiver and satellite system time scale must be introduced:

$$\delta t_r = t_{r_r} - t_{rS}. \tag{1.12}$$

Now Equations (1.11) must be solved for four unknowns in order to define user coordinates x_r, y_r, z_r, δt_r.

A satellite position is a function of time. A control segment provides satellite orbits so its position can be precisely known at any time. To find the satellite position is a reverse task and we consider it later in the book. Note that Equations (1.11) are essentially non-linear because satellite position is a function of time, and time is one of the unknown coordinates.

A GNSS receiver measures range to a satellite r_{S_i} by comparing an incoming signal, which is synchronized to the satellite timeframe, with a locally generated replica, which is aligned with the receiver timeframe. The range to a satellite can be expressed as the time required for signal propagation from the satellite to the receiver multiplied by the speed of light:

$$r_{S_i} = \left(t_{rS} - t_{e_S^i}\right) \cdot c, \tag{1.13}$$

where $t_{e_S^i}$ is an epoch of signal emission from the ith satellite in the satellite timeframe, and t_{rS} is an epoch of signal reception in the satellite timeframe. The range Equation (1.13) can be rewritten using (1.12) as follows:

$$r_{S_i} = (t_{r_r} - t_{e_S^i} - \delta t_r) \cdot c. \tag{1.14}$$

The observables derived from the receiver are a set of r_{S_i} for all observable satellites. The unknowns $x_r(t_{r_r}), y_r(t_{r_r}), z_r(t_{r_r})$, and δt_r can be found from the above set of equations, providing that the number of equations is greater than or equal to the number of unknowns. The number of equations is also equal to the number of signal sources. Then in our three-space-dimensional world, we require four satellites to determine user coordinates and adjust this timeframe to the satellite timeframe.

The receiver measurements of the ranges r_{S_i} include user and satellite clock errors. For that reason these measurements are called pseudoranges, rather than ranges. The corrections to satellite clock errors are transmitted as a part of the navigation message and sequentially compensated by the user, therefore we do not consider them for now. The pseudoranges that are really measured by a receiver are described as follows:

$$\rho_{S_i} = (t_{r_r} - t_{e_S^i}) \cdot c. \tag{1.15}$$

The system (1.11) can now be rewritten as follows:

$$\rho_{S_i} = \sqrt{(x_r(t_r) - x_{S_i}(t_{S_i}))^2 + (y_r(t_r) - y_{S_i}(t_{S_i}))^2 + (z_r(t_r) - z_{S_i}(t_{S_i}))^2} + \delta t_r \cdot c, \quad i = 1, \dots n \tag{1.16}$$

where the first part of the right side of the equation is the distance to the ith satellite and the second is a receiver clock error.

The system (1.16) can be rewritten using matrix form for convenience as follows:

$$\bar{Z} = [A(\bar{X})], \tag{1.17}$$

where $\bar{X}$ is a ***state vector***

$$\bar{X} = \begin{bmatrix} x_r \\ y_r \\ z_r \\ \delta t_r \end{bmatrix}$$

and $\bar{Z}$ is an ***observation vector***

$$\bar{Z} = \begin{bmatrix} \rho_{S_1} \\ \rho_{S_2} \\ \cdots \\ \rho_{S_n} \end{bmatrix}.$$

We have considered only code observables here, i.e. the observables that are derived from comparison of the spreading codes of the incoming signal and the receiver generated replica. We consider other observables in Chapter 8.

In order to solve a positioning task we need to find a fourth-order vector $\overline{X}$ using a vector of pseudorange measurements $\overline{Z}$ of nth order. In order for the system of Equations (1.17) to have a solution, the condition $n \geq 4$ must be satisfied. The system of Equations (1.17) which we need to solve is nonlinear. Methods of solution are considered in this chapter. In order to use most of them, the system should be linearized. The system can be linearized due to the fact that satellites are located very far from a user.

1.7 Taylor theorem and linearization of GNSS equations

We can represent general nonlinear equations by polynomials using the Taylor theorem. The importance of the Taylor theorem is that it allows us to bind a linearization error to a specific required value. Brook Taylor (1685–1731) graduated from St.John's College, Cambridge University in 1709. He published his theorem in 1715, based on work by Newton and Kepler.

The theorem specifies that if a nonlinear function $f(x)$ has $n+1$ continuous derivatives on $[a,b]$ and x, $x_0 \in [a, b]$, then it can be represented as a sum of a polynomial and a remainder as follows:

$$f(x) = p_n(x) + R_n(x), \tag{1.18}$$

where the polynomial is expressed as follows:

$$p_n(x) = \sum_{k=0}^{n} \frac{(x - x_0)^k}{k!} f^{(k)}(x_0), \tag{1.19}$$

and the remainder is

$$R_n(x) = \frac{(x - x_0)^{n+1}}{(n+1)!} f^{(n+1)}(x_\xi), \tag{1.20}$$

where x_ξ is a point between x and x_0.

The remainder can be made arbitrarily small by choosing x_0 close enough to x. If we choose the point in such a way that the second-order remainder can be omitted, then our nonlinear system of equations becomes represented by a linear system. In order to linearize Equations (1.16), we find partial derivatives at the approximate receiver location x_{r0}, y_{r0}, z_{r0}, and $\delta t_{r0} = 0$:

$$\frac{\partial}{\partial x}\rho_{s_i} = \frac{x_{r0} - x_{s_i}}{\sqrt{(x_{r0} - x_{s_i})^2 + (y_{r0} - y_{s_i})^2 + (z_{r0} - z_{s_i})^2}}, i = 1, \ldots n. \tag{1.21}$$

The partial derivatives vector is in fact a unit direction vector from an approximate position x_{r0}, y_{r0}, z_{r0} to a satellite as follows:

$$\bar{e}_i = \left[\frac{\partial}{\partial x}\rho_{s_i}, \frac{\partial}{\partial y}\rho_{s_i}, \frac{\partial}{\partial z}\rho_{s_i}\right]. \tag{1.22}$$

Then the linearized equations can be written in a matrix form as follows:

$$\bar{Z} - \bar{Z}_0 = [H] \times \Delta\bar{X}, \tag{1.23}$$

where matrix $[H]$ called a ***measurement matrix***:

$$[H] = \begin{bmatrix} \frac{\partial \rho_{s1}}{\partial x} & \frac{\partial \rho_{s1}}{\partial y} & \frac{\partial \rho_{s1}}{\partial z} & 1 \\ \frac{\partial \rho_{s2}}{\partial x} & \frac{\partial \rho_{s2}}{\partial y} & \frac{\partial \rho_{s2}}{\partial z} & 1 \\ \dots & \dots & \dots & \dots \\ \frac{\partial \rho_{sn}}{\partial x} & \frac{\partial \rho_{sn}}{\partial y} & \frac{\partial \rho_{sn}}{\partial z} & 1 \end{bmatrix} = \begin{bmatrix} e_{1x} & e_{1y} & e_{1z} & 1 \\ e_{2x} & e_{2y} & e_{2z} & 1 \\ \dots & \dots & \dots & \dots \\ e_{nx} & e_{ny} & e_{nz} & 1 \end{bmatrix},$$

$$\Delta\bar{X} = \begin{bmatrix} x_r - x_{r0} \\ y_r - y_{r0} \\ z_r - z_{r0} \\ \delta t_r \end{bmatrix},$$

and $\bar{Z}_0$ is a vector of measurements predicted for point $x_{r_0}, y_{r_0}, z_{r_0}$.

For a standalone positioning we linearize equations relative to an a-priori estimate of the receiver position. Later we consider differential positioning, using a second receiver as a reference station. In that case linearization can be done in a similar manner but relative to the known position of the reference receiver. Usually a reference receiver is located at a known position. Then differential positioning can be expressed as a task of finding a baseline vector, which is a difference between the unknown position of the receiver and the reference receiver position.

1.8 Least square estimation (LSE)

1.8.1 Theory of LSE

We need to solve the system of Equations (1.23), which can be rewritten as follows without loss of generalization:

$$\bar{Z} = [H] \times \Delta\bar{X}, \tag{1.24}$$

where $\bar{Z}$ is a vector of measurements, consisting of the differences between predicted and measured observables, $\Delta\bar{X}$ is an unknown state vector, which represents an error in positioning, and $[H]$ is the measurement matrix.

In real life the Equations (1.24) incorporate measurement errors and therefore should be written as follows:

$$\bar{Z} = [H] \times \Delta\bar{X} + \bar{\xi}, \tag{1.25}$$

where $\bar{\xi}$ is a vector of measurement errors. Then we can look for the solution $\Delta\hat{\bar{X}}$ of the system (1.25), which minimizes the sum of squared components $\sum_{i=1}^{n} \zeta_i^2$ of the residual vector

$$\bar{\zeta} = [H] \times \Delta\hat{\bar{X}} - \bar{Z}. \tag{1.26}$$

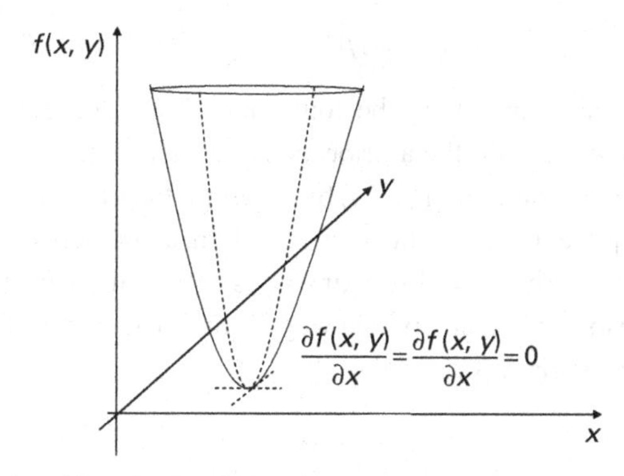

Figure 1.12 Minimum of criterion is achieved when its partial derivatives are zeros.

The minimum value of this criterion is achieved when its partial derivatives are zeros (see Figure 1.12). The equations of partial derivatives which equal zero in matrix form can be written as follows:

$$2 \cdot [H]^T \left| [H] \times \Delta \hat{\bar{X}} - \bar{Z} \right| = 0, \tag{1.27}$$

where $[H]^T$ is the transpose matrix of $[H]$.

Sequentially it leads to what is called ***normal equations***, which we can also derive formally by left multiplication of both sides of Equation (1.25) by matrix $[H]^T$

$$[H]^T \cdot [H] \cdot \Delta \hat{\bar{X}} = [H]^T \cdot \bar{Z}. \tag{1.28}$$

The unique solution of this system,

$$\Delta \hat{\bar{X}} = ([H]^T \times [H])^{-1} \times [H]^T \times \bar{Z}, \tag{1.29}$$

exists if Gramian matrix

$$[G] = [H]^T \times [H] \tag{1.30}$$

is invertible. It is invertible if its rank is equal to or higher than n. If column vectors of the measurement matrix are linearly dependent, then the Gramian matrix becomes singular, i.e. its determinant is equal to zero. The number of linearly independent vectors must be equal to or larger than n. The matrix columns are ***linearly dependent*** if we can get one column vector from another through linear operations, such as addition, subtraction, and multiplication by a constant. In a practical sense the linear dependency means that the correspondent measurements bring no new information. It can happen for example if directional vectors from a receiver to signal sources are parallel or close to parallel.

For practical applications we can assign weight for each observable. For example, we can put low weight on observables from low elevation satellites, because those observables are more corrupt with uncompensated atmospheric errors. Then the weighted LSE solution can be written as follows:

$$\Delta\hat{\bar{X}} = ([H]^T \times [W] \times [H])^{-1} \times [H]^T \times [W] \times \bar{Z}. \tag{1.31}$$

The solution can usually be found in a few iterations. The number of iterations also depends on how far the a-priori position is from the true position. Note that one cannot get a general solution of (1.24) by inverting the measurement matrix, because usually it is not a square matrix. The number of measurements usually exceeds the number of unknowns. When one has extraneous information in these measurements, one would also like to use it in an optimal way to minimize errors. This LSE operation is sometimes called pseudoinverse.

1.8.2 Propagation of covariances and dilution of precision (DOP)

We now consider how the satellite constellation geometry affects position error. To do that we define coordinate vector $\Delta\bar{X}$, which is a difference between assumed and estimated position, in a local horizontal coordinate system instead of Cartesian ECEF (see Figure 1.4) with east, north, up, and time components as follows:

$$\Delta\bar{X} = \begin{bmatrix} \Delta E \\ \Delta N \\ \Delta H \\ \delta t_r \end{bmatrix}. \tag{1.32}$$

Then the components of measurement matrix $[H]$ are unit direction vectors from an a-priori position to a satellite:

$$[H] = \begin{bmatrix} e_{1x} & e_{1y} & e_{1z} & 1 \\ e_{2x} & e_{2y} & e_{2z} & 1 \\ \cdots & \cdots & \cdots & \cdots \\ e_{nx} & e_{ny} & e_{nz} & 1 \end{bmatrix}. \tag{1.33}$$

Measurements are made with errors $\delta\bar{Z}$. Such errors in measurement propagate to the state vector estimate through the same Equation (1.29). Therefore we can write an equation for positioning error caused by measurement error as follows:

$$\delta\hat{\bar{X}} = ([H]^T \times [H])^{-1} \times [H]^T \times \delta\bar{Z} + \bar{\xi}. \tag{1.34}$$

The errors $\delta\hat{\bar{X}}$ are of a stochastic nature, so we are interested in their statistics, rather than specific instances. The first moment, which is mathematical expectation, can be expressed as follows:

$$\mathrm{E}(\delta\Delta\hat{\bar{X}}) = ([H]^T \times [H])^{-1} \times [H]^T \times \mathrm{E}(\delta\bar{Z}) + E(\bar{\xi}). \tag{1.35}$$

The second moment for the positioning error, which is a covariance, can be written

$$\begin{aligned} &\mathrm{cov}(\delta\Delta\hat{\bar{X}}) = \mathrm{E}(\delta\Delta\hat{\bar{X}} \cdot \delta\Delta\hat{\bar{X}}^T) = \\ &([H]^T \times [H])^{-1} \times [H]^T \times \mathrm{cov}(\delta\bar{Z}) \times [H] \times ([H]^T \times [H])^{-1}. \end{aligned} \tag{1.36}$$

Equation (1.36) describes error propagation through geometry, which is called ***dilution of precision (DOP)***. If we make an assumption that all components of the measurement covariance matrix are uncorrelated and have equal variances, then

$$\mathrm{cov}(\delta\bar{Z}) = \sigma^2 \cdot [I], \tag{1.37}$$

where $[I]$ is the identity matrix. Then (1.36) becomes

$$\mathrm{cov}(\delta\Delta\hat{\bar{X}}) = \sigma^2 \cdot ([H]^T \times [H])^{-1} \triangleq \sigma^2 \cdot [D]. \tag{1.38}$$

Diagonal elements of the covariance matrix $\mathrm{cov}(\delta\Delta\hat{\bar{X}})$ define the DOP. Vertical (VDOP) for height error variation, horizontal (HDOP) for horizontal positioning error variation, positional (PDOP), time (TDOP), and geometric (GDOP) dilution of precision are defined as follows:

$$\begin{aligned} d_{\mathrm{VDOP}} &= \sqrt{d_{33}}, \\ d_{\mathrm{HDOP}} &= \sqrt{d_{11} + d_{22}}, \\ d_{\mathrm{PDOP}} &= \sqrt{d_{11} + d_{22} + d_{33}}, \\ d_{\mathrm{TDOP}} &= \sqrt{d_{44}}, \\ d_{\mathrm{GDOP}} &= \sqrt{d_{11} + d_{22} + d_{33} + d_{44}}, \end{aligned} \tag{1.39}$$

where d_{ii} is an ith diagonal element of matrix [D].

It means that errors in pseudorange errors such as those caused for example by atmospheric errors propagate to larger positioning errors. Those positioning errors could be very large in the case of poor satellite geometry. In this case it may be advantageous to have more satellites. Figure 1.13 shows DOP for GPS and GPS+GLONASS constellations calculated for a specified interval of time.

1.8.3 LSE implementation

In order to make a numerical solution more robust towards round-off errors and more economical in terms of memory size and calculation speed, matrixes of Equations (1.23) are transformed to various products of factors. These transformations are called ***decomposition*** or ***factorization***. We consider here only ***QR decomposition*** or ***triangulation***. Gauss first solved Equations (1.23) by elimination using back-substitution.

If matrix [Q] is orthogonal, then by definition

$$[Q]^T[Q] = [I], \tag{1.40}$$

where $[I]$ is the identity matrix.

Matrix [H] can always be expressed as a product of an orthogonal matrix [Q] and an upper triangular matrix [R] in the following way:

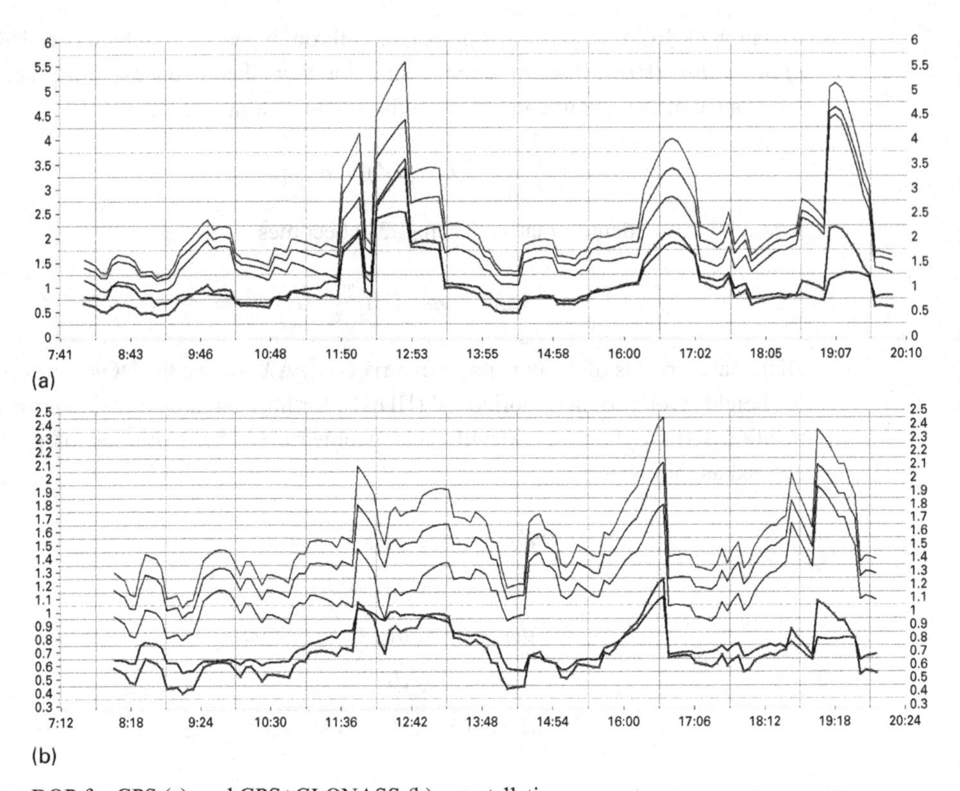

Figure 1.13 DOP for GPS (a), and GPS+GLONASS (b), constellations.

$$[H] = [Q] \times \begin{bmatrix} R_{11} & R_{12} & \cdots & R_{n1} \\ 0 & R_{22} & \cdots & R_{n2} \\ \vdots & \vdots & \ddots & \vdots \\ 0 & 0 & 0 & R_{nn} \end{bmatrix}. \tag{1.41}$$

Then initial Equations (1.24) can be rewritten as follows:

$$[Q][R] \times \Delta\hat{\bar{X}} = \bar{Z}. \tag{1.42}$$

Using (1.40), the equation can be rewritten as

$$[R] \times \Delta\hat{\bar{X}} = [Q]^T \times \bar{Z} \tag{1.43}$$

and then solved by back-substitution. Back-substitution means that we find $\Delta\hat{X}_n$ directly from the last row of (1.43) because $[R]$ is a triangular. Then we put $\Delta\hat{X}_n$ as known to the next row and find $\Delta\hat{X}_{n-1}$, and so on.

The algorithms for calculating QR factorization and back-substitution are well developed and included in most mathematical libraries.

1.8.4 Bancroft method for analytical solution

Depending on the algorithm, we may need an estimate of the initial position $x_{r_0}, y_{r_0}, z_{r_0}$ in order to linearize Equations (1.16). The initial guess in this case can be calculated by a global search or brute force method minimizing residual ranges to a satellite. Such a search is time consuming. (The same method, as we can see below, can be applied to the more general problem, when the epoch of transmission is not known. This problem arises if one has to compute position without retrieving a navigation message.) There are a few analytical methods. The first analytical method for GNSS equations was the Bancroft method [14]. It is widely used to find an a-priori position for linearization.

We use the Lorentz inner product, which is defined as

$$\left\langle \vec{g}, \vec{h} \right\rangle \triangleq \vec{g}^T [M] \vec{h}, \tag{1.44}$$

where $\vec{g}, \vec{h} \in R^4$,

$$[M] = \begin{bmatrix} \|I\|_{3\times3} & 0 \\ 0 & -1 \end{bmatrix}. \tag{1.45}$$

The pseudoranges equations can be rewritten through squaring and grouping to

$$\frac{1}{2}\left\langle \begin{bmatrix} \vec{r}_i \\ \rho_i \end{bmatrix}, \begin{bmatrix} \vec{r}_i \\ \rho_i \end{bmatrix} \right\rangle - \left\langle \begin{bmatrix} \vec{r}_i \\ \rho_i \end{bmatrix}, \begin{bmatrix} \vec{r} \\ b \end{bmatrix} \right\rangle + \frac{1}{2}\left\langle \begin{bmatrix} \vec{r} \\ b \end{bmatrix}, \begin{bmatrix} \vec{r} \\ b \end{bmatrix} \right\rangle = 0. \tag{1.46}$$

Here $\vec{r}_i$ is the radius vector of the ith satellite, $\vec{r}$ is the radius vector of the user position, $b = c \cdot dt$. If we have four such equations for four satellites, we can write them in matrix form:

$$\vec{\alpha} - [B][M]\begin{bmatrix} \vec{r} \\ b \end{bmatrix} + \Lambda \vec{\tau} = 0, \tag{1.47}$$

$$\Lambda = \frac{1}{2}\left\langle \begin{bmatrix} \vec{r} \\ b \end{bmatrix}, \begin{bmatrix} \vec{r} \\ b \end{bmatrix} \right\rangle, \tag{1.48}$$

where $\vec{\tau} = \begin{bmatrix} 1 \\ 1 \\ 1 \\ 1 \end{bmatrix}$

and $\vec{\alpha}$ is 4 x 1 vector with each component described as

$$\alpha_i = \frac{1}{2}\left\langle \begin{bmatrix} \vec{r}_i \\ \rho_i \end{bmatrix}, \begin{bmatrix} \vec{r}_i \\ \rho_i \end{bmatrix} \right\rangle.$$

Solving (1.47) for vector $\begin{bmatrix} \vec{r} \\ b \end{bmatrix}$ yields

$$\begin{bmatrix} \vec{r} \\ b \end{bmatrix} = [M][B]^{-1}\left(\Lambda\vec{\tau} + \vec{\alpha}\right). \tag{1.49}$$

However, Λ is a function $\begin{bmatrix} \vec{r} \\ b \end{bmatrix}$, so finally we can rewrite (1.47) as

$$\left\langle [B]^{-1}\vec{\tau}, [B]^{-1}\vec{\tau} \right\rangle \Lambda^2 + 2\left[\left\langle [B]^{-1}\vec{\tau}, [B]^{-1}\vec{\alpha} \right\rangle - 1\right]\Lambda + \left\langle [B]^{-1}\vec{\alpha}, [B]^{-1}\vec{\alpha} \right\rangle = 0, \tag{1.50}$$

which is a quadratic equation for Λ.

This equation is for the case of four measurements. It can be extended by multiplying Equation (1.47) by $[B]^T$ in the case where we have more than four satellites. Consequently, Equation (1.50) will become:

$$\left\langle [C]\vec{\tau}, [C]\vec{\tau} \right\rangle \Lambda^2 + 2\left[\left\langle [C]\vec{\tau}, [C]\vec{\alpha} \right\rangle - 1\right]\Lambda + \left\langle [C]\vec{\alpha}, [C]\vec{\alpha} \right\rangle = 0, \tag{1.51}$$

where

$$[C] = \left([B]^T[B]\right)^{-1}[B]^T. \tag{1.52}$$

Solving this equation by least squares, we will have two solutions for Λ. We can choose the correct one either by applying some constraints, such as that it should be on the Earth's surface, or by finding the solution for a different set of measurements, which led to only one consistent solution for Λ.

1.9 Integrity monitoring

1.9.1 Usage of redundant measurements in geodetic and navigation tasks

Navigation and geodetic tasks use redundant information differently. For geodetic tasks redundant information comes as measurements from a network of reference stations for a period of a few days. This information can be used to improve accuracy of the estimated station coordinates, either for one station, or for the whole network.

A receiver acquires a signal from a satellite, which is a carrier wave, modulated by a spreading code, and data (see Figure 1.14). In previous equations we have derived observables from code measurements ρ_{s_i} with an accuracy limited by the code tracking loop. In that way we measure distance to a satellite in number of code chips and fractions of a chip.* Observables can also be retrieved from carrier phase measurements, in which case we would measure a distance to a satellite as a number of carrier phase cycles. This will lead to another set of measurement equations with much higher accuracy, but they are ambiguous. The ambiguity numbers are introduced then as part of the state vector. We look at this case later in Chapter 9.

* A chip for GNSS signals can be defined as a spreading code waveform between two transitions. The length of the chip is different for the different signals. For GPS L1 C/A code the chiprate is 1.023 MHz, which means that the chip length is about 300 m.

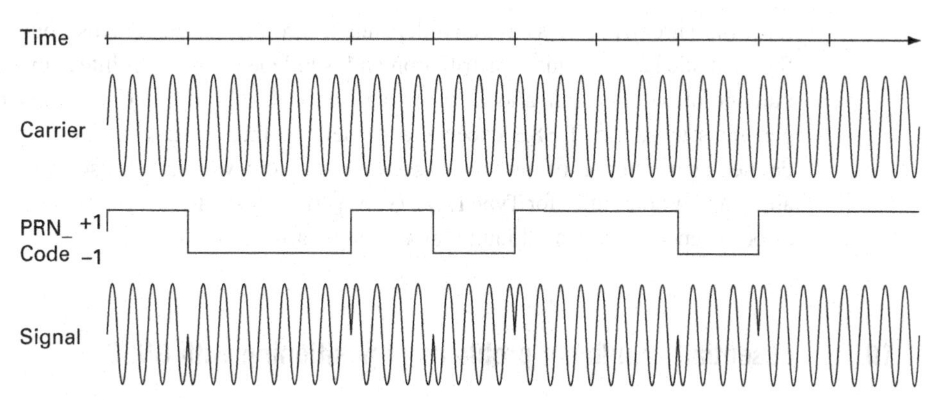

Figure 1.14 Simplified GPS signal.

Extra observables can also be used to include other unknowns in to the state vector. For example, one can estimate orbits of the satellites themselves. The unknown state vector can also include atmospheric parameters, terrain parameters, satellite orbits, and time. In these tasks, however, the coordinates of the receiver, or more often receivers, are assumed to be known. In modern GNSS technology, even the effects which in previous years have been considered only as errors are often used to get useful information. These include atmospheric delays, scintillation, and multipath.

In navigation, redundant information is often used to ensure system integrity. A receiver autonomous integrity monitoring algorithm (RAIM) uses redundant information to find and exclude a failed satellite signal.

1.9.2 Receiver autonomous integrity monitoring (RAIM)

There are a number of methods which allow us to establish that there is a faulty measurement among the group. We consider here a least squares residuals method.

The LSE solution defined by (1.29) gives an error positioning vector $\Delta\hat{\bar{X}}$ relative to an a-priori position. The estimated position

$$\hat{\bar{X}} = \bar{X}_o + \Delta\hat{\bar{X}} \tag{1.53}$$

can be used to recalculate pseudoranges from the estimated position to each satellite:

$$\tilde{\bar{Z}} = [H] \times \Delta\hat{\bar{X}}. \tag{1.54}$$

A vector of residuals between recalculated and measured pseudoranges is constructed as follows:

$$\bar{Y} = \bar{Z} - \tilde{\bar{Z}} = ([I] - [H] \times ([H]^T \times [H])^{-1} \times [H]^T) \times \bar{Z}. \tag{1.55}$$

The sum of the squares of the residuals, called ***sum of squared errors (SSE)*** serves as a test statistic [16]:

$$SSE = \bar{Y}^T \times \bar{Y}. \tag{1.56}$$

If the measurement errors have an independent zero-mean Gaussian distribution, then the SSE statistic has chi-square distribution and is independent of satellite geometry for any number of satellites. It allows us to calculate threshold values in advance. RAIM generates an alarm if SSE exceeds a predefined threshold. The threshold should be defined as a function of number of satellites in view, variance of pseudorange observables, and probabilities for Type I and Type II errors. Type I and Type II errors are defined as occurrences of a missed fault and a wrong alarm.

1.10 Essentially nonlinear problems of positioning with GNSS

1.10.1 Applications which lead to nonlinear problems

There may be a need to solve nonlinear Equations (1.16) or (1.17) in order to find an a-priori position, which may be required for linearization. There are also several problems in GNSS, where these equations cannot be linearized.

One such application is positioning with ***pseudolites***. Pseudolites are sources of a signal similar to a navigation satellite [17],[18]. Initially they transmitted the signal on the same frequency with GPS L1 with especially assigned ***pseudo-random numbers (PRN)*** for codes different from satellites. In order to avoid interference with non-participant receivers, called the near/far problem, the pseudolites transmit the signal in pulses, decreasing the total power of the transmitted signal. Pseudolites were initially proposed as augmentation for aircraft landing systems to provide quick ambiguity resolution for an onboard receiver [19]. The receiver was solving positioning equations, which were using ambiguous carrier phase observables. Currently many countries prohibit any transmissions on GNSS frequencies. Some pseudolite solutions have been moved to other frequencies. Pseudolites are normally located too close to a receiver to linearize Equations (1.16) or (1.17). In order to avoid a nonlinear problem, a receiver initially estimates a position based on GPS only and then involves pseudolites when the initial position guess is available. In this case pseudolite solutions usually use a dead reckoning technique.

Potential applications of pseudolites include social infrastructure [20]. Pseudolites are also used indoors for robotic applications [21],[22]. For such applications it is vitally important to ensure that pseudolites are not broadcasting in the frequency bands allocated to GNSS, which would cause interference. At the least there will be a signal quality degradation, which would affect high-end applications, due to a rise in the background noise level. The common solution in robotic applications is to apply carrier phase dead reckoning calculating rover movement relative to a known initial position. This also allows us to avoid any nonlinear problem, but does not solve the task of initial positioning with pseudolites.

The other important application which requires solution of nonlinear equations is a ***positioning without time mark*** [4],[23],[24]. As we have discussed above, the time to first fix (TTFF) is an important specification parameter for many navigation applications. If satellite orbits can be supplied to the receiver from a source other than a broadcast navigation message, then TTFF will be limited only by the time required to acquire a time mark from the navigation message. Positioning without a time mark allows us to decrease TTFF to around one millisecond.

In the positioning Equations (1.29) we generally need to determine receiver clock error as a part of an unknown state vector. However, the sought clock error should be limited to a certain value. We have to consider the size of the receiver clock error because it directly affects accuracy of the satellite coordinate calculation. Satellite coordinates are assumed to be calculated at the time of transmission. Formally we have linearized the system of Equations (1.23) in the vicinity of point $\bar{X}_0$, or $x_{r_0}, y_{r_0}, z_{r_0}$, and δt_{r_0}. We have assumed

$$\delta t_{r_0} \approx 0. \tag{1.57}$$

If the error δt_{r_0} cannot be assumed to be negligibly small, then the equations cannot be linearized, because errors in satellite positioning can become prohibitive:

$$dx = x_{s_i}(t_{rS}) - x_{s_i}(t_{SS}). \tag{1.58}$$

An average velocity of a GPS satellite is about four kilometers per second and up to about 800 meters per second along the line of sight (LOS). Therefore an error in a receiver clock of 0.1 second will result in satellite position error of 400 meters and 80 meters in LOS direction. These errors will be magnified by constellation geometry (DOP). Usually we consider a situation where a receiver estimates time from a navigation message. The receiver derives time of signal emission from the navigation message and spreading code. Then this time is used to adjust receiver time to an initial value with error δt_{r_0}. Then this value is estimated as a part of the navigation solution. The rule of thumb is that accuracy of the time estimate is at the same level as the other coordinates. We can see by analyzing DOP Equations (1.39), that satellite geometry affects time accuracy slightly differently than positioning accuracy. If the receiver 2DRMS is about 6 meters, then with the speed of light 300 000 kilometers per second the time error will be about 0.00000002 sec. In other words, a standard receiver measures time with about a 20 nanosecond accuracy.

There are various methods which allow us to solve nonlinear equations without knowing time in the user receiver accurately, i.e. without reading a navigation message.

1.10.2 Methods of solving nonlinearizable GNSS tasks

We consider a solution for the original nonlinear GNSS equations (1.16) or in a matrix form (1.17):

$$\bar{Z} = [A(\bar{X}')], \tag{1.59}$$

where $\bar{X}'$ is now an extended state vector,

$$\bar{X}' = \begin{bmatrix} x_r \\ y_r \\ z_r \\ \delta t_r \\ t_r \end{bmatrix}.$$

The only difference is that now we have to introduce an additional variable, which is a time of signal reception t_r, into the state vector. In the Equations (1.17), time of signal reception was assumed to be known as it is derived from the navigation message with certain accuracy.

Note also that $\bar{X}'$ as $\bar{X}$ is a vector of antenna position coordinates rather than of an error in its estimate, as in linearized Equations (1.29).

In searching for the solution one can define an area for the search. The area for the search can be better defined in local horizontal coordinates, because it will impose constraints on the third coordinate, which becomes altitude. The extended state vector can be rewritten as follows:

$$\bar{X}'_{\mathrm{LOC}} = \begin{bmatrix} E_r \\ N_r \\ H_r \\ \delta t_r \\ t_r \end{bmatrix},$$

where E_r, N_r, H_r are correspondingly east, north, and height components in the local horizontal frame.

We can assume that the sought position is in a certain area defined by maximum and minimum values of the state vector components. The area will be covered in a grid of possible $\bar{X}'_{\mathrm{LOC}}$ solutions. The area is located in a five-dimensional time–space continuum with coordinates defined by state vector components. In the brute force method the search is conducted by changing one component of the state vector at a time by a step value:

$$\bar{X}'_{\mathrm{LOC}}(i) = \begin{bmatrix} E_{\mathrm{Min}} + i_E \cdot \Delta E \\ N_{\mathrm{Min}} + i_N \cdot \Delta N \\ H_{\mathrm{Min}} + i_H \cdot \Delta H \\ \delta t_{\mathrm{Min}} + i_\delta \cdot \Delta\delta t \\ t_{\mathrm{Min}} + i_t \cdot \Delta t \end{bmatrix}, \tag{1.60}$$

where i_E, i_N, i_H, i_δ, and i_t are iterators along each dimension.

Each resulting state vector should be evaluated by evaluating a cost function. At each point we can recalculate a vector of measurements as

$$\hat{\bar{Z}}(i) = \left[A(\hat{\bar{X}}'_{LOC}(i))\right]. \tag{1.61}$$

The cost function can be derived as

$$\mathrm{CF} = (\bar{Z} - \hat{\bar{Z}})^T \times (\bar{Z} - \hat{\bar{Z}}). \tag{1.62}$$

The global minimum of cost function CF gives the sought position. It is very important to note that the CF has multiple local minima, so standard optimization algorithms are not applicable for this task.

If we are looking for a position in a two-dimensional plane, with fixed altitude, then the search space can be significantly reduced by removing one of its dimensions. This brute force method, however, currently does not allow real-time implementation even with added constraints. The real-time methods are already available [4], but they are beyond the scope of this book.

1.10.3 Analytical methods of solving nonlinearizable GNSS tasks

There are several analytical methods for solving Equations (1.16), for example [25]. The idea behind these methods is to eliminate unknown variables by polynomial substitution. The method can derive an analytical solution in closed form using modern mathematical packages such as MATLAB, Maple, or Mathematica. It is, however, much more difficult to derive analytical solutions in closed form for Equations (1.59), because they implicitly include additional equations for satellite movements.

References

[1] E. D Kaplan and C. J. Hegarty (editors), *Understanding GPS, Principles and Applications*, 2nd edition, Boston, MA, Artech House, 2006.

[2] Y. Urlichich, V. Subbotin, G. Stupak, *et al.*, GLONASS. Developing strategies for the future, *GPS World*, **22**, (4) 2011, 42–49.

[3] V. Engelsberg, V. Babakov, and I. Petrovski, Expert advice – GLONASS business prospects, *GPS World*, **19**, (3), 2008, 12–15.

[4] I. Petrovski, T. Tsujii, and H. Hojo, First AGPS – now BGPS. Instantaneous precise positioning anywhere, *GPS World*, **19**, (11), 2008, 42–48.

[5] I. Petrovski, *et al.*, LAMOS-BOHSAI ™: LAndslide Monitoring System Based On High-speed Sequential Analysis for Inclination, ION GPS'2000, USA, Salt Lake City, September 2000.

[6] M. Rothacher, W. Gurtner, S. Schaer, R. Weber, and H. Hase, Azimuth- and elevation-dependent phase center corrections for geodetic GPS antennas estimated from GPS calibration campaigns, in IAG Symposium No.115, edited by W. Torge, New York, NY, Springer-Verlag, 1996.

[7] D. D. McCarthy: *IERS Conventions 2000*, Central Bureau of IERS, Observatoire de Paris, IERS Technical Note, 32, Paris, 2004.

[8] V. Galazin, B. Kaplan, M. Lebedev, *et al.*, *Reference Document On Geodetic Parameters System, Parameters of Earth PZ-90*, Moscow, Coordination Scientific Information Center, 1998.

[9] J. M. Dow, R. E. Neilan, and G. Gendt, The international GPS service (IGS): Celebrating the 10th anniversary and looking to the next decade, *Adv. Space Res.* **36**, (3), 2005, 320–326, doi:10.1016/j.asr.2005.05.125.

[10] J. R. Smith, *Introduction to Geodesy, The History and Concepts of Modern Geodesy*, A Wiley-Interscience Publication, John Wiley & Sons, Inc, 1997.

[11] A. Perova and V. Harisov (editors), *GLONASS, Design and Operation Principles*, 4th edition, Moscow, Radiotechnica, 2010 (in Russian).

[12] J. Bennett, Mathematics, instruments and navigation, 1600–1800, in *Mathematics and the Historian's Craft*, M. Kinyon and G. van Brummelen (editors), Berlin/Heidelberg, Springer, 2005.

[13] P. McDonnel, *Introduction to Map Projections*, 2nd edition, Rancho Cordova, CA, Landmark Enterprises, 1991.

[14] C. Audoin and B. Guinot, *The Measurement of Time. Time, Frequency and the Atomic Clock*, Cambridge, Cambridge University Press, 2001.

[15] S. Bancroft, An algebraic solution of the GPS equations, *IEEE Transactions on Aerospace and Electronic Systems*, **21**, 1985, 56–59.

[16] R. Brown, Receiver autonomous integrity monitoring, in *Global Positioning System: Theory, and Applications*, Vol. II, B. W. Parkinson and J. J. Spilker (editors), Washington, DC, American Institute of Aeronautics and Astronautics Inc., 1996.

[17] B. Elrod and A. J. Van Dierendonck, Pseudolites, in *Global Positioning System: Theory, and Applications*, Vol. II, B. W. Parkinson and J. J. Spilker (editors), Washington, DC, American Institute of Aeronautics and Astronautics Inc., 1996.

[18] I. Petrovski, *et al.*, Pedestrian ITS in Japan, *GPS World*, **14**, (3), 2003, 33–37.

[19] C. Cohen, *et al.*, Precision landing of aircraft using integrity beacons, in *Global Positioning System: Theory and Applications*, Vol. II, B. W. Parkinson and J. J. Spilker (editors), Washington, DC, American Institute of Aeronautics and Astronautics Inc., 1996.

[20] I. Petrovski, *et al.*, Pedestrian ITS in Japan, *GPS World*, **14**, (3), 2003, 33–37.

[21] I. Petrovski, *et al.*, *Indoor code and carrier phase positioning with pseudolites and multiple GPS repeaters*. ION GPS'2003, USA, Portland, September 2003.

[22] S. Sugano, Y. Sakamoto, K. Fujii, *et al.*, It's a robot life, *GPS World*, **18**, (9), 48–55.

[23] Frank van Diggelen, *A-GPS: Assisted GPS, GNSS, and SBAS*, Boston, MA, Artech House, 2009.

[24] I. Petrovski, Expert advice – everywhere, without waiting, *GPS World*, **17**, (10), 2006, 12.

[25] J. Awange and E. Grafarend, *Solving Algebraic Computational Problems in Geodesy and Geoinformatics*, Berlin/Heidelberg, Springer, 2005.

Exercise

Use the ReGen signal simulator to generate ground tracks for GLONASS and QZSS. Note that tracks are drawn in green when the satellites are visible from the user location and blue when they are not.

2 Presentations and applications of GNSS orbits

"Therefore it is quite natural, that the Earth, being very heavy in comparison to bodies dropping free on it, stays everywhere motionless under the force from much smaller weights and as if just receives everything, which falls down on her."

Claudius Ptolemy, Almagest, Book 1–7.

Claudius Ptolemy.

In order to use GNSS for almost any practical application or to simulate a GNSS signal, we need to be able to define satellite coordinates at any moment of time. In this chapter we look at GNSS satellite orbits. We consider their mathematical presentations and requirements of constellation design. The subject matter of this chapter in relation to other chapters is shown on Figure 2.1

2.1 Development of models for celestial body movements from Ptolemy to Einstein

Movement of satellites around the Earth is described using mathematical instruments and models which were developed for describing the movement of the planets. The first known applicable mathematical model for orbital movement of the celestial bodies was developed by Claudius Ptolemy, who was working in the first century AD in Alexandria. He created the first accurate working model of a celestial body's motion [1].

Ptolemy's model is centered in immobile Earth. Ptolemy left the description of his model, along with a huge amount of measurement data on which he had built his theory, in ten books called *Mathematical Composition*. The books are known under the name *Almagest*, which came from Arabic al-majisti through Latin almagestum and roughly translates as *Greatest*. Besides making a huge contribution to astronomy, he also authored works on optics and astrology. Even Ptolemy's work in astrology, *Tetrabiblos*, is

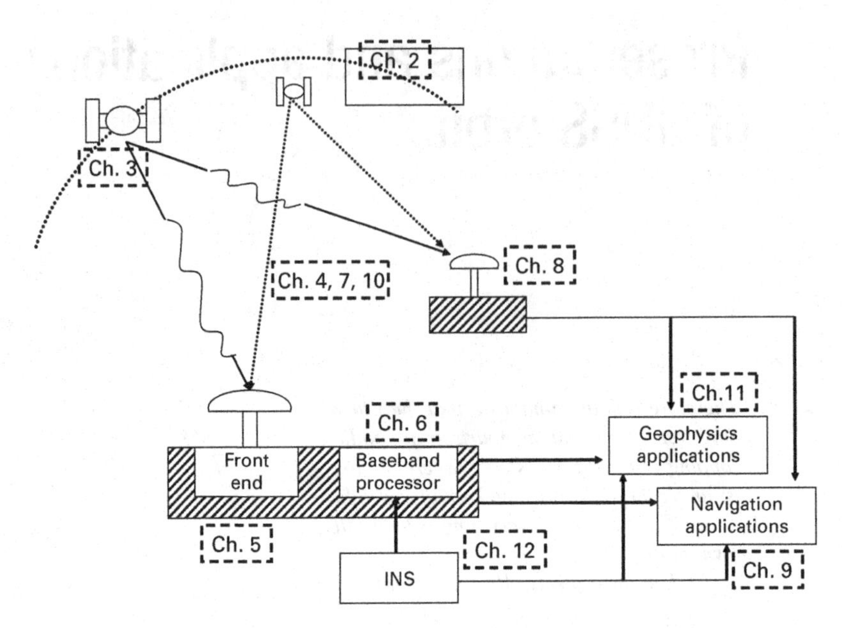

Figure 2.1 Subject of Chapter 2.

considered not as representative of astrology as such, but rather an attempt to rationalize astrology, which was part of life at that time, and to try to explain scientifically some cause-and-effect relations in the background of events [2].

In his *Planisphaerium* Ptolemy refers to probably the first astronomical instrument called a "spider." It is suggested that it is an old astrolabe. *Planisphaerium* is a treatise on stereographic projection, which is a principle of astrolabe operation [3]. It was the most widely used astronomical instrument of the Middle Ages. Figure 2.2 shows a modern replica of an astrolabe. It was used to find the angle of the Sun, the Moon, planets, and stars above the horizon or from the zenith. It was an auxiliary computing device to find latitude, direction of true north, and time, and also it was used in creation of horoscopes. As we discussed in Chapter 1, longitude measurements became possible only with development of accurate chronometers.

Ptolemy's model worked perfectly well in describing the visible motion of the planets. In fact it still works perfectly well today. The main drawback of Ptolemy's model is its complexity in the Ockham razor sense. The fourteenth century Franciscan friar William of Ockham introduced a powerful principle: *Entities should not be multiplied unnecessarily* [4]. Ptolemy had to introduce unnecessary entities in his model in order to account for retrograde motion of the planets. The retrograde motion is a visible backward motion of ***inferior*** planets, the planets which are closer to the Sun than the Earth. The ***Ockham principle*** is very important for science. Sir Isaac Newton restated it as Rule I at the beginning of Book III of the *Principia*: *We are to admit no more causes of natural things than such are both true and sufficient to explain their appearances* [5].

Figure 2.2 The two sides of a modern replica of an astrolabe, which we could use to work with GNSS satellites instead of a GNSS receiver.

Ptolemy's representation of planetary movement was in a way similar to a method developed much later by Fourier for decomposition of a periodical function into harmonics. Any periodical function can be represented by a series of harmonics with any required accuracy. Similarly any planet's periodic motion can be represented by a corresponding series of special entities in the Ptolemy model. Basically any periodic movement which can be described by a continuously differentiable function can be presented by Ptolemy's model.

Copernicus tried to remove those unnecessary entities from the Cosmos model [6]. Copernicus developed a new Sun-centered model, which allowed him to describe planetary movement with a lesser number of entities. It was a huge step in understanding of the physics behind visible events. Even here Ptolemy has played a very significant role, because Copernicus was using Ptolemy's data from Almagest in order to develop his Sun-centered model. Surprisingly, Almagest is in use even today. By overlapping astronomic observations and historical records we can derive a wealth of information to assist historians in researches on chronology, and physicists and astronomers in developing and verifying geophysical models.

Several discrepancies were discovered by comparing astronomic observations and historical records. These discrepancies can be considered as being between our models and measurements. It is interesting that such discrepancies are sometimes explained as errors in measurements, sometimes as faults in models either chronological or geophysical. We briefly return to this again in Chapter 11 when we consider measurements of the Earth's rotation parameters.

To describe planetary movement with the same accuracy as Ptolemy, the Copernicus system initially has required no fewer components than Ptolemy's. This is attributed to the fact that Copernicus used circular orbits instead of elliptical. Further improvement of the Copernicus model had been done by Kepler.

Johannes Kepler analyzed a huge amount of data with respect to a Sun-centered model, which led him to discover his three laws of planetary motion [7]. His first law stated that planets are moving on elliptical orbits with the Sun in one of the focuses. Kepler's work contains musical notes, which can even be played. He was looking for a harmony. It is still a significant scientific criterion, even in modern science. Many scientists feel that a model should be elegant in order to be true. However, it is important to understand that the main criterion for any model is the ability to describe its subject with given accuracy and predict its behavior. The model cannot ever completely describe the subject, because when it does it becomes the subject. In this sense Ptolemy's model was and still is valid, even for planetary movements.

Our choice where to center the system now is just a matter of convenience. In their book *Evolution of Physics*, Einstein and Infeld stated that from the modern physics point of view, Copernicus's and Ptolemy's systems are equally valid, because there is no absolute coordinate frame, fixed in space [8]. Surprisingly, the same view was expressed by the German Cardinal Nicholas of Cusa in the fifteenth century. In his philosophical treatise *Of Learned Ignorance*, he wrote: *The universe is a sphere of which the center is everywhere and the circumference is nowhere, nowhere, for God is its circumference and center and He is everywhere and nowhere* [9]. It is also interesting that the absolute coordinates frame model was a valid scientific model until the late nineteenth century, and played a significant role as the aether in developing modern electro-magnetic theory. Scientific development follows a law of spiral motion, coming back to old ideas on a new level. For instance, the absolute frame model is coming back in quantum mechanics as string theory.

In this book we are mostly interested in Earth-centered models, coordinate frames for which we described in Chapter 1. Ironically, a Sun-centered coordinate system is less in demand today than an Earth-centered system, which is in constant use in numerous applications. An Earth-centered system is obviously more convenient for describing near-Earth satellites, because their movement is governed mostly by the Earth's gravitational field. Understanding of the governing role of gravitation was achieved as a result of work done by Copernicus and Kepler. Isaac Newton discovered the gravitational force behind Kepler's models. Kepler's laws can actually be derived from Newton's laws.

In a sense, all scientists before Newton were giving purely geometrical models. In considering the navigation task, we can work with those models without considering forces which are causing satellite motion. The user is given a geometrical description of satellite motion and can determine satellite position for any moment of time. For a geodetic task, which requires significantly higher accuracy, we need to consider Newton's gravitational model.

A gravitational force between two bodies is defined by Newton:

$$\vec{F}(r) = -\frac{GMm}{r^2}\vec{e}_r, \tag{2.1}$$

where M is the mass of the Earth, m is the mass of a satellite, G is Newton's constant of gravitation, and $\vec{e}_r$ is the unit vector between centers of two bodies:

$$\vec{e}_r = \frac{\vec{r}}{r}. \tag{2.2}$$

For an Earth satellite we can assume that satellite mass is negligibly small with respect to the Earth's mass:

$$m << M. \tag{2.3}$$

Consequently satellite motion can be described in inertial space by a differential equation system:

$$\frac{d^2\vec{r}}{dt^2} = -GM\frac{\vec{e}_r}{r^2}. \tag{2.4}$$

The solutions of this equation can be elliptic, parabolic, or hyperbolic orbits, depending on the initial conditions, which are, in practical terms, the satellite initial velocity.

We use Kepler's laws here in the form derived by Newton and rephrase them for the case of satellites rotating around the Earth. In general Kepler's laws describe any two point masses moving under mutual gravitational attraction. For the GNSS satellites we look only at elliptical orbits, though the orbits are conic sections in general.

Keplers's first law. A satellite is moving around the Earth in an elliptical orbit, with the Earth's center of mass collocated with one of the focuses (see Figure 2.3).

Keplers's second law. A line joining the centers of mass of a satellite and the Earth sweeps out equal areas in equal intervals of time (see Figure 2.3).

Keplers's third law. The sum of the Earth's and satellite's masses multiplied by their period of mutual revolution is proportional to the cube of the mean distance between them:

$$(m + M)T_o^2 = \frac{4\pi^2 a^3}{G}, \tag{2.5}$$

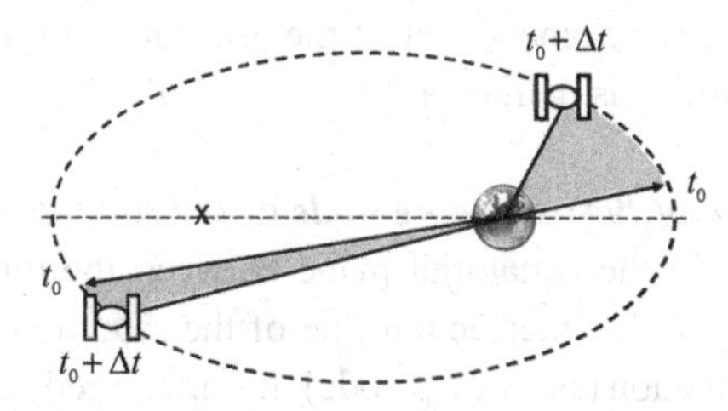

Figure 2.3 A satellite moving around the Earth according to Kepler's laws (not to scale).

where m is satellite mass, M is the Earth mass, T_o period of mutual revolution, G is Newton's gravitational constant, a is the mean distance between the Earth and the satellite centers.

Further development of gravitational theory was carried out by Einstein. His theory of relativity describes specifics of body behavior for the cases when speed and gravitation field are changing significantly. The satellites are actually moving in such conditions that we have to consider the theory of relativity in order for GNSS to work. And that is the case for a navigational task as well.

2.2 Orbital movement description by Keplerian parameters

2.2.1 Keplerian parameters

According to Kepler's first law, the planets are moving around the Sun in elliptical orbits, the same way that satellites are moving around the Earth. Kepler's laws can be derived from Newton's law of gravitation, and satellite orbits can be derived as solutions to Equation (2.4). All satellites we are interested in should be staying near the Earth, and therefore we are not considering satellites with parabolic and hyperbolic orbits which are designed to travel away from the Earth. Consequently, all navigational satellites are moving on elliptical trajectories.

We first examine satellite motion in an Earth Centered Inertial (ECI) frame. A satellite is considered to be a point mass moving along an elliptical orbit. The satellite's position can be defined at any epoch using six Keplerian parameters.

The orbit is fixed in relation to Earth, the center of which coincides with the orbit's focus. Therefore, we need three parameters to describe the orbit position in relation to Earth. Two further parameters describe orbit size and shape. The last parameter describes satellite position in the orbit at the specific epoch. Accordingly, we can define a satellite position at any given epoch using these six parameters. Figure 2.4 shows a satellite orbit and the Keplerian parameters which describe it.

The Keplerian parameters for an orbit are defined as follows:

Eccentricity (e) defines the shape of the orbit ellipse and how close the orbit is to a circular one.

Semi-major axis (a) defines the size of the orbit. It is defined as a half of the sum of the perigee and apogee distances, where the perigee is the closest to the Earth point on the orbit and the apogee is furthest.

The next three elements define the orientation of the orbit in relation to the Earth.

Inclination (i) is defined as the vertical tilt of the orbit with respect to the equatorial plane.

Longitude of the ascending node or ***the right ascension of the ascending node*** (Ω) is the angle in the equatorial plane between the direction to the vernal equinox (see Chapter 1) and the intersection line of the satellite orbit with the equatorial plane in a northerly direction (ascending node). It is measured counterclockwise when viewed from the north side of the equatorial plane.

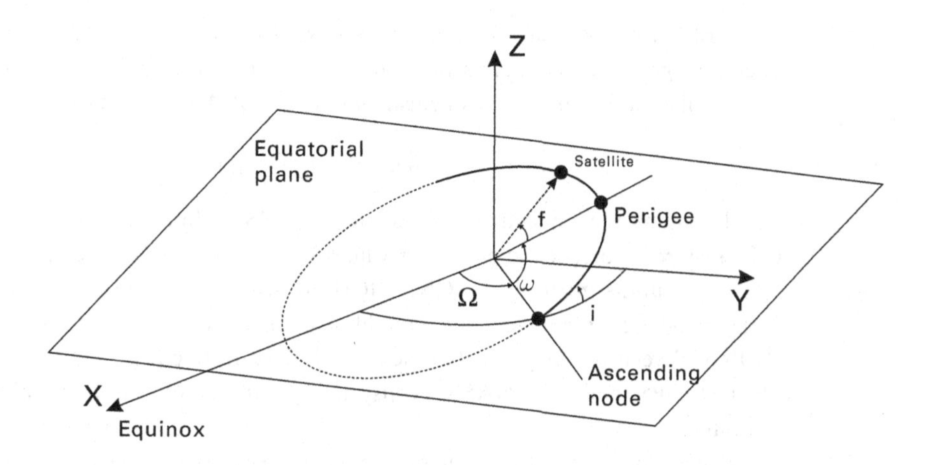

Figure 2.4 Satellite orbit description in Keplerian parameters.

Argument of perigee (ω) is the angle in the orbital plane between the ascending node and the perigee in the direction of motion of the satellite.

The last element defines an epoch, when the satellite is in a certain position on the orbit. Then one can calculate satellite position at any given epoch, by using the time difference between these two epochs.

Time of perigee passage (t_p) is the epoch, when the satellite is in the perigee point.

2.2.2 Satellite position calculation from Keplerian parameters

In the previous section we defined a set of parameters which describes an orbit and a satellite's position on it. In this section we examine how the satellite position is calculated using the Keplerian parameters.

From Kepler's third law, the period of a Keplerian orbit is

$$T_o = 2\pi\sqrt{\frac{a^3}{\mu}}, \tag{2.6}$$

where a is the mean distance between the Earth's and the satellite's centers, μ is the Earth's gravitational constant, which is defined for the Earth as follows:

$$\mu \triangleq GM, \tag{2.7}$$

where M is the Earth's mass, G is Newton's gravitational constant. For GPS the Earth's gravitational constant is defined in GPS ***Interface Control Document (ICD)*** [10] as

$$\mu = 3.986005 \cdot 10^{14}\ [\mathrm{m}^3/\mathrm{sec}^2]. \tag{2.8}$$

The Earth's gravitational constant for GLONASS is defined slightly differently. In general all geodetic parameters may be defined differently for different GNSSs. For example, the Earth's gravitational constant μ for GLONASS is taken as

$$\mu_{\text{GLONASS}} = 3.9860044 \cdot 10^{14}\ [\text{m}^3/\text{sec}^2]. \tag{2.9}$$

The algorithms recommended by a particular GNSS's Interface Control Documents (ICD) may refer to geodetic parameter values slightly different from each other.

The algorithms given in the GNSS ICD are usually adapted to their own values of the geodetic parameters and should use them. Therefore, a combined GPS/GLONASS receiver may use two sets of slightly different geodetic parameters to calculate GPS and GLONASS orbits, though in many cases the difference is negligible.

It is interesting to note that μ can be measured directly rather precisely with astronomical observations. If, however, one uses an explicit notation (2.7) and calculates μ from it, the accuracy will degrade by about 0.06% [11].

The period T_o is called a ***sidereal period***, to stress that it is calculated in respect to an inertial frame and it is defined as a time interval between two epochs of a satellite passing a specific orbit point, for example perigee. It is different from the ***synodic period***, which is defined as a time interval between two epochs of satellite footprint passing the same meridian. The difference is due to Earth rotation. For GPS a sidereal period is 12 hours, which is half of a sidereal day. A sidereal day is 4 minutes shorter than a calendar day; therefore a GPS satellite reappears at the same place in the sky after 11 hours 58 minutes. The GLONASS satellite orbit is lower and the sidereal period is correspondingly shorter. For GLONASS a sidereal period is 8/17 of a sidereal day. It means the GLONASS satellite will reappear at the same place in the sky in 8 days, after it has made 17 revolutions around the Earth.

The ***mean motion*** with which a satellite would be moving on a circular orbit with zero eccentricity can then be expressed as follows:

$$n = \sqrt{\frac{\mu}{a^3}}. \tag{2.10}$$

Correspondingly we can introduce ***mean anomaly***, which defines the position of the satellite on the orbit at any given epoch t:

$$M = n(t - t_p). \tag{2.11}$$

The mean anomaly is an abstraction, which changes linearly with time. It is interesting that equant in Ptolemy's model has a similar function to a mean anomaly and rotates uniformly for a satellite on an elliptical orbit. Satellite movement along the elliptical orbit is not uniform; therefore the real angle to the satellite position is different from the mean anomaly. As a first approximation, we use the ***eccentric anomaly*** E (see Figure 2.5), which describes, similarly to the mean anomaly, the nonuniform movement of the abstract satellite along the real elliptical orbit. A satellite velocity in accordance with Kepler's law reaches its maximum in perigee and minimum in apogee. The eccentric

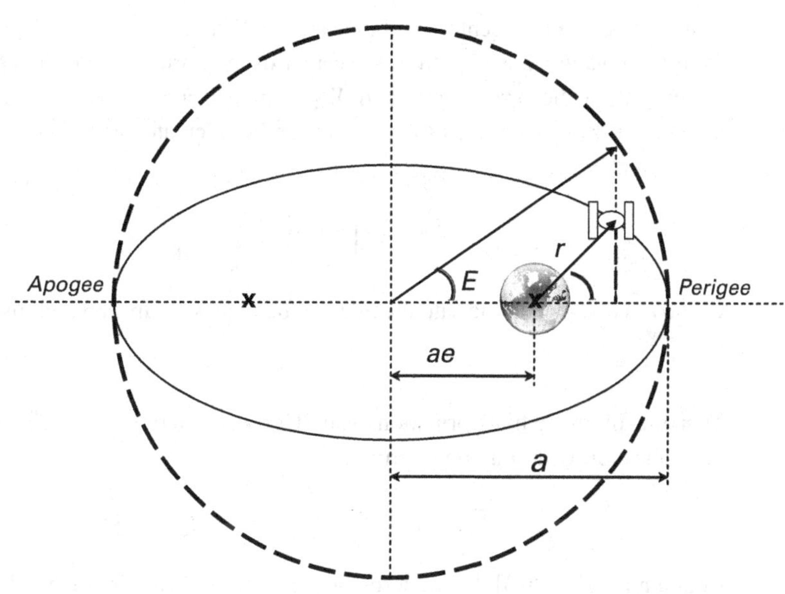

Figure 2.5 True and eccentric anomaly.

anomaly can be defined through the mean anomaly at a given epoch through the Kepler equation:

$$M = E - e \sin E. \tag{2.12}$$

Solution of this equation ultimately gives a position of the satellite. It is a nonlinear equation and can be solved either analytically through expansion to a Fourier series or Bessel function series or numerically in a few iterations [12]. We present here a method of iterations developed by Newton.

An initial approximation for the eccentric anomaly is given by

$$E_0 = M + \frac{e \sin(M)}{1 - \sin(M + e) + \sin(M)}, \tag{2.13}$$

with sequential iterations defined as

$$E_{n+1} = E_n - \frac{E_n - e \sin(E_n) - M}{1 - e \cos(E_n)}, \tag{2.14}$$

where $n = 1,2, \ldots$

The ***true anomaly*** is the angle to the real satellite position on the real elliptical orbit from the Earth. It can be derived from the eccentric anomaly as follows [12]:

$$\nu = \cos^{-1}[(e - \cos E)/(e \cos E - 1)], \tag{2.15}$$

where ν is defined in the same quadrant with E.

We define an Earth-centered orbital coordinate frame with axes X_{orb} and Y_{orb} located in the orbital plane. Axis X_{orb} from the center of the Earth to the orbit's perigee, axis Z_{orb} orthogonal to the orbit plane, and Y_{orb} complete a right-handed coordinate system. Satellite coordinates in the orbital frame can be calculated as follows:

$$\vec{X}_{orb} = \begin{bmatrix} x_{orb} \\ y_{orb} \\ z_{orb} \end{bmatrix} = \begin{bmatrix} r\cos(\nu) \\ r\sin(\nu) \\ 0 \end{bmatrix}, \tag{2.16}$$

where r is distance to the satellite and can be expressed from the ellipse geometry as

$$r = a(1 - e\cos E). \tag{2.17}$$

Then the following transformation to an ECI frame can be described by sequential matrix multiplication with rotation matrixes:

$$\vec{X}_{ECI} = [R_3(-\Omega)][R_1(-i)][R_3(-\omega)]\vec{X}_{orb}. \tag{2.18}$$

Rotation matrix $[R(\alpha)]$ defines a rotation of an orthogonal frame in relation to one of its axes on angle α.The rotation matrixes are generally defined as follows:

$$[R_1(\alpha)] = \begin{bmatrix} 1 & 0 & 0 \\ 0 & \cos(\alpha) & \sin(\alpha) \\ 0 & -\sin(\alpha) & \cos(\alpha) \end{bmatrix},$$

$$[R_2(\alpha)] = \begin{bmatrix} \cos(\alpha) & 0 & -\sin(\alpha) \\ 0 & 1 & 0 \\ \sin(\alpha) & 0 & \cos(\alpha) \end{bmatrix},$$

$$[R_3(\alpha)] = \begin{bmatrix} \cos(\alpha) & \sin(\alpha) & 0 \\ -\sin(\alpha) & \cos(\alpha) & 0 \\ 0 & 0 & 1 \end{bmatrix}.$$

Multiplication of a vector by a rotation matrix does not change the vector magnitude. Correspondingly, satellite coordinates in ECI can be calculated as follows:

$$\vec{X}_{ECI} = \begin{bmatrix} x_{ECI} \\ y_{ECI} \\ z_{ECI} \end{bmatrix} = \begin{bmatrix} r[\cos(\nu+\omega)\cos\Omega - \sin(\nu+\omega)\sin\Omega\cos i] \\ r[\cos(\nu+\omega)\sin\Omega + \sin(\nu+\omega)\cos\Omega\cos i] \\ r\sin(\nu+\omega)\sin i \end{bmatrix}. \tag{2.19}$$

The sum of the argument of perigee and the true anomaly is defined as ***the argument of latitude***

$$u = \nu + \omega. \tag{2.20}$$

The argument of latitude therefore is defined as the angle between the ascending node and the position of the satellite at the current epoch. Sometimes it is convenient to use this value as the sixth Keplerian parameter.

2.3 Orbit implementation in GPS navigation message

When we described satellite position in the previous sections, we assumed it to coincide with the satellite center of mass. In GNSS navigation messages and GNSS algorithms the satellite coordinates are assumed to coincide with the satellite antenna phase center, because it defines the propagation distance and therefore measured observations. When it comes to describing satellite movement in a gravitation field as is required in geodetic tasks, the corresponding adjustments should be made.

The GPS navigation message embedded in the transmitted signal provides a user with orbital parameters. These parameters allow the user to calculate satellite position in ECEF. Keplerian parameters given in a GPS navigation message for almanac and ephemeris are not strictly speaking Keplerian parameters for a GPS satellite orbit. These parameters are defined in the ECEF frame by ***control segments*** by fitting these parameters to a set of measurements from a control segment reference station network. If we want to apply the Equations (2.19) to GPS satellites, we need to take that into account. The Keplerian parameters in that case are calculated not in an inertial ECI frame, which is fixed relative to the stars, but in a special "non-rotating" with the Earth ECEF frame. This frame is different from an ECI frame by small rotations caused by polar motion, nutation, and precession. We can transfer satellite coordinates in (2.19) to ECEF just by multiplying them by a matrix, which describes Earth rotation. That provides us with satellite coordinates accurate enough for navigational tasks.

When we consider geodetic tasks, it becomes necessary to account for polar motion, nutation, and precession. In this case we need to consider a transformation between ECI and ECEF frames. Satellite orbits and corresponding coordinates are defined by respective control segments in ECEF frames, as described above. That is because a control segment reference network, which defines these coordinates, is in fact the ECEF frame itself. The GPS ICD recommended algorithm for satellite position calculation gives the satellite coordinates in an ECEF frame. Then satellite positions should be transferred to an ECI frame, in which receiver antenna position should be calculated. Then user position can be transferred back to the constellation ECEF frame (WGS-84, PZ-90, ITRF) or to a local frame. This is done in order to make all calculations in an inertial frame, thus avoiding complications related to the dynamics of ECEF and the time difference between signal transmission and reception. For navigational tasks, however, we can omit this transformation. As described in Chapter 1, the ECI to ECEF transformation for such tasks can be simplified to account only for Earth rotation, which is implicitly embedded in the GPS ICD algorithm for satellite position calculation. In this case we calculate GPS satellite coordinates in ECEF at the epoch of reception according to the GPS ICD algorithm and correct satellite positions for the distance they have traveled during signal propagation. Then applying, for example, a least square estimation (LSE) method (see Chapter 1), we can derive receiver position at the time of signal reception in ECEF frame.

For more demanding tasks a transformation between ECI and ECEF would be required. As a first step we need to specify a particular implementation of the ECI frame. As described in the previous chapter, the vernal equinox, defined as the intersection line of the equatorial and the ecliptic planes is not fixed in space due to precession and nutation. Therefore it is necessary to specify a reference epoch for equator and equinox in order to make the inertial frame unique. One of the options is to use J2000.0 reference systems [13]. The J2000.0 reference system is the most common. Users working on geodetic tasks can apply ***Earth orientation parameters*** (EOP), which are freely available from International GNSS Service (IGS) services. A navigation message for an L2C GPS signal also contains EOP. The GPS ICD recommends the following algorithm for their implementation [10]. The coordinate transformation of a satellite position from ECEF to ECI is described by a series of rotations as follows:

$$\vec{X}_{\mathrm{ECEF}} = [R_{\mathrm{PM}}][R_{\mathrm{ER}}][R_{\mathrm{N}}][R_{\mathrm{P}}]\vec{X}_{\mathrm{ECI}}, \tag{2.21}$$

where the $[R_{\mathrm{PM}}]$, $[R_{\mathrm{ER}}]$, $[R_{\mathrm{N}}]$, $[R_{\mathrm{P}}]$ are rotation matrices for polar motion, Earth rotation, nutation, and precession, respectively.

2.4 Ephemerides of GNSS satellites

2.4.1 Osculating Keplerian parameters

In order to describe satellite orbits with enough precision for positioning, the Keplerian model must be refined. A satellite orbit is not perfectly elliptical due to numerous forces, which are affecting the satellite movement. In Equation (2.4) we have considered basically a one-body problem, by taking into account only the Earth's mass. Now we consider other forces affecting satellite movement as well, following [14]–[16].

Equation (2.4) can be rewritten in general form as follows:

$$\frac{d^2\vec{r}}{dt^2} = -GM\frac{\vec{r}}{r^3} + \vec{f}(t, \vec{r}, \frac{d\vec{r}}{dt}, p_0, p_1, \ldots), \tag{2.22}$$

where nonlinear function $\vec{f}(t, \ldots)$ includes all forces that are functions of satellite position at a particular epoch, which can be sufficiently well modeled, and unknown parameters p_i, which are associated with radiation pressure. These parameters should be estimated along with satellite position $\vec{r}$ if included in the consideration. All these forces, including other harmonics of Earth gravitational field, have significantly less, but still significant impact on satellite movement periodical in nature. Therefore all Keplerian parameters, semi-major axis, eccentricity, inclination, ascending node, and argument of perigee are oscillating. Consequently the orbits cannot be precisely described as ellipses.

Let us define ***osculating Keplerian parameters*** for a particular epoch as Keplerian parameters, which give correct coordinates and velocities for a given epoch, based on unperturbed Equations (2.4). Consequently the real orbit can be presented as an envelope of a set of osculating Keplerian orbits, each corresponding to the position of the satellite in the specific epoch.

Satellite movement for a specific epoch can be equally described in terms of six Keplerian parameters and in terms of six coordinates (three coordinate and three velocity projections) in an ECI frame. This implies one-to-one correspondence between the orbit description in ***Descartes (Cartesian) coordinates*** $(X, Y, Z, \dot{X}, \dot{Y}, \dot{Z})$ and in Keplerian parameters. Satellite coordinates and velocity can be precisely calculated by numerical integration of the vector equation given by (2.22). We can also calculate Keplerian parameters which correspond to this set of coordinates and velocities. These Keplerian parameters, called ***osculated Keplerian parameters***, are different for each point at which we make these calculations. The whole orbit therefore can be viewed as an envelope of osculating parameters.

2.4.2 GPS ephemerides

In the case of GPS, a 12 hour satellite orbit is divided into two-hour segments. Osculating Keplerian parameters are calculated for each segment. In order to account for changes of the Keplerian parameters from one segment to another, and provide more accurate orbit presentation, Keplerian parameters in each segment are presented as time series (see Table 2.1). The almanac parameters in the GPS are defined in the same way as ephemeris. GPS almanac parameters in this sense can be viewed as a subset of ephemeris.

Orbital parameters of a satellite are estimated using measurements of distance to the satellite provided by numerous stations, the same type of measurements that are used by a GNSS receiver user for positioning. In the case of broadcast ephemerides, these stations belong to a control segment network. In order to be delivered to a user, these orbital parameters are embedded into the satellite signal as a navigation message.

The accuracy of broadcast ephemeris is constantly increased. It can be assumed to be on the meter level. Broadcast ephemerides are valid only for the specified period of two hours due to the way they are constructed, which was described above. They can be propagated forward, using accurate force models. This technique may be useful when a receiver has acquired a satellite signal only for short time, which was not enough for

Table 2.1 Ephemeris parameters.

"Keplerian" parameter	Rate	Sine harmonic	Cosine harmonic
Mean anomaly, M	Mean motion Δn difference from computed	–	–
Eccentricity, e	–	–	–
Square root of semi-major axis, $\sqrt{a}$	–	Correction term to the orbit radius, C_{RS}	Correction term to the orbit radius, C_{RC}
Longitude of ascending node, Ω	Rate of right ascension, $\dot{\Omega}$	–	–
Inclination, i	Rate of inclination angle, $IDOT$	Correction term to the inclination C_{IS}	Correction term to the inclination, C_{IC}
Argument of perigee, ω	–	Correction term to the argument of latitude, $(u)\ C_{US}$	Correction term to the argument of latitude, $(u)\ C_{UC}$

decoding ephemeris information from the navigation message. In this case the receiver may try to calculate satellite coordinates from the previously decoded set of ephemerides. This would require, besides a set of algorithms, a set of geophysical parameters to be kept in the receiver. Such a set would require also to be updated approximately once a year for navigational applications.

We follow GPS ICD [10] in describing an algorithm of satellite position calculation from GNSS ephemeris. It is recommended always to start with algorithms as they are given in the official documents, such as ICD, which are normally freely available to the public. For the purpose of presentation the notations here are slightly changed from those in ICD in order to be consistent with notations in the book.

Ephemerides for a satellite are calculated for a specific reference epoch t_{OE}, called the time of ephemeris. The time interval Δt between the reference epoch and time of signal transmission is

$$\Delta t = t - t_{OE}. \tag{2.23}$$

Because GPS time count starts each week at Saturday night from 0, and the reference epoch should be specified in a navigation message within a week, a user should account for the end of week crossover as follows:

$$\begin{cases} \text{if } \Delta t > 302,400 \text{ [sec] then } \Delta t = \Delta t - 604,800, \\ \text{else if } \Delta t < -302,400 \text{ [sec] then } \Delta t = \Delta t + 604,800. \end{cases} \tag{2.24}$$

Now we apply corrections from the navigation message as they are specified in the second, third, and fourth columns of Table 2.1 to the nominal values of Keplerian parameters given in the first column. Corrected mean motion is calculated as

$$n = n_0 + \Delta n, \tag{2.25}$$

where mean motion n_0 is calculated from (2.10). Mean anomaly is calculated as

$$M = M_0 + n \cdot \Delta t. \tag{2.26}$$

The Keplerian Equation (2.12) may be solved interactively for eccentric anomaly E. The true anomaly should be calculated using the following formula:

$$\nu = \tan^{-1}\left\{\frac{\sqrt{1-e^2}\sin E/(1-e\cos E)}{(\cos E - e)/(1-e\cos E)}\right\}. \tag{2.27}$$

The numerator and denominator cannot be simplified by reducing $(1 - e\cos E)$ because the arctangent should be defined using the signs of both arguments to determine the quadrant of the return value.

The corrected argument of latitude is calculated correspondingly to its definition (2.20) and by applying corrections as follows:

$$u = \nu + \omega + C_{US}\sin(2u_0) + C_{UC}\cos(2u_0). \tag{2.28}$$

The corrected radius is calculated as

$$r = a(1 - e\cos E) + C_{RS}\sin(2u_0) + C_{RC}\cos(2u_0), \tag{2.29}$$

and corrected inclination as follows:

$$i = i_0 + +C_{\mathrm{IS}}\sin(2u_0) + C_{\mathrm{IC}}\cos(2u_0) + \dot{i}\Delta t. \tag{2.30}$$

Correspondingly the satellite position in the orbital plane will be given by an equation similar to (2.16):

$$\vec{X}_{\mathrm{orb}} = \begin{bmatrix} x_{\mathrm{orb}} \\ y_{\mathrm{orb}} \\ z_{\mathrm{orb}} \end{bmatrix} = \begin{bmatrix} r\cos(\nu) \\ r\sin(\nu) \\ 0 \end{bmatrix}. \tag{2.31}$$

In the ICD all the parameters which we described earlier in Section 2.3 for ECI are calculated in ECEF, because all GPS control station network stations basically define ECEF. Therefore if we account only for Earth rotation we get equations for satellite position in ECEF instead of ECI as in (2.16), where we assumed that all Keplerian parameters are defined in ECI. If that were the case we would need, besides Earth rotation, also to account for pole rotation, precession, and nutation as they are given by (2.21). In general we need to use these transformations only if we want to transfer coordinates to ECI, for example for some accuracy demanding applications.

Corrected longitude of an ascending node with Earth rotation taken into account can be written as follows:

$$\Omega = \Omega_0 + (\dot{\Omega} - \dot{\omega}_{\mathrm{E}})\Delta t - \dot{\omega}_{\mathrm{E}} t_{\mathrm{OE}}. \tag{2.32}$$

Accordingly, satellite position in an ECEF frame can be described as follows:

$$\vec{X}_{\mathrm{ECEF}} = \begin{bmatrix} x_{\mathrm{ECEF}} \\ y_{\mathrm{ECEF}} \\ z_{\mathrm{ECEF}} \end{bmatrix} = \begin{bmatrix} r[\cos u\cos\Omega - \sin u\sin\Omega\cos i] \\ r[\cos u\sin\Omega + \sin u\cos\Omega\cos i] \\ r\sin u\sin i \end{bmatrix}. \tag{2.33}$$

which looks exactly the same as (2.19), with all parameters corrected and Earth rotation hidden in the corrected longitude of an ascending node.

Broadcast ephemerides for post-processing are available through the Internet from the IGS (International GNSS Service) network, as well as from some national and regional networks, such as the GSI (Geographical Survey Institute) in Japan, usually as RINEX navigation files. A snapshot of a RINEX navigation file is given in Figure 2.6. The file gives satellite PRN, epoch, and then a set of ephemeris parameters for this epoch. Then it continues to another satellite/epoch record.

2.5 Tabular orbits

In Section 2.4.1 we described correspondence between an orbit description in Descartes coordinates and in Keplerian parameters. Consequently, this gives us an alternative method of providing orbit information as tabular orbits. In this case satellite coordinates and velocities are provided for the boundaries of each segment.

One may need to use force models when working with tabular orbits. The requirements on these models depend on the interval between the tabular points and the required orbit

```
     2.10           N: GPS NAV DATA                         RINEX VERSION / TYPE
teqc  2002Mar14     GSI, JAPAN          20090522 04:44:18UTCPGM / RUN BY / DATE
Linux 2.0.36|Pentium II|gcc -static|Linux|486/DX+           COMMENT
teqc  2002Mar14     GSI, JAPAN          20090520 03:15:31UTCCOMMENT
     2              NAVIGATION DATA                         COMMENT
DAT2RIN 3.53        GSI, JAPAN          20MAY09 10:03:55    COMMENT
                                                            COMMENT
    7.4510D-09  2.2350D-08 -5.9600D-08 -1.1920D-07          ION ALPHA
    8.6020D+04  8.1920D+04 -1.3110D+05 -5.2430D+05          ION BETA
    7.450580596920D-09 2.753353101070D-14   503808     1532 DELTA-UTC: A0,A1,T,W
    15                                                      LEAP SECONDS
                                                            END OF HEADER
 2 09  5 20  0  0  0.0 1.544686965640E-04 5.684341886080E-13 0.000000000000E+00
    6.900000000000E+01 8.003125000000E+01 4.917704909960E-09-1.452650735020E-02
    4.025176167490E-06 9.066797443670E-03 9.216368198390E-06 5.153657329560E+03
    2.592000000000E+05 2.216547727580E-07-2.839270790000E-01-6.891787052150E-08
    9.416867444840E-01 1.944375000000E+02 2.816546211150E+00-8.417136321270E-09
    1.071473187950E-10 1.000000000000E+00 1.532000000000E+03 0.000000000000E+00
    2.400000000000E+00 0.000000000000E+00-1.722946763040E-08 6.900000000000E+01
    2.520180000000E+05
 2 09  5 20  2  0  0.0 1.544733531770E-04 5.684341886080E-13 0.000000000000E+00
    8.300000000000E+01 8.228125000000E+01 4.865917002660E-09 1.035608460950E+00
    4.287809133530E-06 9.067845880050E-03 8.540228009220E-06 5.153659574510E+03
    2.664000000000E+05 2.589076757430E-07-2.839869050340E-01 1.769512891770E-07
    9.416879791870E-01 2.023437500000E+02 2.816603926200E+00-8.258558281680E-09
    3.821587837850E-11 1.000000000000E+00 1.532000000000E+03 0.000000000000E+00
    2.400000000000E+00 0.000000000000E+00-1.722946763040E-08 8.300000000000E+01
    2.592180000000E+05
```

Figure 2.6 RINEX navigation file.

accuracy between them. It may be necessary to use some force models if the interval between points is large, such as in the case of GLONASS broadcast orbits. It is necessary to use a force model also if the requirements for orbit definition accuracy are high, such as in the case of geodetic applications.

2.5.1 Broadcast GLONASS ephemerides

The coordinates of GLONASS satellites are estimated through a boundary value problem solution. Ephemerides in such a format are provided in the GLONASS navigation message. Satellite position in this case should be found using a fourth-order Runge–Kutta method, as recommended in the GLONASS ICD [17]:

$$\begin{cases} \frac{dx_{\mathrm{ECEF}}}{dt} = V_{x_{\mathrm{ECEF}}} \\ \frac{dy_{\mathrm{ECEF}}}{dt} = V_{y_{\mathrm{ECEF}}} \\ \frac{dz_{\mathrm{ECEF}}}{dt} = V_{z_{\mathrm{ECEF}}} \\ \frac{dV_{x_{\mathrm{ECEF}}}}{dt} = -\frac{\mu}{r^3} x_{\mathrm{ECEF}} - \frac{3}{2} J_2 \frac{\mu a_E^2}{r^5} x_{\mathrm{ECEF}} \left(1 - \frac{5 z_{\mathrm{ECEF}}^2}{r^2}\right) + \omega^2 x_{\mathrm{ECEF}} + 2\omega V_{y_{\mathrm{ECEF}}} + \ddot{x}_{\mathrm{ECEF}} \\ \frac{dV_{y_{\mathrm{ECEF}}}}{dt} = -\frac{\mu}{r^3} y_{\mathrm{ECEF}} - \frac{3}{2} J_2 \frac{\mu a_E^2}{r^5} y_{\mathrm{ECEF}} \left(1 - \frac{5 z_{\mathrm{ECEF}}^2}{r^2}\right) + \omega^2 y_{\mathrm{ECEF}} + 2\omega V_{x_{\mathrm{ECEF}}} + \ddot{y}_{\mathrm{ECEF}} \\ \frac{dV_{z_{\mathrm{ECEF}}}}{dt} = -\frac{\mu}{r^3} z_{\mathrm{ECEF}} - \frac{3}{2} J_2 \frac{\mu a_E^2}{r^5} z_{\mathrm{ECEF}} \left(1 - \frac{5 z_{\mathrm{ECEF}}^2}{r^2}\right) + \ddot{z}_{\mathrm{ECEF}}, \end{cases} \quad (*) \tag{2.34}$$

(*) *Note that for the sake of clarity we have changed notation here from that given in the ICD, which shows* J_2 *as* J_0^2 *in the formulas.*

where $r = \sqrt{x_{ECEF}^2 + y_{ECEF}^2 + z_{ECEF}^2}$, J_2 is the second spherical harmonic of Earth geopotential, which represents the oblateness of Earth (2.8.1). In these equations the ECEF frame is defined as PZ-90.

2.5.2 Tabular orbits in geodetic algorithms

When it comes to more precise coordinate estimation, we have to consider gravitational and other forces affecting satellites. Here we have to move from a geometric description to physical models based on Newton's equations (2.22).

Broadcast GPS and GLONASS ephemerides are estimated by corresponding ground segments, which belong to the USA and Russian authorities respectively. The international geodetic community also calculates ephemerides for both constellations. These ephemerides are available freely from the Internet from IGS. The common format for ephemeris is the SP3 format. The ephemerides are free for all applications, including commercial [18]. There are two main bodies, IGS and CODE (Center for Orbit Determination in Europe, located at AIUB (Astronomical Institute, University of Bern)). Initially, the IGS was designed as an orbit determination service for GPS satellites. Some of the available ephemerides products are summarized in Table 2.2.

Figure 2.7 shows an SP3 file from CODE with tabular orbits for GPS and GLONASS. The file gives an epoch and then lists tabular parameters for each satellite in one row, which contains satellite PRN, three coordinates and clock. Letter G denotes GPS and R GLONASS satellites.

2.5.3 Tabular orbits in navigation

We can use predicted ephemeris in tabular format for navigation. It allows us to avoid the necessity to wait until the navigation message is decoded by a receiver. It also allows a receiver to calculate a position, using short chunks of GNSS signal, as may be required in snapshot positioning in particular for indoor applications. We also may use precise tabular ephemeris for signal simulation purposes. In order to use these ephemerides for navigation solutions, or for simulation, we can consider geometrical interpolation within points of tabular orbits without applying the numerical integration of (2.22) and considering force models. We can use various types of interpolation for this purpose. Here we consider an interpolation using a Fourier series.

Table 2.2 Some of the available GNSS ephemerides.

Ephemerides	Orbit accuracy	Conditions	Availability	Source
Broadcast	$\sim$1 m	none	real-time	GNSS signal
Real-time	$\sim$10 cm	contract	real-time	JPL or proprietary
Predicted 24 hours	$<$10 cm	free	predicted	CODE or IGS
Predicted 5 days	$<$20 cm	free	predicted	CODE or IGS
Final orbits	$<$5 cm	free	after 5 days	CODE or IGS

```
 1 #cP2009  2 23  0  0  0.00000000     480 d+D    IGS05 EXT AIUB
 2 ## 1520  86400.00000000   900.00000000 54885 0.000000000000
 3 +   50   G02G03G04G05G06G07G08G09G10G11G12G13G14G15G16G17G18
 4 +        G19G20G21G22G23G24G25G26G27G28G29G30G31G32R02R03R04
 5 +        R06R07R08R10R11R13R14R15R17R18R19R20R21R22R23R24  0
 6 +          0  0  0  0  0  0  0  0  0  0  0  0  0  0  0  0  0
 7 +          0  0  0  0  0  0  0  0  0  0  0  0  0  0  0  0  0
 8 ++         5  8  5  5  7  5  7  5  5  5  5  7  7  7  5  7  5
 9 ++         7  5  5  5  7  5  5  7  5  5  7  5  5  5 36 36  6
10 ++         6  6 36  6  6  6  6  6  7  7  7  7  7  7  7  7  0
11 ++         0  0  0  0  0  0  0  0  0  0  0  0  0  0  0  0  0
12 ++         0  0  0  0  0  0  0  0  0  0  0  0  0  0  0  0  0
13 %c M  cc GPS ccc cccc cccc cccc cccc ccccc ccccc ccccc ccccc
14 %c cc cc ccc ccc cccc cccc cccc cccc ccccc ccccc ccccc ccccc
15 %f  1.2500000  1.025000000  0.00000000000  0.000000000000000
16 %f  0.0000000  0.000000000  0.00000000000  0.000000000000000
17 %i    0    0    0    0      0      0      0      0         0
18 %i    0    0    0    0      0      0      0      0         0
19 /* CENTER FOR ORBIT DETERMINATION IN EUROPE (CODE)
20 /* GNSS ORBIT PREDICTION (0-5 DAYS) STARTING YEAR-DAY 09054
21 /* THESE ORBITS ARE DERIVED FROM CODE NRT ORBIT RESULTS
22 /* PCV:IGS05_1515 OL/AL:FES2004  NONE     YN ORB:CoN CLK:BRD
23 *  2009  2 23  0  0  0.00000000
24 PG02  13558.194340  22183.148230  -6493.377924    156.553389
25 PG03 -20837.446641 -10952.709005 -12936.842720    356.685418
26 PG04   4561.915085  25256.800892   5903.447542   -346.633111
27 PG05  19571.441450 -10847.447005  13916.526428     -0.827906
28 PG06 -16745.401827 -13558.590832 -15364.612255     46.068435
29 PG07 -10204.040892  13373.839166 -20543.468553     22.399578
30 PG08  -1781.465093  23021.069785 -12718.420545   -191.804306
31 PG09  16419.826107   4870.788917  19687.309257     41.913539
32 PG10  12079.353764   9447.561781 -21849.349893    -11.581265
```

Figure 2.7 SP3 tabular orbit file.

The sidereal day as a base period in this interpolation is taken in accordance with the behavior of oscillating Keplerian parameters (Section 2.4.1). From the behavior of the parameters we can see that these parameters have corresponding harmonics. Therefore the trigonometric series can be expressed:

$$\begin{aligned} x_i =& A_{i0} + A_{i1}\sin\left(\frac{2\pi}{T_{\mathrm{SD}}}t\right) + B_{i1}\cos\left(\frac{2\pi}{T_{\mathrm{SD}}}t\right) + A_{i2}\sin\left(\frac{4\pi}{T_{\mathrm{SD}}}t\right)\ldots \\ &+ B_{in}\sin\left(\frac{2n\pi}{T_{\mathrm{SD}}}t\right), \end{aligned} \tag{2.35}$$

where x_i are orbit coordinates, $T_{\mathrm{SD}} = 0.99726956634$ is the sidereal day [10].

Given $2n+1$ measurements, which are coordinate values in the tabular orbit table, we can estimate A_{ij}, B_{ij} coefficients. If we are not sure about what the harmonics in these series are, we can estimate them from the data (see an algorithm in [19]).

There are various more complex techniques for such interpolation. A few methods of polynomial interpolation were given by Isaac Newton. A comparison and accuracy analysis for polynomial and trigonometric interpolation are given in [20]. It is shown that implementation of five terms in an interpolation series gives a standard deviation error less than 0.4 cm and maximum error less than 70 cm.

We can assume that three terms give us approximately the same accuracy as broadcast ephemeris. The optimal number of seven members gives a standard deviation error less than 0.1 cm and maximum error less than 8.2 cm. That gives limits to the orbit accuracy achievable by pure geometric approximation of precise ephemeris. Using precise ephemeris with force models allows us to achieve millimeter level accuracy.

As we can see there are underlying trigonometric series in broadcast GPS ephemeris representation as well (see Table 2.1). Accuracy analysis also implies that broadcast GPS ephemerides are limited in accuracy due to a limited number of members in their series rather than anything else.

2.5.4 Calculation of Keplerian parameters from Cartesian coordinates

As described in previous sections, the osculating Keplerian parameters are calculated from the orbit Cartesian coordinates by a control segment. There is one-to-one correspondence between six Keplerian parameters and six Cartesian coordinates, which may be coordinates and their first derivatives, at the beginning, middle, or end of a segment, or coordinates at the ends of a segment. In the case of a control segment or for geodetic solutions there are many more data, so the Keplerian parameters are defined through a solution of an overdetermined problem, for example by using LSE.

We may use a similar algorithm in order to equip a GNSS signal simulator with an ephemeris propagator. Generally the standard high-end RF GNSS simulator simulates an orbit using almanac or ephemeris parameters propagating with equations similar to those given in the corresponding ICD. Only special simulators, developed for the special needs of generating geodetic grade observations, use Equation (2.22), which explicitly includes force models. In such a case orbits are generated by numerical integration. These orbit propagators can be used indefinitely. However, a GPS simulator needs to recalculate a set of ephemeris for every two hours in order to update the navigation message realistically in a way similar to a control segment.

2.6 Satellite clocks

The same interpolation technique as we have described in the previous section can be used to interpolate clocks. However, if we want to use predicted clocks, we may encounter some difficulties. We may want to use a predicted orbit and clocks for various real-time applications, in particular when a broadcast navigation message is unavailable. For navigation purposes clock values can be interpolated from predicted ephemerides. However, clocks are much more difficult to predict. Satellite orbits are very predictable, especially for GPS, for which the most unpredictable factor, solar radiation pressure, is still well modeled. Clocks have less stable parameters. Clock drift can change significantly over a short period.

A satellite with a less predictable clock may significantly degrade the quality of a position solution. In such cases, special modification of the integrity monitoring algorithm in a receiver may be able to exclude a clock with maximum drift.

Ephemeris information in broadcast navigation messages includes a satellite clock. With broadcast parameters clock error is calculated using the drift parameters. The clock error is calculated using

$$\Delta t_{S_i} = (k_{2_i} dt + k_{1_i}) dt + k_{0_i}, \tag{2.36}$$

where $k_{1_i}, k_{2_i}, k_{3_i}$ are clock corrections transmitted in a navigation message for the ith satellite, dt is the difference between the transmission epoch and reference time for which corrections were calculated.

2.7 Application of Keplerian almanac for constellation analysis

2.7.1 GNSS almanac implementation

Six Keplerian parameters are enough to describe an approximate position of a satellite. These parameters given for each satellite constitute a GNSS almanac. The almanac can be decoded from a satellite signal and downloaded from the Internet in the ASCII (text) format. The most common almanac format is the Yuma format, named after the GPS testing facilities in Yuma, USA. The Yuma format includes six Keplerian parameters, health status, and two clock error coefficients for each satellite. An example of an almanac is shown in Figure 2.8. The almanac parameters do not provide information required for

```
 1 ******** Week 611 almanac for PRN-02 ********
 2 ID:                         02
 3 Health:                     000
 4 Eccentricity:               0.1023864746E-001
 5 Time of Applicability(s):  319488.0000
 6 Orbital Inclination(rad):   0.9391288757
 7 Rate of Right Ascen(r/s):  -0.8018105291E-008
 8 SQRT(A)  (m 1/2):           5153.699219
 9 Right Ascen at Week(rad):  -0.6342332363E+000
10 Argument of Perigee(rad):  -3.002793193
11 Mean Anom(rad):             0.2672459364E+001
12 Af0(s):                     0.3356933594E-003
13 Af1(s/s):                   0.3637978807E-011
14 week:                        611
15
16 ******** Week 611 almanac for PRN-03 ********
17 ID:                         03
18 Health:                     000
19 Eccentricity:               0.1417636871E-001
20 Time of Applicability(s):  319488.0000
21 Orbital Inclination(rad):   0.9283962250
22 Rate of Right Ascen(r/s):  -0.8276401786E-008
23 SQRT(A)  (m 1/2):           5153.594727
24 Right Ascen at Week(rad):  -0.1792258859E+001
25 Argument of Perigee(rad):   1.086403370
26 Mean Anom(rad):             0.2760034442E+001
27 Af0(s):                     0.7133483887E-003
28 Af1(s/s):                   0.3637978807E-011
29 week:                        611
30
```

Figure 2.8 GPS almanac in Yuma format.

satellite position calculation in positioning algorithms, because the real orbits cannot be described without taking into account various perturbations, as described in the previous section.

However, these parameters are useful enough to be embedded into the satellite signal. The almanac can be used for the purposes of defining an approximate position of the satellite in order for a receiver to look only for satellites visible from the user location. It allows the user to omit other satellites from the search. This approximate satellite position can also be used to speed up a satellite signal acquisition in a receiver, because satellite velocity along the line-of-sight can be calculated in advance. When the navigation message is decoded from a satellite signal, a receiver can use almanac parameters in order to find an approximate position of the satellite at any epoch. For these purposes, an almanac can be used for weeks, so the receiver can use an almanac from the previous session to assist an acquisition. For many receivers, knowing an almanac significantly decreases time to first fix (TTFF), i.e. the time required for a receiver to make a position fix.

We can see that different approaches to satellite position calculation from an almanac in GLONASS and GPS are based on the system specifics. If for a GPS satellite position from an almanac is calculated based only on geometrical considerations, then for a GLONASS satellite position from an almanac calculation involves Earth gravitational field modeling. This difference has a basis in the corresponding almanac parameter calculations.

A control segment network for GPS is global, whereas for GLONASS it is more regional, so GPS data incorporate implicitly the force models in the data, whereas in the case of GLONASS these models are included explicitly to work with less data.

Almanac parameters can also be used in order to analyze satellite visibility, availability, and geometry. Therefore they can also be used for mission planning and constellation planning. Mission planning is important if for some reason the number of satellites is limited. In this case we can find the period of time at which more satellites are visible and DOP is smaller. Using an almanac we can also find out how a choice of constellation parameters may affect satellite visibility and possibly accuracy.

The accuracy of a position derived with an almanac rapidly degrades with time; however, one can use an almanac in many tasks, such as mission planning, for months.

2.7.2 Case study: types of orbit and navigation satellite constellation analysis based on an almanac

Keplerian parameters provide us with enough mathematical apparatus to describe the main features of a GNSS constellation in terms of satellite coverage and achievable accuracy due to the constellation geometry.

Despite a huge possible range of satellite orbits, for a satellite rotating near the Earth, all the orbits can actually be classified into four main groups. Their classification is done on the basis of their ***mean altitude***.

The orbits with mean altitude below 1500 km are called ***low Earth orbits (LEO)***. The orbits with an altitude around 20 000 km are called ***medium Earth orbits (MEO)***. The orbits with altitude of 36 000 km are on a ***geostationary Earth orbit (GEO)*** or **Clarke**

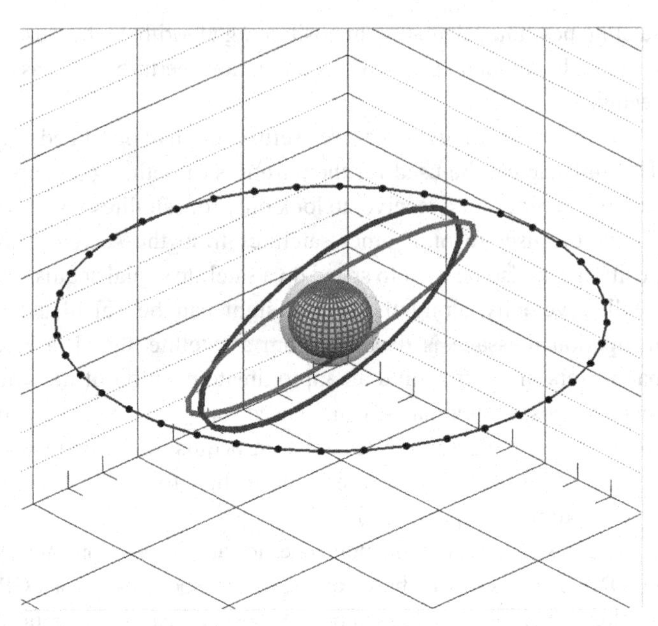

Figure 2.9 GPS, GLONASS, and GEO in ECI frame (screenshot of ReGen panel). (Tinted shell around the Earth depicts the ionosphere.)

orbit named after Sir Arthur Clarke, SF writer, who suggested them in 1945. GEO satellites are rotating synchronously with the Earth.

GNSS satellites are located usually on an MEO. GNSS also widely use GEO satellites to provide regional corrections and an integrity service as ***space-based augmentation systems (SBAS)***, which we cover in more detail in Chapter 8. Figure 2.9 shows orbits for GPS, GLONASS, and geostationary satellites in ECI. We can see that MEO and GEO satellites have low eccentricity and their orbits are rather close to circular orbits. The same is true for LEO satellites. GEO satellites are restricted to strictly defined slots and should undergo periodical orbit corrections to keep their position. LEO satellites are used together with GNSS satellites for geophysical measurements by methods of ionosphere probing with eclipsing satellites. LEO satellites are subjected to forces of air-drag, which are very difficult to account for. Therefore their ephemeris can be valid for significantly less time and their calculation is a much more difficult task.

Using the almanac we can analyze the main features of a particular orbit. Let us look at the satellite ground tracks, which should be created in Exercise 2.2. The revolution period of a GPS satellite is approximately 11 hours 58 minutes and a GLONASS satellite approximately 11 hours 16 minutes. As a result, a GPS satellite ground track repeats itself every sidereal day with about 4 minutes difference. Looking at the GPS and GLONASS orbits in an ECI frame we can see only a difference in the orbit size and inclination. Figure 2.10 shows a GPS orbit over 12 hours in an ECEF frame. The GPS orbit is closed. The GLONASS orbit in ECEF is open (Figure 2.11), the GLONASS recurrence cycle is about eight sidereal days. Within that period the GLONASS ground track does not repeat itself. GPS satellites are therefore periodically affected by Earth gravitational field

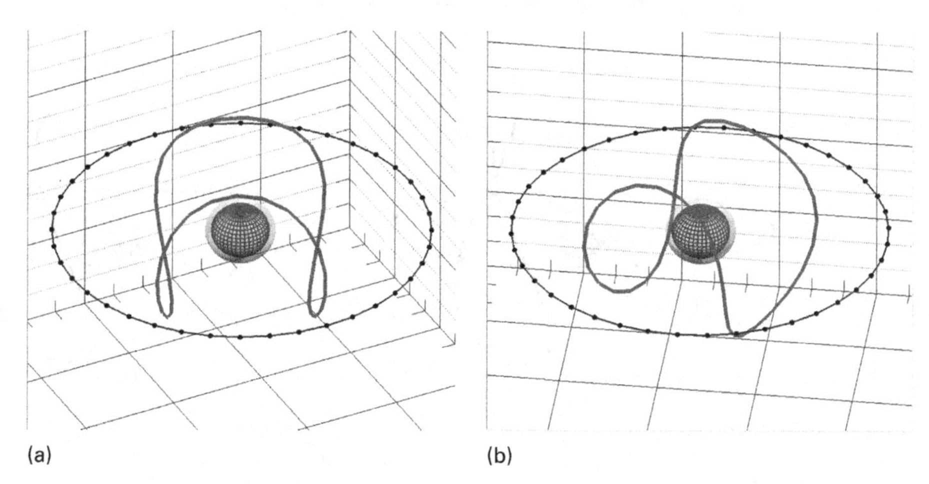

Figure 2.10 GPS and GEO in ECEF and ECI (screenshot of ReGen panel); (a) and (b) show the same 3D plot from different angles.

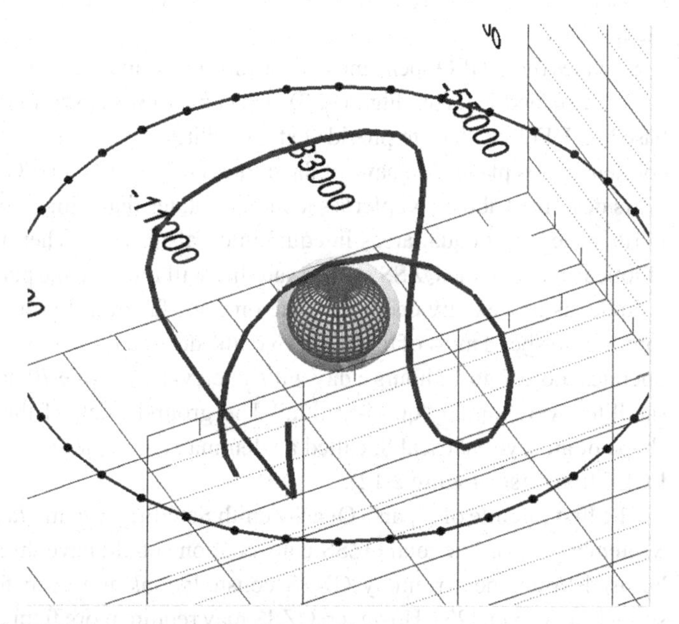

Figure 2.11 GLONASS in ECEF (screenshot of ReGen panel).

irregularities [21],[22]. The GLONASS constellation is much less affected by this factor. This does not significantly affect accuracy on the end-user side, because GPS ephemerides are estimated regularly and the network for GPS ephemerides estimation has good distribution. However, it affects the frequency of necessary correction maneuvers for GPS satellites.

There are also orbits with high eccentricity, called a highly eccentric orbit, or a ***highly elliptical orbit (HEO)***. Typical examples of HEO are ***Molniya*** type orbits, named after

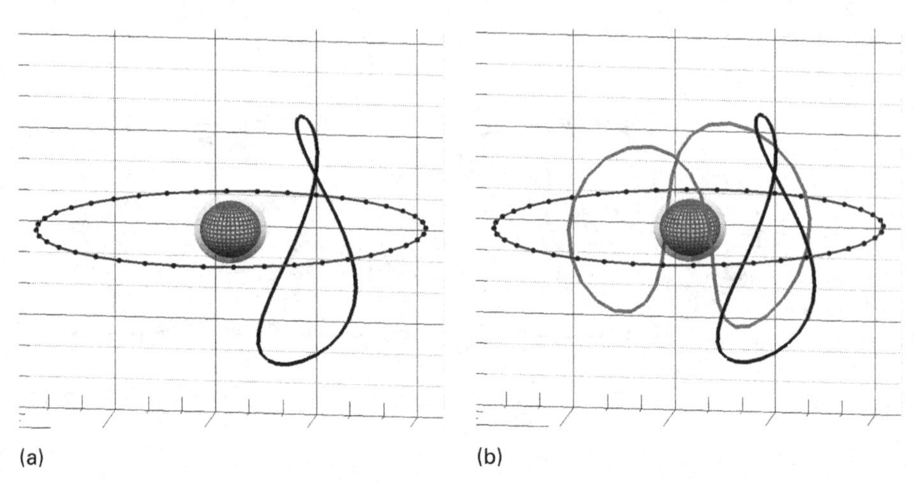

Figure 2.12 QZSS (a), QZSS and GPS (b), in ECEF (screenshot of ReGen panel).

Russian communication satellites, which were designed to cover mostly Siberian areas of the former USSR.

Satellites on an HEO spend most of the time in the apogee area. This comes directly from Kepler's second law (see Figure 2.3). Japan's ***Quasi Zenith Satellite System (QZSS)*** is based on HEO satellites to provide extra satellites for users in urban canyons. The QZSS constellation is planned to have of three satellites moving in HEO over the Asia region.

As described above, Kepler's second law states that a line joining a satellite and the Earth sweeps out equal areas in equal intervals of time. Therefore, due to the highly elliptical shape of the QZSS orbit, a satellite will linger in the part of the orbit with high altitude, as its velocity decreases when it goes far from Earth. This allows the QZSS satellites to spend most of their time over the desired region. A QZSS satellite typically operates more than 12 hours a day with an elevation above 70 degrees. This is why the satellites were called "quasi-zenith." The ground track of the QZSS satellite has a shape of an asymmetrical 8, called analemma. The same shape is clearly visible in an ECEF frame (see Figure 2.12).

The first satellite of Japan's Quasi-Zenith Satellite System *Michibiki* was launched in September 2010. The total QZSS constellation should have three satellites, which can be used to augment primary GNSS constellations and even for a basic positioning service on its own [23]. However, QZSS may require more frequent satellite maneuvers than, for example, GPS satellites, for station keeping and collision avoidance. This may have an impact on the accuracy of ephemeris and depends on how QZSS timekeeping is organized and on the two way satellite time and frequency transfer (TWSTFT) system [24].

The ReGen simulator can be used to visualize a GNSS constellation. Figure 2.13 shows GPS satellites in ECEF coordinates and satellite orbital deployment. For example, Figure 2.14 shows orbital deployment of Galileo satellites. We can see that GPS satellites are located on six orbital planes, whereas Galileo is planned to have three orbital planes.

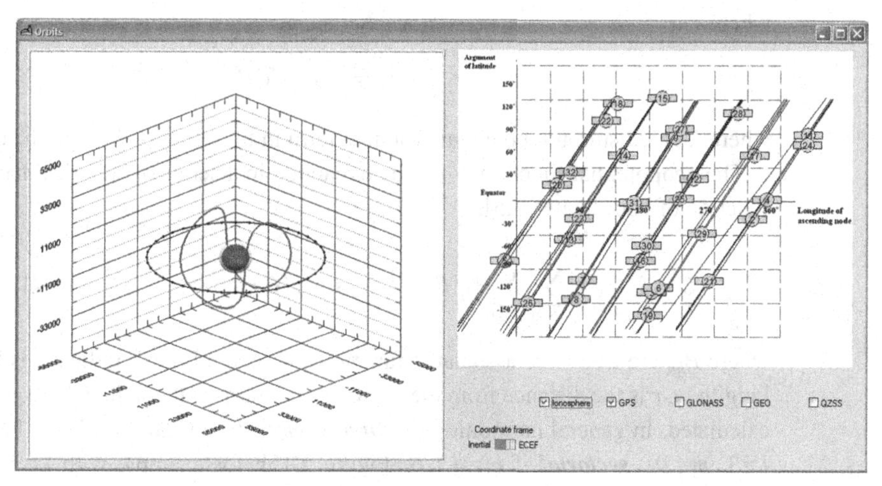

Figure 2.13 GPS orbits in ECEF and satellite orbital deployment (screenshot of ReGen panel).

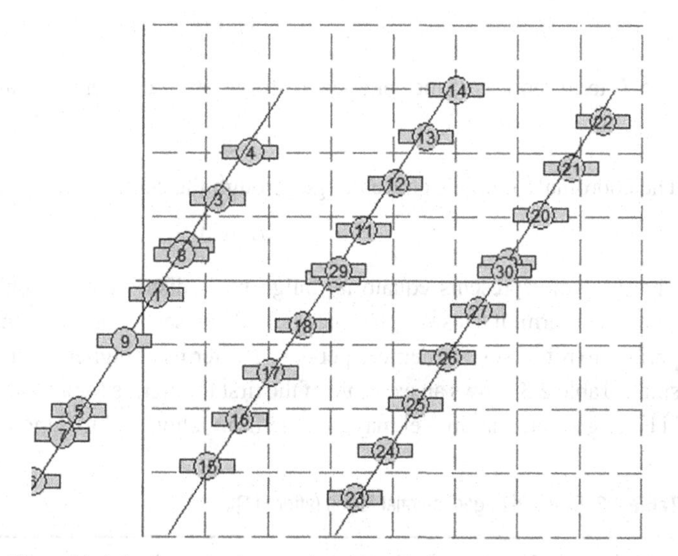

Figure 2.14 Galileo satellite orbital deployment (screenshot of ReGen panel).

2.8 Force models

2.8.1 Spherical harmonics of Earth geopotential

Acceleration in Newton's equations (2.4) can be expressed as a gradient of a potential energy of the orbit ∇U,

$$\vec{a} = -\nabla U, \tag{2.37}$$

where

$$\nabla U = \frac{\mu}{r^3}\vec{r} - \nabla R, \tag{2.38}$$

where R is a disturbing potential. The first term in the equation is from the central body.

The geopotential function U can be expanded in a series of spherical harmonics as a function of coordinates [25]:

$$U(r, \lambda, \varphi) = \sum_{l=0}^{\infty}\sum_{m=0}^{l} (r/R_e)^{l+1} P_{lm} \sin \varphi (C_{lm} \cos m\lambda + S_{lm} \sin m\lambda), \tag{2.39}$$

where P_{lm} are Legendre associated functions, R_e is the Earth's radius, λ is latitude, φ is longitude, r is the distance from the center of the Earth to the point where geopotential is calculated. In general harmonics are ***zonal*** if $m = 0$ (see Figure 2.15a for harmonic with $l = 3$, $m = 0$), ***sectorial*** if $m = l$ (see Figure 2.15b for example with $l = 2$, $m = 2$), and ***tesserial*** for $m \neq 0$, $m \neq l$ (see Figure 2.15c).

A geopotential model of the Earth is defined as a set of coefficients in (2.39) series expansion. The zonal coefficients are usually denoted as follows:

$$J_l \triangleq C_{l0}. \tag{2.40}$$

The J_0 term represents the spherical distribution of a potential changed by a *1*/*r* law,

$$J_0 \triangleq 1. \tag{2.41}$$

The coordinate frame is chosen to go through the center of mass, therefore

$$J_1 \triangleq 0. \tag{2.42}$$

The J_2 term represents equatorial bulge mass distribution or oblateness. It is a most significant term. It causes right ascension of an ascending node, and the argument of the perigee to rotate several degrees per day. The terms of higher order than J_2 become rather small. Table 2.3 shows an example of the first four terms for a *Goddard Earth Model 10B* [11]. A geopotential model may include from a hundred to a thousand terms.

Table 2.3 Example of geopotential terms (after [11]).

J_0	J_1	J_2	J_3	J_4
1	0	0.001 082 63	–0.000 002 54	–0.000 001 61

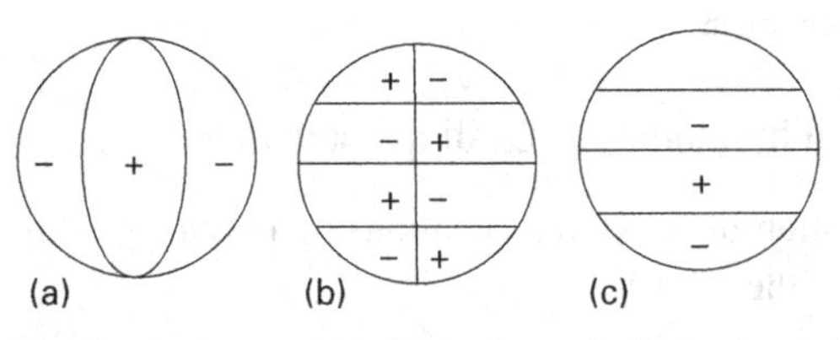

Figure 2.15 Spherical harmonics of the Earth geopotential. (a) $l = 3$, $m = 0$, (b) $l = 4$, $m = 1$, (c) $l = 2$, $m = 2$ (after [25],[26]).

Table 2.4 Contributions from various forces to satellite movement.

Force	Satellite acceleration caused by force (m/s^2)	Error in satellite position after one day, if the force is not accounted for (m).
Oblateness of the Earth	$5 \cdot 10^{-5}$	10 000
Lunar gravitation	$5 \cdot 10^{-6}$	3000
Sun gravitation	$2 \cdot 10^{-6}$	800
Other harmonics of the Earth gravitational field	$3 \cdot 10^{-7}$	200
Sun radiation pressure	$9 \cdot 10^{-8}$	200
Y-bias	$5 \cdot 10^{-10}$	2
Solid Earth tides	$1 \cdot 10^{-9}$	0.3

This spherical harmonics model can also account for elastic properties of the Earth. In that case coefficients should be expressed as functions of time. Solid Earth tides can be taken into account by considering the lowest terms. The coefficients in (2.39) are determined by analyzing tracking data for a large number of satellites.

2.8.2 Other forces in the force budget

The models for various forces in (2.22) which affect a satellite are well standardized, verified, and generally available. These models are generally not required for the user side of navigation applications, though they may be required in order to provide various data and services to a user, such as ephemerides. They also may be required on the user side to extend the validity of broadcast ephemeris by extrapolating them.

Table 2.4 summarizes contributions to satellite movement from various forces. It combines results from [14]–[16] and [27],[28]. The forces in the first and third rows of this table were considered in the previous section. An interested reader can learn more about these models from these references.

2.9 Where do all the satellites go? A satellite life cycle and space garbage hazards

Where do all the calculators go after they die? – asked Kryten from Red Dwarf. A philosophical question in his case has very serious implications in real life. GNSS satellites do not go anywhere. They all stay in space. This is a part of a much bigger and more general problem of satellite decommissioning and space garbage.

We consider what is happening to MEO decommissioning satellites by looking at a GPS satellite as an example. After a GPS satellite has been decommissioned, it is placed on a disposal orbit with altitude slightly higher than the operational orbit. The degraded GPS orbit becomes more eccentrically deformed with the years and may pose a threat first to other MEO satellites, and then even to GEO and LEO satellites [29]. The threat to

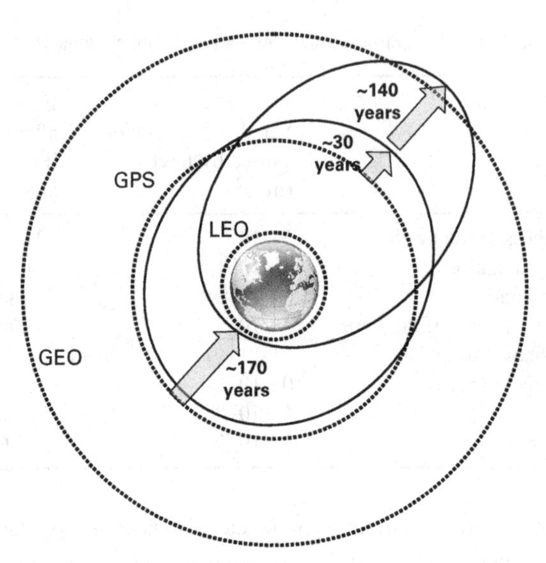

Figure 2.16 Degradation of GNSS orbits with time (after [29]).

other MEO satellites may become significant after a few decades, whereas the threat to GEO and LEO satellites arises after more than a hundred years (see Figure 2.16).

Initially decommissioned satellites are placed on an almost circular orbit. The source of its eccentric deformation is the same periodical resonance effects which we considered in the previous section.

GLONASS have more than a hundred decommissioned satellites in orbit. Though GLONASS orbits are less prone to resonance effects, they also become a collision threat in about 30 years. Galileo satellites may have even stronger resonance effects, a rising from Sun and Moon gravitational forces, due to their higher altitude.

This space junk problem, however, is not specific to GNSS satellites. Collisions between satellites are extremely dangerous because they produce a large number of pieces of debris. Each piece moves on its own trajectory and can cause a new collision, thus multiplying the number of scattered fragments, which in their turn can cause new collisions with higher probability [21]. Most of these scattered fragments of the collided satellites are continuously tracked by specially designated ground stations. However, there is a gap there. The size of the smallest fragment which can be tracked is larger than the size of the smallest fragment which can cause a problem on impact with another satellite or launch vehicle. This poses a problem, which should be addressed sooner rather than later in order not to jeopardize all human space activity.

References

[1] *Ptolemy's Almagest*, translated and annotated by G. J. Toomer, Princeton, NJ, Princeton University Press, 1998.

[2] S. Heilen, Ptolemy's doctrine of the terms and its reception, in *Ptolemy in Perspective*, A. Jones (editor), Springer Science + Business Media B.V., 2010.

[3] J. D. North, *Stars, Minds and Fate, Essays in Ancient and Medieval Cosmology*, London, The Hambledon Press, 1989.

[4] William of Ockham, *Philosophical Writings*, Cambridge, MA, Hackett Publishing Company, 1990.

[5] Isaac Newton, *Mathematical Principles of Natural Philosophy and His System of the World*, Berkeley, CA, University of California Press, 1934.

[6] Nicolaus Copernicus, *On the Revolutions of Heavenly Spheres*, Amherst, NY, Prometheus Books, 1995.

[7] I. Kepler, *New Astronomy*, Cambridge, Cambridge University Press, 1992.

[8] A. Einstein and L. Infeld, *The Evolution of Physics*, New York, Simon and Schuster, 1938.

[9] Nicholas Cusanus, *Of Learned Ignorance*, G. Heron (Translator), D. J. B. Hawkins (Introduction), Eugene, OR, Wipf & Stock Publishers, 2007.

[10] GPS IS, *Navstar GPS Space Segment/Navigation User Interfaces*, GPS Interface Specification IS-GPS-200, Rev D, GPS Joint Program Office, and ARINC Engineering Services, March, 2006.

[11] J. R. Wertz, *Mission Geometry: Orbit and Constellation Design and Management*, El Segundo, CA, Microcosm Press, and Dordrecht, The Netherlands, Kluwer Academic Publishers, 2001.

[12] J. J. Spilker Jr., *Satellite Constellation and Geometric Dilution of Precision*, in *Global Positioning System: Theory, and Applications*, Vol. I, B. W. Parkinson and J. J. Spilker (editors), Washington, DC: American Institute of Aeronautics and Astronautics Inc., 1996.

[13] D. D. McCarthy: IERS Conventions 2000, Central Bureau of IERS, Observatoire de Paris, IERS Technical Note, 32, Paris, 2004.

[14] G. Beutler, *Methods of Celestial Mechanics,* Vol. I*: Physical, Mathematical, and Numerical Principles*, Berlin/Heidelberg, Springer-Verlag, 2005.

[15] G. Beutler, *Methods of Celestial Mechanics,* Vol. II*: Application to Planetary System, Geodynamics and Satellite Geodesy*, Berlin/Heidelberg, Springer-Verlag, 2005.

[16] R. Dach, U. Hugentobler, P. Fridez, and M. Meindl (editors), *User Manual of the Bernese GPS Software Version 5.0*, Bern, Astronomical Institute, University of Bern, 2007.

[17] *Global Navigation Satellite System GLONASS, Interface Control Document, Navigational Radiosignal in Bands L1, L2*, Edition 5.1, Moscow, Russian Institute of Space Device Engineering, 2008.

[18] J. M. Dow, R. E. Neilan, and G. Gendt, The International GPS Service (IGS): Celebrating the 10th anniversary and looking to the next decade, *Adv. Space Res.*, **36**, (3), 2005, 320–326, doi:10.1016/j.asr.2005.05.125.

[19] Cornelius Lanczos, *Applied Analysis*, Englewood Cliffs, NJ, Prentice-Hall, Inc., 1956; republication by Mineola, NJ, Dover, 1988.

[20] M. Schenewerk, *A Brief Review of Basic GPS Orbit Interpolation Strategies*, GPS Solutions 6:265–267, Berlin/Heidelberg, Springer-Verlag, 2003.

[21] U. Hugentobler, *Astrometry and Satellite Orbits: Theoretical Considerations and Typical Applications*. Volume 57 of Geodätisch-geophysikalische Arbeiten in der Schweiz, Schweizerische Geodätische Kommission, Institut für Geodäsie und Photogrammetrie, Zürich, Switzerland, Eidg. Technische Hochschule Zürich, 1998.

[22] D. Ineichen, G. Beutler, and U. Hugentobler, Sensitivity of GPS and GLONASS orbits with respect to resonant geopotential parameters, *Journal of Geodesy*, **77**, 2003, 478–486, Springer-Verlag.

[23] I. Petrovski, *et al.*, QZSS – Japan's new integrated communication and positioning service for mobile users, *GPS World*, **14**, (6), 2006, 24–29.

[24] I. Petrovski and U. Hugentobler, Analysis of impact of onboard time scale instrumental error accuracy of Earth satellite ephemerides estimation with one-way measurements, In Proceedings of the European Navigation Conference 2006, London, Royal Institute of Navigation, 2006.
[25] W. M. Kaula, *Theory of Satellite Geodesy – Applications of Satellites to Geodesy*, Waltham, MA, Blaisdell Publishing Company, 1966.
[26] *GPS for Geodesy*, P. Teunissen and A. Kleusberg (editors), 2nd edition, Berlin/Heidelberg, Springer, 1998.
[27] O. Montenbruck and E. Gill, *Satellite Orbits, Models, Methods, Applications*, Berlin/Heidelberg, Springer-Verlag, 2000.
[28] M. Capderou, *Satellites, Orbits and Missions*, France, Springer-Verlag, 2005.
[29] Collision prevention for GPS, *Crosslink*, The Aerospace Corporation Magazine, Summer, 2002, **3**, (N2), 2002, The Aerospace Corporation, LA, USA.

Exercises

Exercise 2.1. This is a thought experiment. Based on the information in this chapter, consider how an astrolabe can be changed in order to be useful for getting information from GNSS satellites. We need of course, actually to be able to see satellites in order to use an astrolabe. This is feasible today. Most GNSS satellites are equipped with laser reflectors for two-way measurements by laser tracking stations. Therefore we could make a system which would combine lasers with an astrolabe, and this system is fairly easy to make if we already have all the components, such as satellites and a laser tracking station. What conditions should be met? How will a new astrolabe look? What information will be available? How long can an astrolabe serve?

Exercise 2.2. Draw GPS and GLONASS ground tracks using ReGen signal simulator software. Estimate an approximate sidereal period from a ground track. Ground tracks for GPS and Galileo satellites are shown in Figure 1.1.

Exercise 2.3. Use ephemeris in SP3 format from IGS for predicted and precise orbits to estimate the accuracy of predicted clocks versus a true clock. We assume that clocks from precise ephemeris can be used as a true clock. See if the satellites demonstrate different accuracy of clock prediction.

Exercise 2.4. Using ReGen analyze the difference between GPS and GLONASS visibility in high latitudes, using GPS and GLONASS almanacs. GLONASS inclination is about 64.8° and GPS inclination is about 55°. Find out whether the difference in coverage is significant.

Exercise 2.5. Plot orbital deployment for GLONASS satellites. How many orbital planes does GLONASS use?

3 GNSS signal generation in transmitters and simulators

In this chapter we examine how GNSS signals are created. We mostly concentrate our presentation on the signals of currently operating GPS and GLONASS. We also consider the main specific features of future GNSS signals including modernized GPS, modernized GLONASS, and GALILEO. We also describe in this chapter how GNSS signals are generated in simulators and how those signals differ from the signals generated by satellite transmitters and pseudolites. The particular place which this chapter occupies in the book is schematically depicted on Figure 3.1. In this book we describe GNSS using some help from a GNSS signal simulator. GNSS simulation today has become an important and significant part of the GNSS technology, which enjoys a high demand. Further, any presentation is based on models. These models can be mathematical, empirical, or speculative. A simulator-based model provides one with the best combination of those models, being as close to the real system from a receiver point of view as possible.

3.1 Spread-spectrum radio signals for satellite navigation

3.1.1 Spread-spectrum concept and benefits

The spread-spectrum concept lies at the core of all GNSS signals. As described in the previous chapter, GNSS satellites are located on medium Earth orbits (MEO) at a distance of about 20 000 km from the user receiver. Even spaceborne receivers located on low Earth orbit (LEO) spacecraft are almost at the same distance to GNSS satellites as a user on the surface of the Earth. Signal energy decreases inversely as the square of distance between signal source and a receiver. We have already encountered the inverse square law in relation to gravitational forces. The signal power as well as gravitational force follows an inverse square law because energy is equally spread on the spherical surface at the receiver distance. Satellite payload should be minimized in order to make its delivery to an orbit cheaper, and the transmission energy also should be minimized in order for a transmitter to consume less energy. As a result of these constraints, the satellite signal is not very powerful, and when it reaches the Earth it is well below noise level. The GNSS signal power levels are given in Table 3.1 based on GPS ICD for L1/L2/L2C signals [1], GPS L5 ICD [2], GPS L1C ICD [3], GLONASS ICD [4] , GALILEO ICD [5], and QZSS ICD [6].

Table 3.1 Minimum power level of GNSS signals.

System	Signal	Minimum power level (dBW)
GPS	L1 C/A	−158.5
	L2 C (satellite type II/IIA/IIR)	−164.5
	L2 C (satellite type IIR-M/IIF)	−160.0
	L5 I and Q	−157.9
GLONASS	L1 and L2 (all satellite types)	−161
GALILEO	E1(L1)	−157
	All other than E1(L1) frequencies	−155

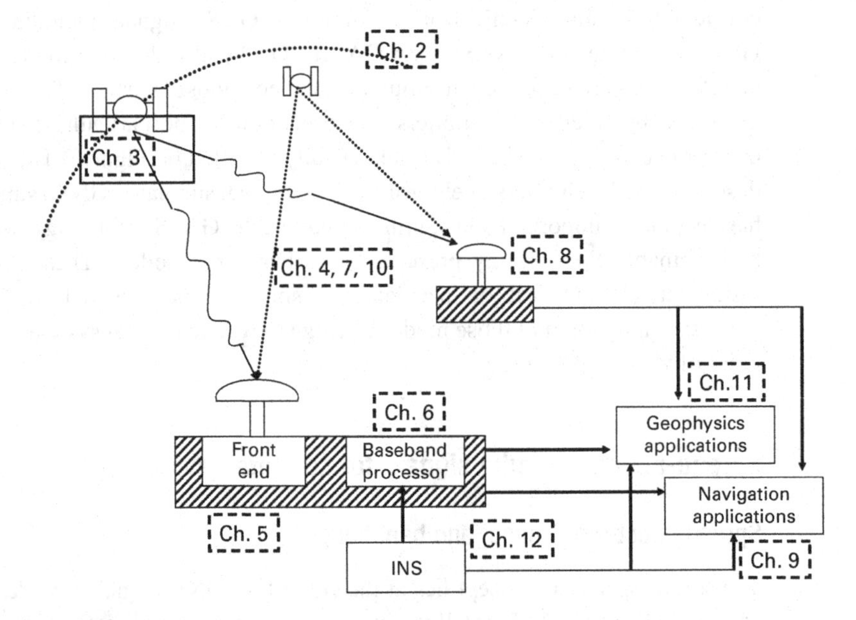

Figure 3.1 Subject of Chapter 3.

Due to such low signal power, satellite navigation systems have to apply spread-spectrum signals. In the case of spread-spectrum signals, the carrier phase of the signal is modulated by a spread code. In general it allows a very low power signal to be detected due to the fact that a receiver knows the signal pattern. However, that is not the only advantage of the spread spectrum. Spread-spectrum technology allows us to achieve several goals simultaneously:

1. It allows a receiver to acquire and track a low power signal, which is under the noise level.
2. The spread code allows a receiver to derive signal delay information, which in the case of satellite navigation systems translates into a distance to a satellite.
3. For satellite systems, in which a number of satellites transmit on the same frequency (such as GPS), it allows us to distinguish between signals from different satellites

which are using the same carrier frequency. Because of this feature, spread-spectrum signals which facilitate this function are in general also called code division multiple access (CDMA) signals. The CDMA principle is used not only in satellite navigation, but in communication, for example in cellular phones.
4. It allows us to derive useful geophysical information, for example about an atmosphere using the difference in code and carrier propagation delays.

3.1.2 Central frequencies of GNSS signals

GPS, GLONASS, and GALILEO are transmitting signals on the central frequencies listed in Table 3.2.

Figure 3.2 shows the central carrier frequencies and bands allocated to GPS and GLONASS signals.

A GNSS signal carrier can be expressed as follows:

$$A(t) = A_0 \sin(\omega t + \varphi). \tag{3.1}$$

The signal spectrum represents a distribution of signal energy as a function of frequency. The spectrum of the sine wave is a line at frequency $f = \frac{\omega}{2\pi}$.

GPS signals are generated using a common satellite clock and baseline clock frequency. The clock frequency is 10.23 MHz. The GPS frequencies are as follows:

$$\begin{aligned} \mathrm{L1} &= 1575.42 = 154 \times 10.23, \\ \mathrm{L2} &= 1227.6 = 120 \times 10.23, \\ \mathrm{L5} &= 1176.45 = 115 \times 10.23. \end{aligned} \tag{3.2}$$

Table 3.2 Central frequencies of GNSS signals.

System	Signal	Central frequency (MHz)	Currently available for civilian user
GPS	L1	1575.42 with 20.46 band	C/A
	L2	1227.6 with 20.46 band	L2C*
	L5	1176.45 with 24 band	
GLONASS	L1	1602 (from 1598.0625 to 1605.375)	SP
	L2	1246 (from 1242.9375 to 1248.625)	SP
	L3	1202.025	CDMA**
QZSS	L1, L2, L5	same as GPS	C/A, L1C, L2C, L5***
GALILEO	E1(L1)	1575.420	
	E5	1191.795	
	E5a	1176.450	
	E5b	1207.140)	
	E6	1278.750	

* The signal is transmitted from some of the satellites.
** The signal is transmitted from GLONASS-K satellites.
*** Transmitted in test mode.

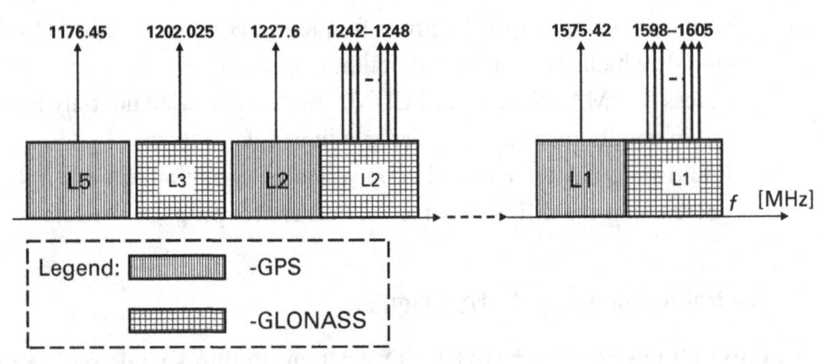

Figure 3.2 GPS and GLONASS current frequency allocation.

Though the GPS L2 signal is not intended for the civilian community and in general cannot be used for real-time stand-alone positioning by an unauthorized user, it has become essential for all geodetic applications today. Geodetic receivers are able to get measurements on both frequencies, though measurements on L2 are more noisy due to the fact that they are not currently designed for civilian applications. The L2C GPS signal is intended to provide civilian users with dual frequency capabilities to eliminate ionospheric error for most stand-alone positioning applications. The L2C signal is already transmitted from some GPS satellites. The L1C and L5 signals are currently transmitted by the QZSS satellite in test mode. Further GPS modernization will provide more demanding civil applications, such as aviation, with the L5 signal.

The recovery of a spread-spectrum signal in a receiver has a mechanical analogy in a resonance. When a receiver multiplies the incoming signal repeatedly with the same spread code, the result of such multiplication is much higher when there is no shift between the code in the incoming signal and the replica code. As a result of this resonance type recovery the receiver can distinguish between signals from different satellites which are using the same carrier. That is why GPS and GALILEO are sometimes related to CDMA systems.

This feature of the spread code is also starting to be utilized in GLONASS. Currently GLONASS transmit civilian signals on L1 and L2 frequencies. All GLONASS satellites are distinguished not by a different spreading code but by a shift in frequencies. The GLONASS frequencies are defined for L1 as follows:

$$L1_k = L1_0 + k\Delta L_1. \tag{3.3}$$

where $L1_0 = 1602$ MHz, $\Delta L_1 = 562.5$ kHz. For L2, GLONASS frequencies are defined as follows:

$$L2_k = L2_0 + k\Delta L_2. \tag{3.4}$$

where $L2_0 = 1246$ MHz, $\Delta L_2 = 437.5$ kHz. GLONASS utilizes a single spread code for all satellites. This feature sometimes leads to GLONASS being described as a FDMA (frequency division multiple access) system. This is, however, not exactly correct, because GLONASS essentially uses a spread code for two other reasons: to be able to work with low power signals, and to recover a pseudorange to a satellite. The spread code is an essential component of CDMA, but is not required for FDMA. Therefore GLONASS can rather be described as FDMA on top of CDMA. Initially FDMA was envisioned to provide more resistance against interference. A single carrier system is less resisant to intentional or unintentional interference than a system in which carriers are spread over some range. In new GPS signals and planned GALILEO and GLONASS signals, similar resistance should be achieved by modulating the signal with a subcarrier. GLONASS is beginning to implement signals on the single frequency with CDMA in L3, which is transmitted from GLONASS-K satellites. The first GLONASS-K satellite started to transmit CDMA on L3 in 2011.

3.1.3 Generating GPS signals

An ideal spread code would be an infinite sequence of independent, identically distributed, random binary variables taking one of two values (0 or 1) with equal probability. Such a random code would not repeat itself, which would make such code difficult to implement in a way that it can be generated both in a transmitter and a receiver in an economic implementation. It is possible though to implement not random but chaotic codes using nonlinear systems and chaos theory. These chaotic codes would have the same properties as random codes and have infinite length.

Currently, however, a more deterministic approach has been implemented in GNSS. The approximations to binary random sequences are generated by linear feedback shift registers (see Figure 3.3). Mathematically such a shift register can be represented as follows:

$$y(n) = \sum_{k=1}^{m} h_k y(n-k), \tag{3.5}$$

where m is the number of adders, h_k can be either 1 or 0, where 0 would indicate an absence of the corresponding adder in the shift register. Given the initial state of the shift register, a $[y_0(n-1), y_0(n-1), \ldots y_0(n-m)]$ sequentially pseudorandom noise (PRN) sequence can be generated.

Because of its deterministic nature, the sequence will not be random and at some point the shift register will start to repeat it. A shift register can generate sequences with period no greater than $N = 2m - 1$. The sequences with such a maximum period are called maximal length sequences (m-sequences).

These codes have a specific feature that they are almost orthogonal to each other [7]. That means they are almost not correlated between each other, the feature which we

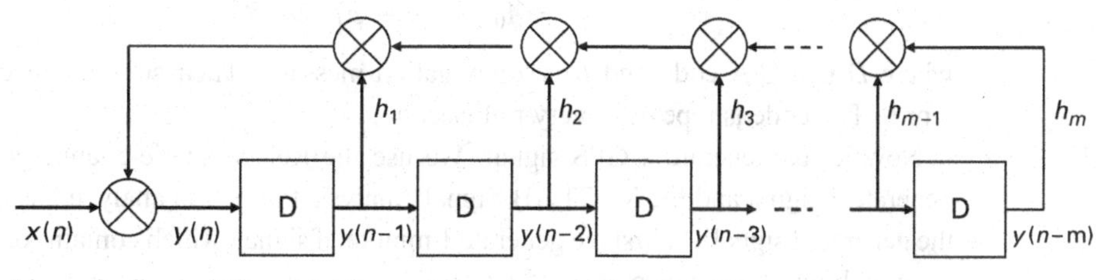

Figure 3.3 Linear feedback shift register for m-sequence generation.

would expect from a really random code. The orthogonal codes have an autocorrelation function with a single correlation peak. The autocorrelation function can be found as the result of multiplication of the sequence on its own shifted version:

$$R(n) = \frac{1}{N}\sum_{k=0}^{N-1} y(k)y(n+k), \tag{3.6}$$

where N is the sequence period. Here we should use a sequence of $\{-1,1\}$ random values, rather then $\{0,1\}$. The autocorrelation function then has a peak corresponding to 0 lag, where n is a multiple of N. When the lag is not 0, the autocorrelation function has $-1/N$ value.

The shift register functionality (3.5) is usually represented in the form of a polynomial as follows:

$$G(X) = \sum_{k=0}^{m} X^k. \tag{3.7}$$

The GPS C/A (coarse/acquisition) signal for civilian users on the L1 frequency implements codes taken from the Gold Codes family [8]. Gold Codes PRN have a number of specific features. They are short with a maximum length of 1023 chips. They are generated at a rate of 1023 chips per millisecond, using a 10.23 MHz onboard satellite clock.

GPS C/A codes are described as a product of two polynomials:

$$\begin{aligned} G1(X) &= 1 + X^3 + X^{10}, \\ G2(X) &= 1 + X^2 + X^3 + X^6 + X^8 + X^9 + X^{10}. \end{aligned} \tag{3.8}$$

The initial states of both polynomials define the code for a specific satellite. The resulting Gold codes are non-orthogonal. As a result their autocorrelation functions are not minimal for non-zero lag as for orthogonal codes. The Gold Codes, on the other hand, guarantee uniformly low cross-correlation properties within their own family. This function is not essential for GLONASS civilian signals on L1, L2 frequencies, which implement an m-sequence, because these signals currently use only one code sequence for all satellites. Figure 3.4 depicts autocorrelation and cross-correlation functions for GPS Gold codes PRN = 2 and PRN = 3 and a GLONASS m-sequence. The shift of the g2 code for PRN = 2 and PRN = 3 is 6 and 7 respectively.

Now we can simulate a GPS L1 C/A signal. If we use the ReGen single-channel simulator (Figure 3.5), the GUI allows one to create a signal out of components, such as carrier, code, navigation message. The GPS signal at each epoch is simulated as an L1 carrier wave modulated by a code and navigation message,

$$A = A_0 \sin(\omega t + \varphi) \cdot D \cdot B, \tag{3.9}$$

where D is a C/A code and B is a navigation message. Their summation gives either 1 or 0. The code is repeated every millisecond.

Now let us generate a GPS signal. We use the ReGen single-channel simulator to generate a signal and the MATLAB Signal Analysis toolbox to analyze the spectrum of the generated signal. At first we generate 1 minute of signal, which contains only a spread code with minimum noise.

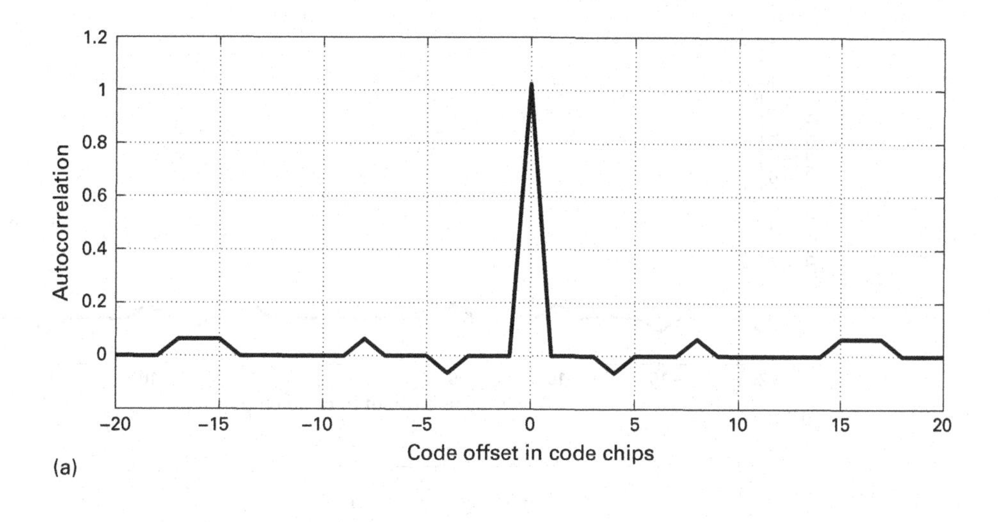

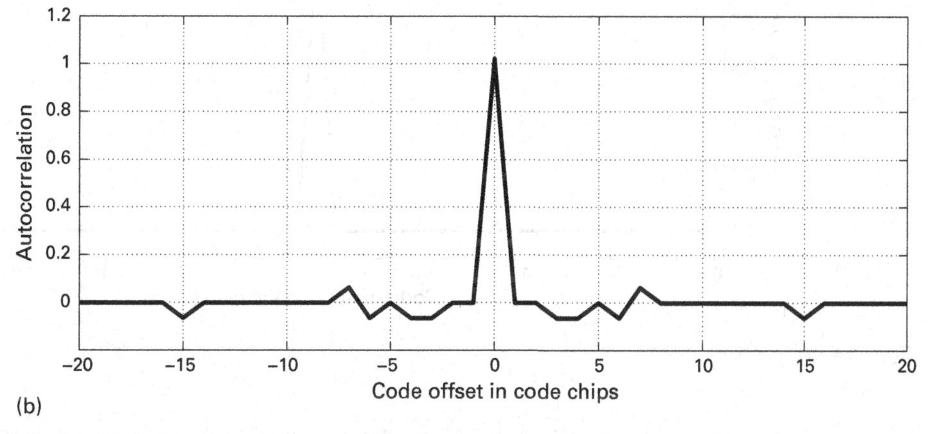

Figure 3.4 Normalized correlation functions: (a) Autocorrelation function for GPS Gold code PRN= 2, (b) Autocorrelation function for GPS Gold code PRN= 3, (c) Cross-correlation function for GPS Gold code PRN= 2 and PRN = 3, (d) Autocorrelation function GLONASS m-sequence.

The instantaneous spectrum of this signal is presented in Figure 3.6. It can be viewed as a development of the square wave spectrum sin(x)/x. If we sum up spread code with a carrier, a spectrum of the result signal is a result of convolution of the pulse and carrier spectra (Figure 3.7). Therefore, when a spread code is applied, it spreads spectrum lines to a wider area, which is defined by the chip length of the code. A code is mixed with a GPS navigation message which is issued with a rate of 50 bits per second and repeats itself every 12.5 minutes.

Now we can add a carrier to the signal. The signal now corresponds to (3.9) and its spectrum is presented in Figure 3.8. We are now generating a ***digitized intermediate frequency (DIF)*** signal, as viewed at the output of a receiver front end. The L1 radio frequency (RF) signal on the front-end input, which came from an antenna, is down-converted to ***intermediate frequency (IF)***. Therefore the carrier moves the code spectrum

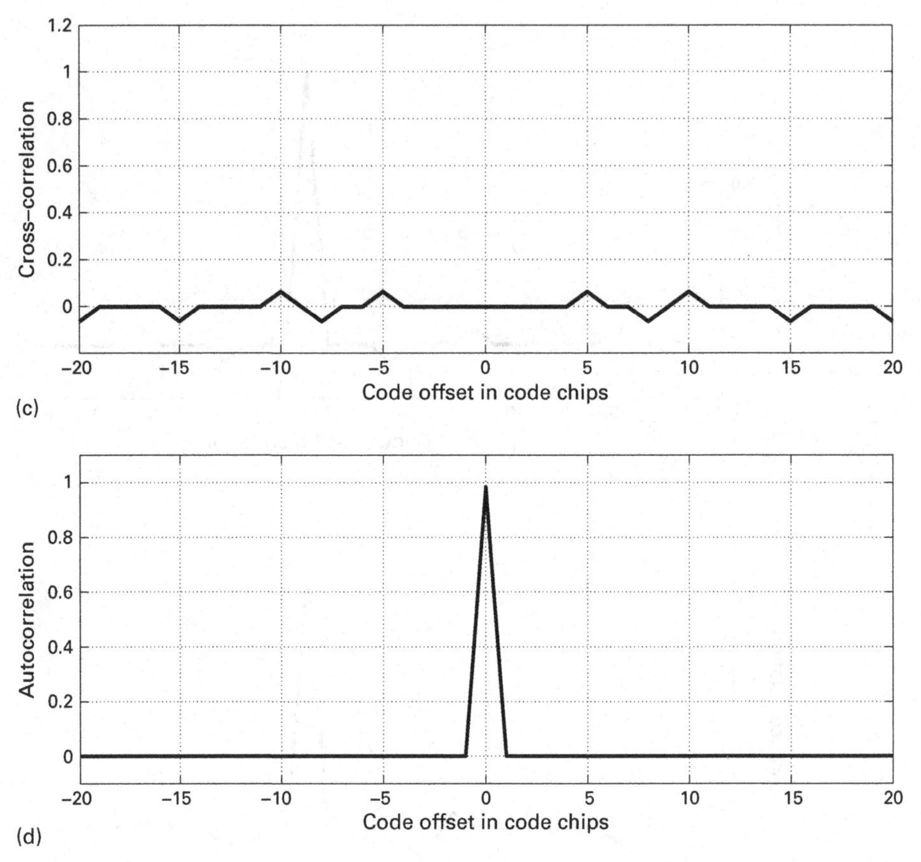

Figure 3.4 (cont).

to the central IF, which is chosen now to be about 4 MHz. Parameters such as IF, sampling rate, and others can be specified from the GUI (Figure 3.5). By simulating a Doppler shift, we can see that it is not changed during the down-conversion process.

Next we generate a signal taking into account a front end as a dynamic system, simulated by a passband filter. The spectrum of such a signal is presented in Figure 3.9. We simulate a narrow band front end and all lobes except the main lobe are suppressed.

Finally we add a noise to the signal (see Figure 3.10). In order to compare the results with those from the real front end, we simulate a GPS signal with off-the-shelf high-end Spirent simulators (Figure 3.11), which are default standards for the industry. We have also used Spirent simulators along with a ReGen simulator to verify our software receiver design. A spectrum from a generated signal, simulated by a Spirent RF GNSS simulator is shown in Figure 3.12. Using the RF signal from a simulator rather than from a real antenna, we can ensure that the signal is not corrupted with multipath.

The type of signal generation described above is called binary phase shift keying (BPSK). This signal is summed up with a restricted access P code,

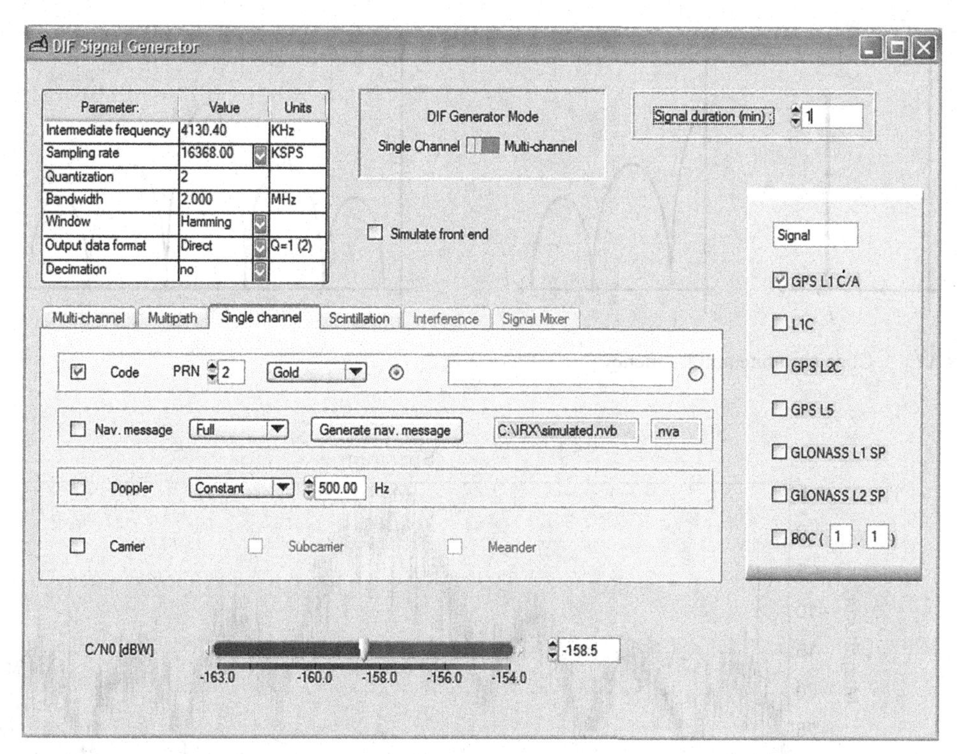

Figure 3.5 ReGen single-channel simulator.

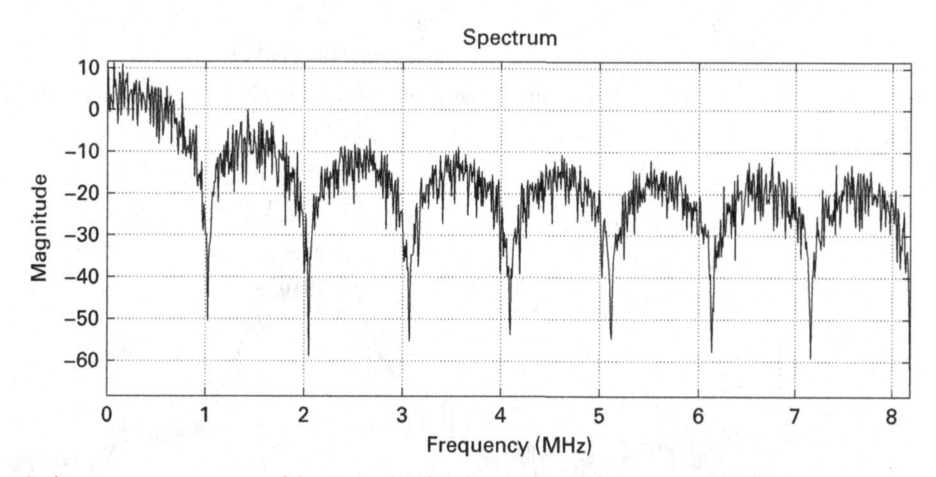

Figure 3.6 An instantaneous spectrum of the GPS code signal generated with a single-channel ReGen simulator.

$$A_P = A_{P0} \sin(\omega t + \varphi + \frac{\pi}{2}) \cdot D_P \cdot B_P, \tag{3.10}$$

where D_P is a P code and B_P is a navigation message embedded into the P code. The carrier for the P code has the same L1 frequency, but shifted by $\pi/2$ relative to the C/A carrier. Two quadrature carrier components then form what is called a quadrature phase

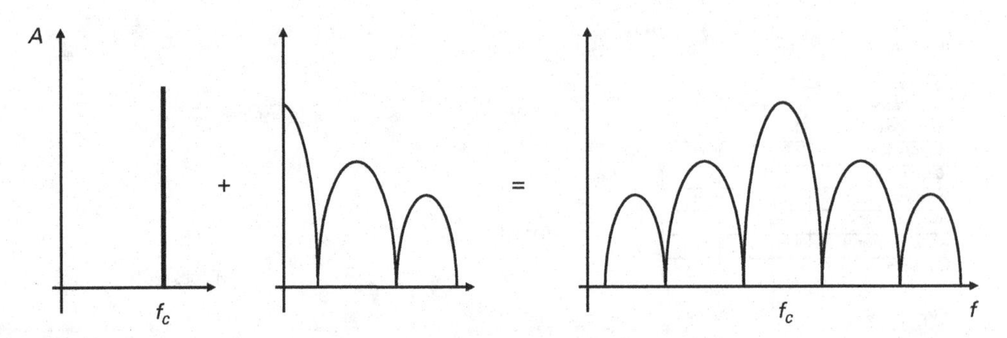

Figure 3.7 Code chip rate and frequency.

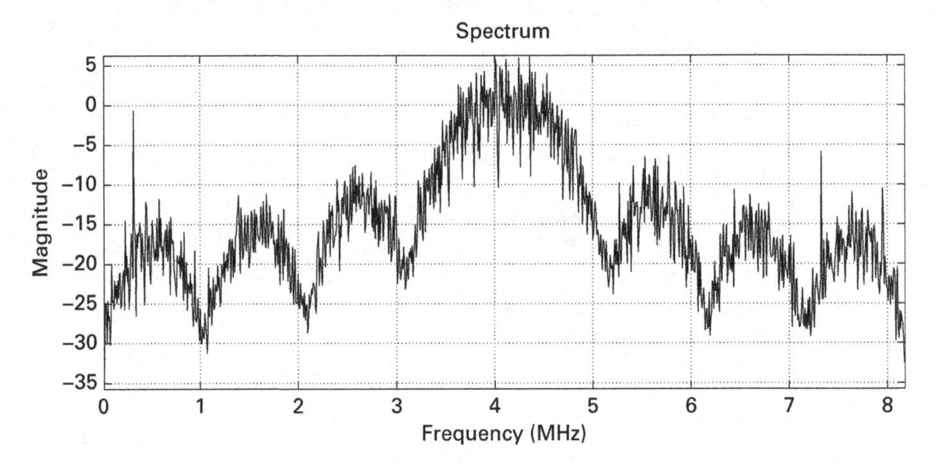

Figure 3.8 A spectrum of the "code+carrier" signal generated with a single-channel ReGen simulator.

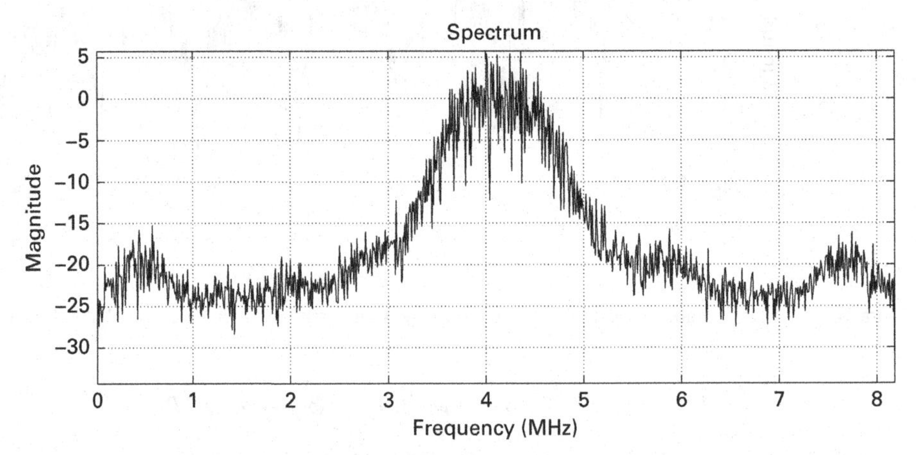

Figure 3.9 A spectrum of the "code + carrier + front end" signal simulated with a single-channel ReGen simulator.

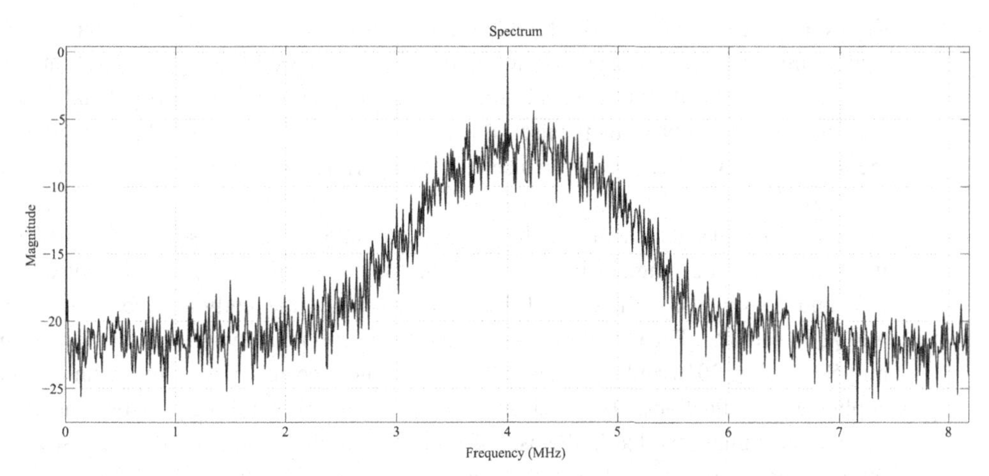

Figure 3.10 A spectrum of the "code + carrier + front end + noise" signal simulated with single-channel ReGen simulator.

Figure 3.11 Spirent GSS6300 Multi-GNSS generator and GSS6700 Multi-GNSS simulator generating a signal for iP-Solutions GNSS receivers.

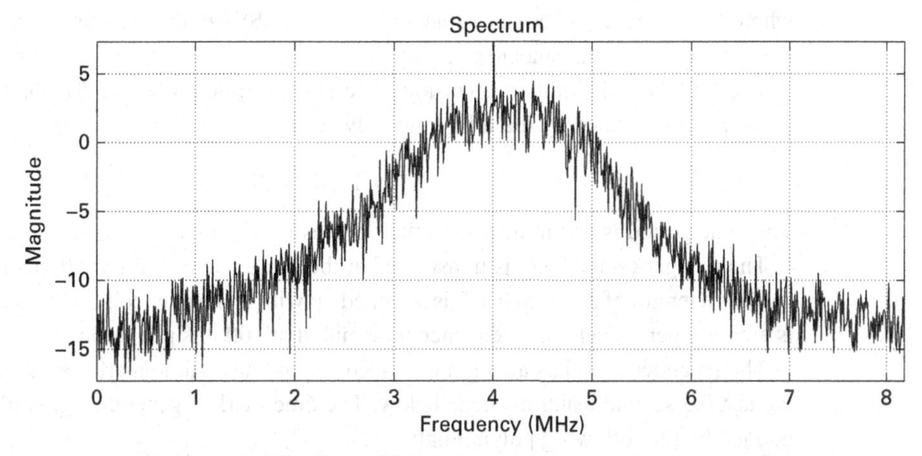

Figure 3.12 A spectrum of the recorded live signal (iP-Solutions recorder with Rakon module).

shift keyed signal (QPSK). The P code is also transmitted on L2 frequency, but P code signals are not generally accessible for civil users in the case of navigation applications.

To have signals on two frequencies from the same satellite would be beneficial for the user, because it would allow the user to eliminate errors related to signal propagation in the ionosphere. Geodetic receivers use special algorithms to partly retrieve information from P-codes.

New GPS civil signals will include L1C, L2C, and L5. A signal description is available from ICDs. The L2C signals have been broadcast from some of the GPS satellites. The L1C and L5 signals, at the time of the writing, were permanently available only from the QZSS satellite. The QZSS satellite began to transmit these signals in test mode from mid-September 2010 and in regular mode from mid-December 2010 for technical and application verifications. QZSS currently has one satellite in the constellation. Therefore transmitted signals can hardly be used for positioning. It is possible to use one satellite for positioning if the satellite timeframe can be connected to the timeframe of the constellation, which provides other satellites with sufficient accuracy.

3.1.4 Generating GLONASS open access signals

Ideally the GLONASS signal spectrum should be represented by a set of non-overlapped narrow beams for each frequency. The frequencies in fact overlap, which introduces some signal degradation on the level of 54 dB. Similarly, the GPS L1 C/A Gold code has signal degradation in comparison with pure orthogonal code sequences on the level of 21.6 dB. On L1 and L2, GLONASS currently use only one m-sequence for all satellites. A receiver distinguishes between signals from different GLONASS satellites by frequency.

GLONASS currently transmits the same open standard precision (SP) code on L1 and L2 frequencies. The code is generated in a similar way to the GPS C/A code described by (3.9), with the difference that there is an extra binary sequence used as a time mark. The GLONASS signal is generated as follows:

$$A = A_0 \sin(\omega t + \varphi) \cdot D \cdot B \cdot M, \tag{3.11}$$

where D is an m-sequence (the same for all GLONASS satellites) code, B is a navigation message and M is a meander sequence.

The GLONASS standard precision code is transmitted at a rate 511 bits per millisecond and generated by the following polynomial:

$$\mathrm{G(X)} = 1 + \mathrm{X}^5 + \mathrm{X}^9. \tag{3.12}$$

The code period is one millisecond, the same as for GPS L1 C/A.

The navigation message is transmitted at the same rate as GPS at 50 bits per second and has a length of 2.5 minutes. It is summed up with a meander code. The meander code is a binary periodical {0,1} sequence transmitted at 100 bits per second.

The meander code has a length of 2 seconds and has a time mark transmitted during the last 0.3 second using the code below. The time mark is generated by shift a register defined by the following polynomial:

$$G(X) = 1 + X^3 + X^5, \tag{3.13}$$

and cut down to 30 symbols. The length of each symbol is 10 milliseconds.

GLONASS has also started to transmit a civilian CDMA signal on L3. The signal uses a Kasami sequence and is described in detail in [9].

3.1.5 GPS and GLONASS navigation messages

All essential information related to satellite ephemerides, satellite clocks, and message time mark are encoded in a binary message, denoted as B in Equations (3.9) and (3.11). The GPS navigation message contains ionospheric parameters to compensate for variable signal delay in the ionosphere. The GLONASS navigation message does not contain ionospheric information.

Both GPS and GLONASS transmit an L1 civil signal navigation message with bit rate 50 bits per second. The complete navigation message for GPS is transmitted within 12.5 minutes, and for GLONASS within 2.5 minutes. Any part of the GPS L1 navigation message with length of 36 seconds contains enough information for a receiver to make a positioning. A GPS L1 receiver can in general make a positioning within 36 seconds after it is switched on. As we saw in the previous chapter, GLONASS and GPS epehemerides are provided in different formats. GPS ephemerides are provided as a set of Keplerian osculating elements and GLONASS ephemerides are provided in tabular format. Each navigation message also provides a user with a time mark to pin down satellite position to the specific moment of time. A GPS L1 C/A user gets a time mark within 6 seconds of a navigation message, which is the length of one frame. A GLONASS user gets a time mark within 2 seconds of a navigation message.

A GLONASS navigation message contains parameters to calculate the difference $\Delta\tau$ between GLONASS and GPS time. The parameters allow a GPS/GLONASS receiver to connect time scales with an accuracy of 30 nanoseconds, which roughly translates to about 9 meters error. A much more accurate way to estimate the time difference between two scales is by introducing it as an extra unknown into the Equations (1.19),

$$\bar{Z} = [A(\bar{X})], \tag{3.14}$$

with the state vector $\bar{X}$ now defined by

$$\bar{X} = \begin{bmatrix} x_r \\ y_r \\ z_r \\ \delta t_r \\ \Delta\tau \end{bmatrix}. \tag{3.15}$$

This allows us to find $\Delta\tau$ with the same accuracy level as other coordinates. The downside of this is that this method requires use of an extra satellite to provide an extra measurement for the extra unknown.

3.1.6 New features of future GALILEO and modernized GPS, GLONASS signals

It is planned that GALILEO will utilize a slightly different signal structure. We consider here general new characteristics of these signals. The main feature of the new signals is that the majority of the signal energy is no longer concentrated around the carrier frequency, but is split. The main reasons behind the development of such signals were on the one hand, the need to improve traditional GNSS signal properties for better resistance to multipath and interferences, and on the other hand, the need for improved spectral sharing between various GNSSs.

Binary offset carrier (BOC) modulation is implemented by additionally adding to (3.9) a binary periodical {0,1} sequence called a subcarrier [10]. It is generated as follows:

$$S(t) = \text{sign}(\cos(\omega_S t)), \tag{3.16}$$

or alternatively, as follows:

$$S(t) = \text{sign}(\sin(\omega_S t)). \tag{3.17}$$

The subcarrier is generated synchronously with the code. A BOC(n,m) is defined as a binary offset carrier signal with chipping code rate of $1023 \cdot m$ chips per millisecond and a subcarrier frequency of $1.023 \cdot n$ MHz. Thus,

$$A = A_0 \sin(\omega t + \varphi) \cdot D \cdot B \cdot S, \tag{3.18}$$

where S is a subcarrier binary sequence $S = \{01\}$. The subcarrier can be aligned with the code or shifted on some phase.

Currently it is planned that GALILEO will use BOC(1,1) on E1 frequency, AltBOC(15,10) on E5 frequency and BPSK(5) on E6. The L1C signal and GPS Military M-code are also designed as BOC type signals.

Some of the new codes cannot be generated by shift registers and therefore can only be set in the memory.

Some of the signals (GPS L5, GLONASS L3) will have two channels, one data channel and one pilot channel. The data channel has code and navigation messages. The pilot channel has a secondary code instead of a navigation message. A receiver knows a secondary code, and therefore, as we see in the next chapter, can have a higher sensitivity, due to the possibility of integrating the signal longer.

A navigation message in GALILEO should basically contain the ephemerides, satellite clock drift parameters, ionospheric model parameters, and time mark information. The new signal design did not leave navigation messages aside. The GALILEO AltBOC signal is intended to be useful for a user who accesses it on either or both sidelobes. The navigation message is provided on two channels in such a way that a user who accesses both channels can receive some parts (in particular the almanac) of the navigation message twice as quickly.

The other new feature of navigation message design is interleaving [11]. The interleaving on the transmitter side performs permutations of the bits of navigation message. A deinterleaving process on the receiver side restores the original order of bits. If parts of a navigation message have been corrupted at certain localized points, for example as a result of multipath, obstructions, or interference, then the burst of corrupted bits is dispersed over a number of words. A small number of errors can then be restored through redundancy provided in the implemented error correction algorithms.

3.2 GNSS signal simulator

3.2.1 Why does one need a simulator?

Why are simulators so important? Everything in the GNSS business can be related either to signal generation or to signal processing. Simulators are widely used in GNSS-related fields for testing. However, the methodological and research benefits of GNSS signal simulators are often overlooked. We can view at a simulator as a model of a real GNSS. When we are trying to learn about and understand something in the outside world, we are constructing a model of it inside our minds. A simulator facilitates our understanding of GNSS including the physical world in which this system exists, i.e. effects of Earth and planetary gravitational forces, effects of the Sun and its radiation, effects of the atmosphere on radio propagation, and so on. Throughout this book we use a ReGen signal simulator, a free version of which is available from the book website, for generating signals, making tests, and providing examples in our GNSS study. One also can use any other off-the-shelf simulators. In particular, we also use a Spirent GSS6300 Multi-GNSS generator and a GSS6700 Multi-GNSS simulator for testing iPRx receivers in parallel with ReGen (these simulators and the receivers under test are shown in Figure 3.11). The purpose of a RF simulator is to generate a GNSS signal as it would be received by a GNSS receiver from a number of satellites. A standard RF simulator generates a signal as it would be after the receiver antenna and before its front end. A ReGen DIF signal simulator generates a signal as it would be after the receiver front end. A basic version of a DIF signal simulator is bundled with this book, and we use it, as well as a professional version, for demonstrations throughout the book.

As we are pursuing academic purposes in this book, we are mostly interested in how a simulator can be used for GNSS study. A study of GNSS through a simulator not only helps us to understand the GNSS and its environment, but also to validate our models and understanding through a GNSS receiver, which works both with simulated and real signals. The basic principle is that if one can simulate a GNSS signal successfully, then one has a successful model and therefore a correct perception. In a way the GNSS itself, which is the object of simulation, is a model of the perception of its authors. We need also to appreciate a model's limitations. Sometimes these limitations may not play a significant role. For example, a white noise model can be used without limitations if the system under test has a bandwidth within which there is no difference between limited power noise and white noise, which has unlimited power (see Figure 3.13).

An understanding of how to create a GNSS signal in a simulator gives not only an understanding of the real signal, but also an understanding of all related processing of this signal in a receiver and post-processing software, which allow us to gain a variety of information, not only on the receiver position.

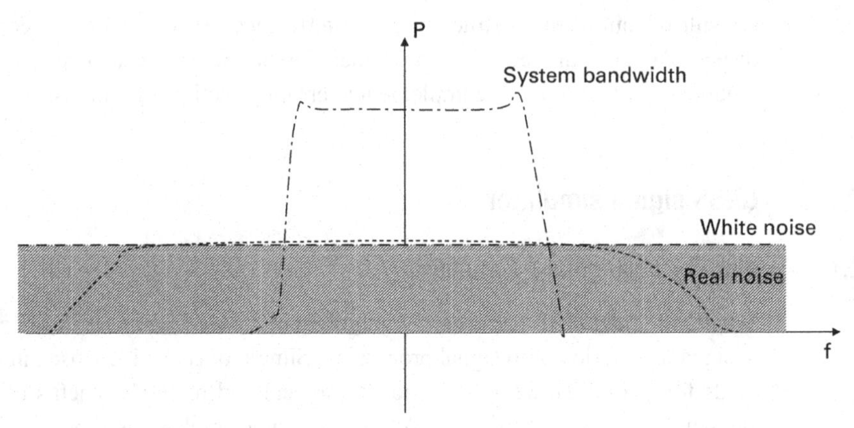

Figure 3.13 Noise model application.

In this chapter we generate GNSS signals in such way that they will be indistinguishable to a certain extent from those acquired from real satellites. The difference between a simulated signal and an acquired signal should be in where the range information contained in the signal comes from. In the case of the real signal this range information comes from a delay in propagation media simulated signal, and in the case of the simulated signal it comes from our models. When these models describe signal propagation precisely enough, the simulated signal is close to the real one. The models affecting propagation of the signal are described in Chapter 4. In this chapter we discuss the GNSS signal, its reconstruction in a simulator, and what the signal looks like when it reaches a receiver antenna.

Simulators provide a universal, high-level capability to test GNSS equipment at every stage of its life cycle. GNSS equipment, in particular a GNSS receiver life cycle, can go through the following stages [12]:

1. Research and development. This is a stage at which new designs and solutions are created.
2. Design and validation. At this stage a specific receiver design is created and tested for its ability to meet a given specification.
3. Hierarchical element production. This stage encapsulates production of chips, modules, OEMs, and user devices. At each sub-stage a particular set of tests is required to ensure quality on input and output of the production line.
4. Certification.
5. Consumer testing.
6. Repair and maintenance.

Quite often live signals from satellites are used for testing GNSS devices. However, when we use live signals we have no control over the environment and we lack repeatability in our tests, which makes it difficult to specify test requirements. For example, as we have seen earlier, the number of satellites and constellation geometry affect accuracy. In the case of many receiver tests at the development stage, we would like to remove the geometry factor from consideration and make sure that we consider only those effects due to the parameters of the receiver which we are working on. We also cannot test many situations that happen only rarely in real life, such as for example a satellite failure, which we need to simulate in order to test RAIM algorithms (Chapter 1). For these reasons, a GNSS simulator is a tool of choice for GNSS device development at all stages.

3.2.2 Main designs of GNSS simulator

Here we look at the main types of simulator design following [13] and [14].

Analog simulator

An analog simulator was the first type to be developed. It was designed in a way similar to a satellite transmitter. In this simulator a scenario is generated in a host computer. If we put a borderline between hardware including a digital part (FPGA) and a PC part, then all the information coming from the PC constitutes a scenario. Doppler, spread code, and navigation messages are generated in the digital hardware (FPGA). The radio frequency analog part is responsible for generating the signal. The generated baseband carrier is adjusted by Doppler and then modulated by a spread code, which is already mixed with the navigation message and also adjusted by Doppler. The baseband signals from all satellites are then mixed together. The resulting signal is up-converted to the L-band. Figure 3.14 shows a flowchart of this type of simulator.

Digital simulator

Digital simulators have appeared within probably a decade. Most GNSS off-the-shelf simulators today are of this type. Figure 3.15 shows a flowchart of a digital simulator. A signal is completely generated on an intermediate frequency (IF) in the digital part

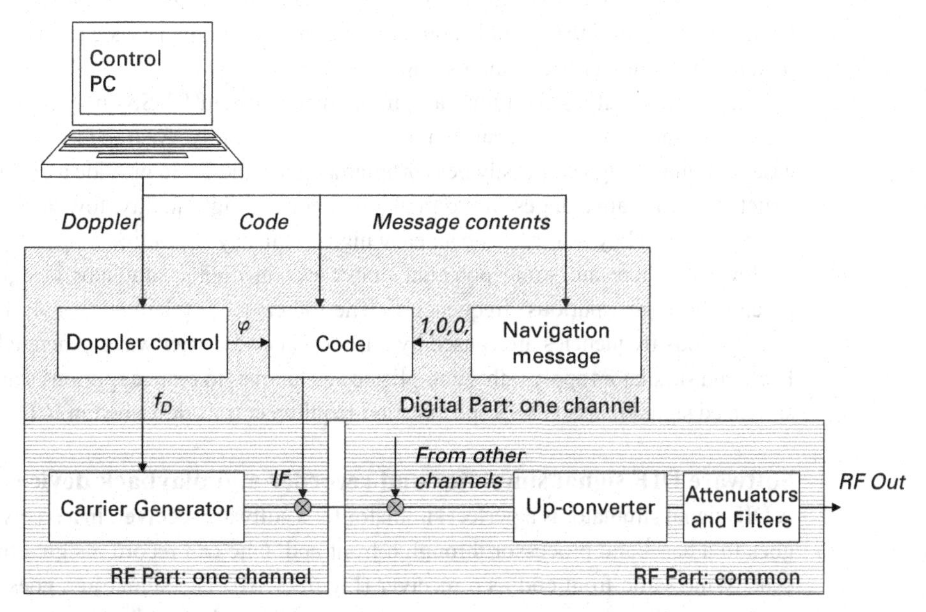

Figure 3.14 Analog simulator.

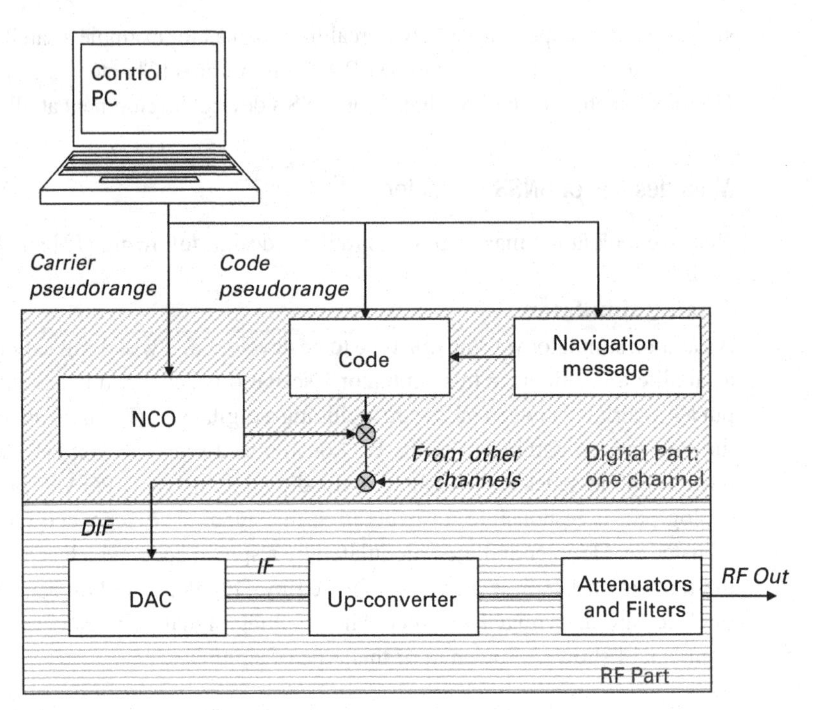

Figure 3.15 Digital simulator.

(FPGA), and then it is converted to analog signal and up-converted to an RF. The carrier is generated by a numerically controlled oscillator (NCO). Together with a code and navigation message, it creates a digitized intermediate frequency signal (DIF).

It is summed with DIF signals from other channels and then goes through digital to analog converters (DAC) and become an intermediate frequency signal with its further path no different from an analog simulator.

Satellite transmitters for future and modernized current GNSSs may move to digital signal generation as well, because it allows more flexibility. Also the transmitters in the case of digital design can easily be reprogrammed on-the-fly to provide a different signal structure. Other advantages of a digital simulator are high quality, low aging, and high predictability. They also provide an easy high-frequency resolution.

However, there are some potential drawbacks in digital simulator designs. One is potentially high spurious frequencies. The other is a relatively small bandwidth. The spurious frequencies are caused by a ladder-shaped carrier wave generated by NCO. High-end simulators apply rather complicated techniques to reduce spurious signals, so the simulated signal meets GNSS specification requirements as described in ICD.

Software DIF signal simulator and recorder and playback device

A DIF signal simulator is a perfect simulator for a software receiver. In a receiver a signal goes to a baseband processor through a front end. One can record a signal by utilizing exactly the same front end as a receiver (Figure 3.16). The signal just goes to storage media instead of the receiver, and can be processed by the receiver in exactly the same way later. Figure 3.16 shows that the same receiver/recorder front end can serve both as a front end for a standard receiver and as a recorder to record DIF to a file in memory. A software receiver can read such a DIF signal file.

This technology allows the same test to be processed over and over with different settings in the same way as with a simulator. A DIF signal thus is a signal taken after a receiver front end and stored in memory for future use.

A DIF signal simulator generates a signal similar to one which can be recorded. The difference is that the simulator allows full control over the environment and the signal. The recorded or generated DIF signal can then also be played back as an RF signal by a playback device (Figure 3.17). Such a front-end device basically works to reverse the processes in a receiver front end. It consists of a digital-to-analog converter (DAC), which converts the DIF signal to IF, and an up-converter, which converts the IF signal to RF.

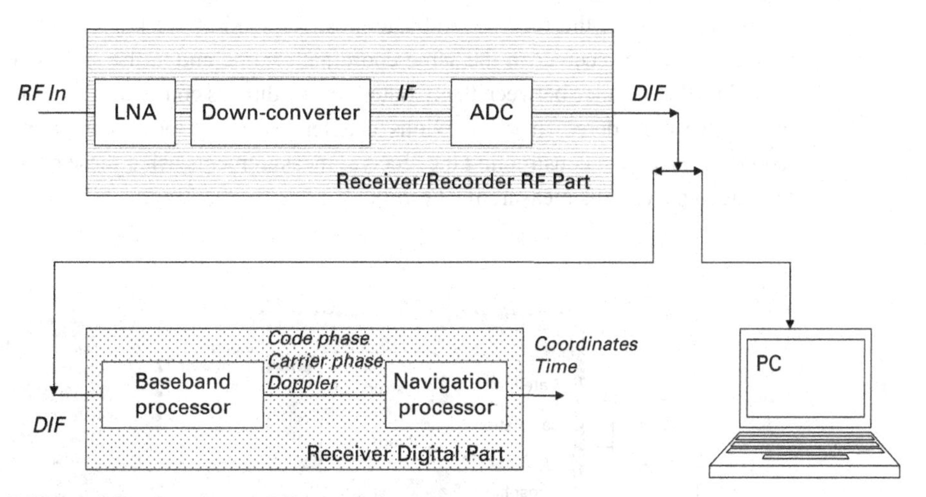

Figure 3.16 DIF signal from receiver or RF recorder.

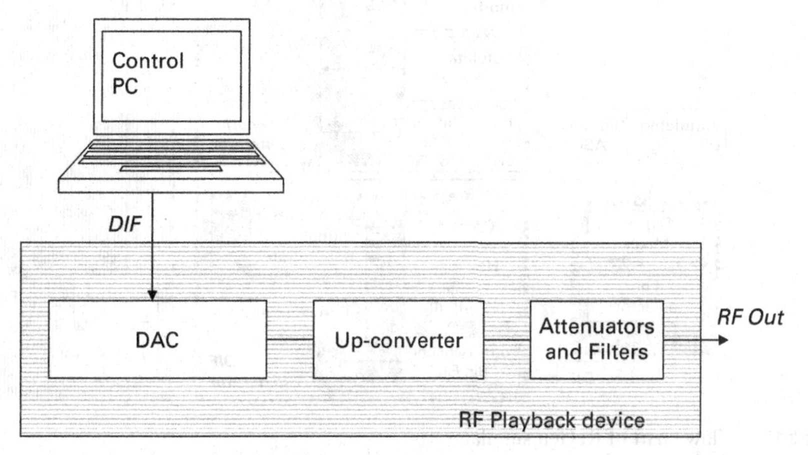

Figure 3.17 DIF signal simulator with playback front end.

3.3 Case study: scenario data generation for airborne receiver

In the next few sections we describe creation of a GNSS simulated signal. In this section we prepare a scenario, which should provide information for creation of all components of the signal. We start with creation of the receiver observables, which will be used in the next paragraph to construct a GNSS signal. All software components which we will be using are shown in Figure 3.18. We start here with trajectory generation (right top box), and a GNSS scenario generator (left top box) which creates a scenario for the given trajectory. The scenario then provides an input for a DIF generator. The generated DIF signal can be mixed with other DIF signals, such as noise, interference, or even a recorded live satellite signal, in a mixer. The flowchart also shows an INS data simulator, which is used to generate INS aiding for the simulated trajectory.

Receiver measurements could be constructed on the basis of line-of-sight (LOS) distances between the receiver and satellites. As soon as we know coordinates of the satellite and the receiver in the same coordinate frame and at the same epoch, we can calculate the distance between them. Satellite coordinates come from satellite ephemeris, which we considered in the previous chapter. Now we look at getting the receiver coordinates, and at constructing the receiver measurements using these distances, taking into account various measurement errors.

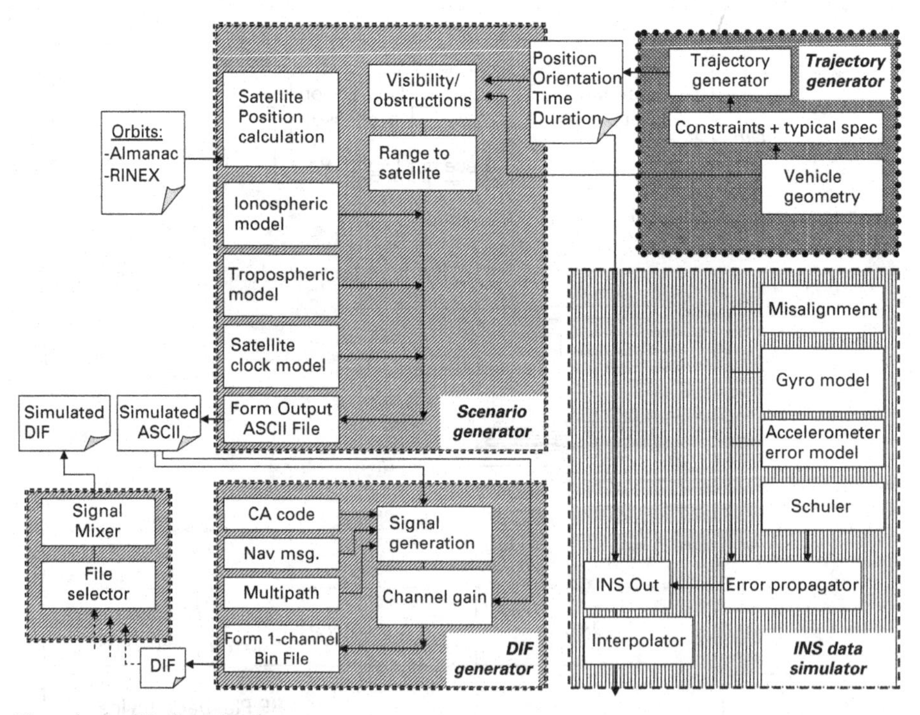

Figure 3.18 Flow chart of ReGen simulator.

3.3.1 Aircraft trajectory generation for GNSS and INS simulation

The simplest way to implement a receiver trajectory is to make it static, i.e. fixed on the Earth surface. However, for many practical applications, a receiver is moving. The movement will make a big difference: if a receiver is on a high dynamics vehicle, then the signal processing in the receiver will be affected. We look at these effects in detail in Chapters 6 and 12. To address a user with significantly different dynamics we look at aircraft movement and simulate a relevant trajectory. At the first stage we generate a simple aircraft trajectory, which includes take-off, straight flight, turn, and landing.

In Chapter 12 we look at tightly integrated GNSS and INS. For that and similar applications we need to generate two trajectories. One is a true trajectory, which is generated for all GNSS simulations with a moving user. This is a trajectory used to generate a GNSS signal, and it is an analog of real user movement in the case of live satellites. The other trajectory is a measured trajectory, which is a trajectory as it is estimated by an onboard INS. This trajectory will be considered in Chapter 12, when we use it to simulate INS assistance to a tightly coupled onboard receiver.

A trajectory in general should describe a movement of a rigid 3D body, in this case an aircraft, in 3D space, and therefore six coordinates are required to describe a position, and their derivatives of the first and second order to describe the dynamics. We will need these derivatives only if we need to create INS measurements, otherwise six coordinates for each epoch will be enough to create a GNSS signal.

Here we are describing creation of a GNSS signal as it is acquired by a receiver, and not as it is transmitted from a satellite. The six coordinates include three coordinates describing a receiver position, and three coordinates describing a vehicle attitude. It is useful to recall that when we talk about a receiver position we in fact mean a receiver antenna phase center, because those are the coordinates, which are actually measured. The coordinates which describe the vehicle attitude are important in two respects. Firstly, we use them to create strapdown INS measurements, which measure acceleration in the body frame. Secondly, vehicle attitude defines which satellites are visible to the receiver at any particular moment.

In terms of creating INS measurements in particular, and a realistic trajectory in general, it is important to ensure that a trajectory is continuous and its derivatives up to the second order are also continuous. The derivatives are also constrained by the dynamics of the vehicle chosen for the simulation.

The trajectory consists of segments, which we can describe in two-dimensional space. The first type of segment is a change in flight echelon. This segment can also describe take-off and landing. The echelon change segment is located in a vertical 2D plane. The same type of segment located in a horizontal plane describes a lane change and is adaptable to describe car motion as well. We construct this segment using parabolic blends following [15]. In a linear aircraft movement model we can consider vertical and horizontal movements independently.

A horizontal movement can be described by two intervals. The first interval of aircraft movement on a runway has constant acceleration and increasing horizontal velocity, and the second interval has constant velocity.

In vertical movement we need to achieve height change H during the interval T. These two parameters are defined by the aircraft personality file. The vertical movement in this segment is divided into three sub-segments, acceleration, constant velocity and deceleration. We can take an interval of acceleration equal to the interval of deceleration. At the end of the deceleration interval the vertical velocity is equal to zero. Vertical coordinate, velocity, and acceleration during the acceleration interval can be described as follows:

$$\begin{aligned} y(t) &= y_0 + \dot{y}_0 \cdot t + \ddot{y}_0 \cdot t^2/2, \\ \frac{\partial y(t)}{\partial t} &= \dot{y}_0 + \ddot{y}_0 \cdot t, \\ \frac{\partial^2 y(t)}{\partial t^2} &= \ddot{y}_0, \end{aligned} \tag{3.19}$$

where in the case of take-off, initial altitude and vertical velocity can both be set to zero: $y_0 = 0, \cdot y_0 = 0$. Here we assign index a to the acceleration interval, index c to the constant velocity interval, and index d to the deceleration interval.

Acceleration is found from the desired velocity at the end of the acceleration interval and its length:

$$\ddot{y}_0 = \dot{y}_{t_a} / t_a \tag{3.20}$$

Vertical coordinate, velocity, and acceleration during the constant velocity interval can be described as follows:

$$\begin{aligned} y(t) &= y_{t_a} + \dot{y}_{t_a} \cdot t, \\ \frac{\partial y(t)}{\partial t} &= \dot{y}_{t_a}, \\ \frac{\partial^2 y(t)}{\partial t^2} &= 0, \end{aligned} \tag{3.21}$$

where $y_{t_a} = y_0 + \dot{y}_0 \cdot t_a + \ddot{y}_0 \cdot t_a^2/2$. Vertical coordinate, velocity, and acceleration during the deceleration interval can be described by the Equations (3.19) with the acceleration $\ddot{y}_{t_d} = -\ddot{y}_{t_a}$ and $y_0 = y_{t_a} + \dot{y}_{t_a} \cdot t_c, \dot{y}_c = \dot{y}_{t_a}$.

In the case of descending, the height difference is negative and the first sub-segment becomes deceleration. In the case of a lane changing maneuver in the horizontal plane, we consider a lateral component of horizontal velocity instead of vertical velocity. We can describe a turn as a circular maneuver in a horizontal plane.

Figure 3.19 shows a screenshot of a ReGen add-on trajectory generator with an example of the trajectory. The software allows the user to specify aircraft parameters which may be used to provide realistic dynamics according to aircraft maneuverability. The trajectory is generated based on a set of standard trajectory segments with user specified parameters.

3.3.2 Making a simulation scenario

An antenna coordinates can be introduced either as static data or as a trajectory. If the antenna is not static, then the coordinates are provided versus time. Satellite coordinates are also calculated for each epoch based on satellite ephemerides. The ephemerides can

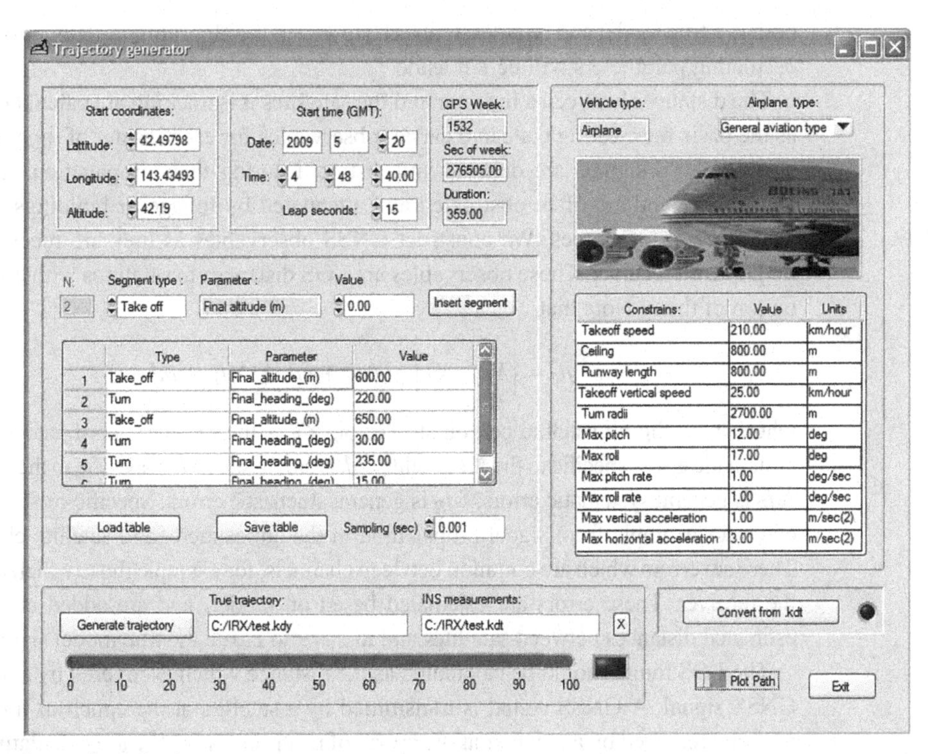

Figure 3.19 Screenshot of trajectory generator.

be introduced to a simulator from a recorded broadcast navigation message, satellite almanac, or user input.

We can describe satellite ephemerides at several levels of increasing precision (see the previous chapter). The least precise formalization is given by six main Keplerian parameters that essentially describe an orbital ellipsoid, its shape, and its position in relation to the Earth. The current epoch specifies the satellite position on the ellipsoid. These parameters basically constitute a GPS almanac. In a real environment, a GPS almanac gives just enough information to calculate an approximate satellite position with sufficient accuracy to enable a receiver to "guess" which satellite to look for and where, as well as to estimate the expected Doppler shift to assist in acquiring the signal.

In order to make an initial position fix, however, the receiver needs more precise orbits that are described by the main Keplerian parameters plus osculating parameters. These osculating parameters describe a deviation of the satellite orbit from an ideal ellipsoid, similar to the one contained in the almanac. Such a complete set of the orbital parameters can come, for example, from a receiver in a RINEX (receiver independent exchange) format file. RINEX files are also available for IGS stations from the IGS websites.

A simulator, however, might use only almanac parameters and leave all osculating parameters set to zero. In a sense, that creates a special case, but the receiver undergoing test will calculate a position all the same, because those parameters are presented to the receiver in a simulated navigation message as a satellite ephemeris. The accuracy of a tested receiver's positioning will not suffer, because only almanac parameters are used both in a true model and broadcast model. However, the algorithm functionality related to osculating parameters will be left aside.

The distances between a receiver and the satellites it is tracking are calculated to serve as the basis for creating code and carrier observables for each instant of time. In the real world, GNSS signals are delayed and distorted during their propagation through the atmosphere and are often obstructed and attenuated by foliage or buildings. There are also other error sources. We construct GNSS observables as they are received on the output from receiver. These observables are LOS distances to satellites with errors added on top of them. Note that

$$\rho_i = \left|\vec{X}_i - \vec{X}_R\right| + \Delta d_i + \delta d_i + \Delta d_R + \delta d_R, \tag{3.22}$$

where $\vec{X}_i$ is the ith satellite coordinate vector, $\vec{X}_R$ is the receiver coordinate vector, d_i is systematic errors specific to the ith satellite, δd_i is stochastic errors specific to the ith satellite, Δd_R is general systematic errors, δd_R is general stochastic errors. Specific-to-satellite errors come for example from signal propagation in the atmosphere, and satellite clock errors. There are errors which are common between channels, for example those which come from the receiver. These errors are calculated based on models and are added on top of the estimated distances between satellites and the user to create the true model signals.

The LOS range should be calculated as the distance which is covered by a transmitted GNSS signal. A GNSS signal is transmitted by a satellite at the epoch of transmission t_{TOT} and received by a receiver at the epoch of reception t_{TOR}. We are calculating a range (3.22) for t_{TOR} epochs. We also take receiver antenna coordinates $\vec{X}_R$ at t_{TOR} epochs. The satellite coordinates $\vec{X}_i$ should be taken at t_{TOT} epochs. Otherwise the difference in time between these two moments is

$$\Delta t = t_{\text{TOR}} - t_{\text{TOT}} \approx \frac{20\,000 \text{ km}}{300\,000 \text{ km/ sec}} \approx 0.7 \text{ sec.} \tag{3.23}$$

An average GPS satellite velocity about V=4 km/sec causes an error in satellite coordinates due to this time difference,

$$\Delta r = \Delta t \cdot V \approx 0.28 \text{ km.} \tag{3.24}$$

Therefore we need at least two iterations in order to calculate range to the satellite, which corresponds to the signal propagation pass (see Figure 3.20).

The navigation message transmitted by a GNSS satellite carries satellite orbits, satellite clock parameters, and ionospheric model parameters (GPS and GALILEO). The parameters embedded into the navigation message represent our supposed knowledge about a true model, but with an element of inaccuracy. The true model parameters are used for signal creation and may in general differ from the corresponding parameters embedded in the navigation message.

The broadcast ionospheric-corrections model is implemented either in a shape of a Klobuchar model in the case of GPS or in the form of the NeQuick model for GALILEO. If the simulator is simulating both systems at the same time, then it should use the same underlying true model for both broadcast models.

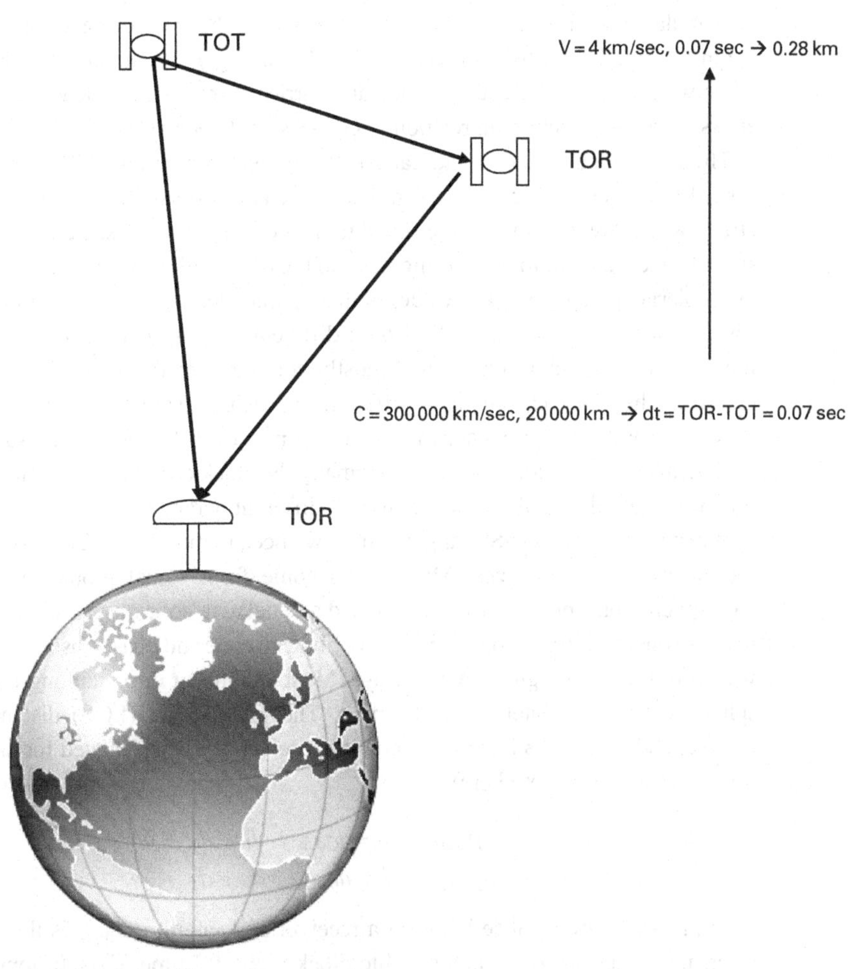

Figure 3.20 Two iterations in satellite position calculation.

The purpose of the software simulator described in this chapter is to create data for a GNSS signal simulation and not to simulate output of a receiver as such. In this respect, there is a difference between all errors described above and multipath and scintillation errors. We look at these errors in detail in the next chapters. Here we need to mention them, because these errors are different from others in Equation (3.22). Multipath is an error which caused by satellite signal reflections, and scintillation is an error caused by signal propagation through the atmosphere. If the purpose of the software simulator was to simulate output of a receiver, then it would be required to simulate an effect of multipath and ionospheric scintillation errors on range measurements, as for example a bias and stochastic component in Equation (3.22). However, we are designing a signal simulator. Therefore both scintillation and multipath errors are implemented in simulation differently from other errors because they do not affect code and carrier measurements in the same way as propagation delays. The multipath and scintillation errors are transformed to range errors only after they go through a receiver baseband processor. Therefore these errors are not simulated at this stage, unlike those errors which change code delay, and therefore they are in fact excluded from Equation (3.22). Accordingly, we should not account for them when we prepare range data for code and carrier observables, instead we consider these errors in the next section, in relation to GNSS signal generation.

The range data for code and carrier observables for various GPS, GLONASS, and GALILEO satellites are different mostly in relation to satellite orbits and ionospheric errors, which are functions of the signal frequency. Some stochastic errors also depend on spread code chip length. Regarding the satellite orbits, all constellations can be described by Keplerian parameters. The values of these parameters should be specified accordingly. The various GNSS signals will also be different in noise components, which are also functions of code chip length, and mostly coming after the signal is processed in the receiver. The clock error is also different for each constellation. The model could be different as well, because it depends not only on clock drift model, but also on the way an analytical compensation is formed. Normally the clock errors can be simulated using the broadcast model with the same or slightly different parameters.

When we simulate GNSS ranging errors we need to modify (3.22) to account explicitly for the main error sources. Main errors come from signal propagation through the atmosphere. Ionospheric error is calculated as a delay in code observables and an advance in phase observables. All other errors contribute to code and phase observables in the same way with the same sign, only as delays. Most significant errors among these are tropospheric delays and satellite clock errors. The line-of-sight (LOS) distance between a receiver and satellite as it is measured by a receiver can be calculated for each satellite as follows (we omit index i here):

$$\begin{aligned} \rho_{\text{code}} &= d + d_{\text{clock}} + d_{\text{I}} + d_{\text{T}} + \delta d, \\ \rho_{\text{carrier}} &= d + d_{\text{clock}} - d_{\text{I}} + d_{\text{T}} + \delta d, \end{aligned} \tag{3.25}$$

where d is a LOS distance between a receiver and satellite, d_{clock} is the satellite clock error, i.e. offset between each satellite clock and GPS time, d_{I} is an ionospheric error along the signal LOS path (strictly speaking it also can be simulated with a stochastic component, which will be included in code and carrier observables with the opposite sign), and d_{T} is a tropospheric error along the signal LOS path.

Proper generation of ionospheric error is very important, because this error influences many basic positioning algorithms, and we consider a particular example later in this book. A simulator should be able to output ρ_{code} and ρ_{carrier} data. In this case we can check if a simulator generates data for code and carrier correctly, by generating two scenarios as follows:

1. A scenario without errors, in which case only the LOS distance between a receiver and satellite (d) in Equations (3.25) is simulated.
2. A scenario with an error only due to the ionosphere (d_{I}).

By comparing the two data sets we should see that these values are calculated correctly as in the equations.

These ρ_{code} and ρ_{carrier} values are transferred unambiguously specifically to the code delays and carrier phase values in the signal generated by the simulator. These values

Table 3.3 GNSS error budget.

Error	Ionospheric	Tropospheric	Clock and ephemeris	Receiver noise	Multipath
1σ SPS range error [m]	45*, 7**	20*, 0.7**	3.6	1.5	1.2

* Uncompensated maximum value at low elevation.
** After compensation in a receiver.

need to be calculated with a rate no more than once every 10 milliseconds in the case of a static user. Inside of a 10 millisecond interval, the data can be interpolated to the DIF sampling rate, usually about $16 \cdot 10^6$ samples per second in the case of GPS. It is important to note that ρ_{code} values are not pseudoranges. A pseudorange by definition includes a receiver clock error, which is not simulated in the signal simulator.

An approximate estimation of error component shares in the total error budget is presented in Table 3.3 [16],[24].

3.3.3 GNSS observables at a receiver

The ρ_{code} and ρ_{carrier} values calculated in the scenario are basically the same observables, which we have at the output of the receiver. The only difference is that they do not include receiver clock error, other receiver induced errors, and also multipath and scintillation errors, which are introduced at the later stages of signal generation.

A GPS satellite moves with a velocity of approximately four kilometers per second. The projection of this velocity to a line-of-sight between the satellite and a ground-based receiver is usually about 800 meters per second or less. This relative movement changes the signal frequency as observed by the receiver. This shift in frequency, called the Doppler effect, can be expressed as follows:

$$f_D = f_0 \frac{v_{\text{LOS}}}{c}, \tag{3.26}$$

where f_0 is the transmitted signal frequency, c is the speed of light, and v_{LOS} is the relative velocity between the satellite and the receiver along the line-of-sight. The shift increases the frequency if receiver and satellite are converging. Then Equation (3.9) describing the signal at the input of the receiver can be written as follows:

$$A = A_0 \sin((\omega_0 + \omega_D)t + \varphi) \cdot D \cdot B, \tag{3.27}$$

where ω_0 is the signal central angular frequency, and ω_D is the Doppler angular frequency. Equations describing GLONASS and GALILEO signals can be expressed in a similar form. For a low-dynamic vehicle the Doppler shift is limited to 6 kHz. For a high-dynamic user, Equation (3.26) describing a Doppler shift is generally not adequate because it fails to incorporate derivatives of higher order necessary for a correct solution. If the simulator creates a signal for each moment from the geometrical considerations, as is done in the ReGen, it automatically accounts for the Doppler effect regardless of the simulated vehicle's dynamics.

```
     2.11           OBSERVATION DATA    G (GPS)             RINEX VERSION / TYPE
 iPRx v.3           ???                 3-APR-96 00:10      PGM / RUN BY / DATE
                                                            COMMENT
iPRx 01                                                     MARKER NAME
001                                                         MARKER NUMBER
USER                ???                                     OBSERVER / AGENCY
X1234A123           XX                  ZZZ                 REC # / TYPE / VERS
234                 YY                                      ANT # / TYPE
  4375274.        587466.       4589095.                    APPROX POSITION XYZ
        .9030           .0000           .0000               ANTENNA: DELTA H/E/N
     1     1                                                WAVELENGTH FACT L1/2
     3    C1    L1    D1                                    # / TYPES OF OBSERV
    1                                                       INTERVAL
  1990     3    24    13    10   36.000000                  TIME OF FIRST OBS
                                                            END OF HEADER
 09 02 24 04 12 42.4457680  0  1G 2
  19476944.137     41733803.616 7      66745.487
 09 02 24 04 12 44.4457680  0  1G 2
  19475936.213     43303333.650 7      66740.906
 09 02 24 04 12 52.4457680  0  1G 2
  19471916.084     49581453.788 7      66740.727
 09 02 24 04 12 58.4457680  0  1G 2
  19468904.647     54290043.891 7      66744.288
 09 02 24 04 13  4.4457680  0  1G 2
  19465900.697     58998633.995 7      66736.002
 09 02 24 04 13  9.4457680  0  1G 2
  19463387.569     62922459.081 7      66731.170
 09 02 24 04 13 13.4457680  0  1G 2
  19461384.345     66061519.150 7      66732.109
 09 02 24 04 13 18.4457680  0  1G 2
  19458885.046     69985344.236 7      66732.175
 09 02 24 04 13 21.4457680  0  1G 2
```

Figure 3.21 RINEX observation file generated as part of ReGen scenario.

In order to make these data accessible to the positioning algorithm, which is handy if we want, for example, to check the correctness of these algorithms, the ReGen simulator outputs these data in RINEX format. The RINEX format was introduced by the geodetic community to provide a standard for basically all the information which can be retrieved from a receiver. In such a case a receiver is used as a sensor to collect information and its own navigation processor is not engaged. An output of the ReGen scenario in RINEX format is given in Figure 3.21. The file shown contains a header and lines with time marks, the number of satellites, and observables generated for each satellite. In this case code phase, carrier phase, and Doppler measurements are generated for one satellite with PRN = 2.

RINEX also provides a standard for writing GNSS satellite ephemerides information (RINEX navigation file) and meteo information (RINEX meteo file). Most off-the-shelf simulators can use RINEX navigation files to introduce orbits into the simulator.

The ReGen orbit parameters panel allows us to introduce orbital parameters into ***true*** and ***broadcast models*** (see Figure 3.22). The true model orbits define the calculation of satellite orbits in the scenario generator and consequently observables, the broadcast model defines orbit parameters in the navigation message. The orbital parameters on the left are for a broadcast model and on the right for a true model. The true model can use ephemerides in Keplerian or tabular format. If tabular ephemerides are used in the true model, then they must correspond to the ephemeris in the broadcast model, which should be in Keplerian form.

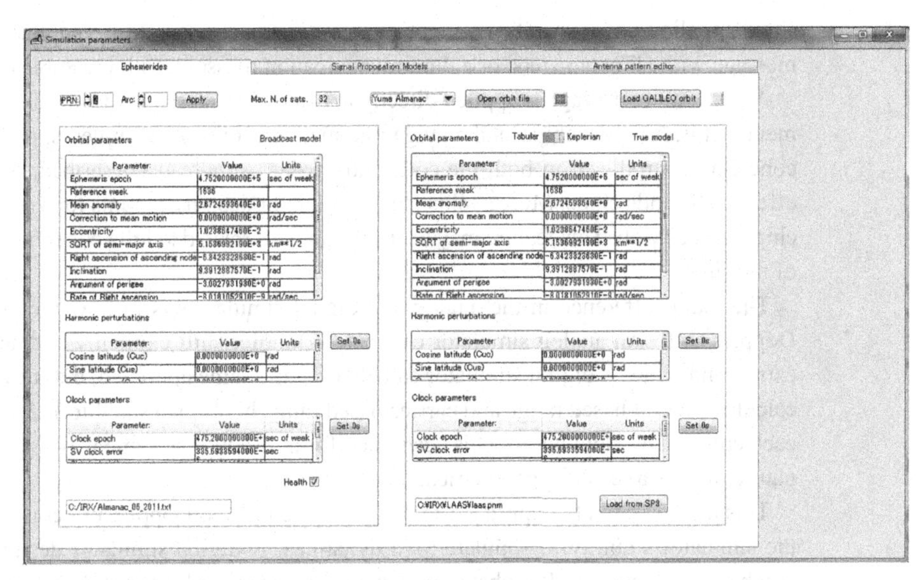

Figure 3.22 ReGen orbit parameters panel.

3.4 DIF signal generation in GNSS simulator

In the previous section we described how to create a simulator scenario which contains all data necessary for creation of a GNSS signal. Here we look at how such scenarios are used in off-the-shelf RF signal simulators of various types.

We start with a single-channel simulator. Single-channel simulators can simulate one satellite, but can give a user more control over simulation parameters. For multi-channel simulators all the parameters are defined by the scenario. A single-channel simulator does not use a scenario as such. It allows us to control a signal's Doppler profile with much more flexibility. This can be very useful, for example, for tuning receiver tracking loops. Single-channel simulators are often used for production and R&D tests. A single-channel simulator should be able to provide:

- control over signal power within a required range,
- control of the Doppler shift,
- an editable navigation message.

The ReGen simulator bundled with this book provides all the functionality of the digital signal simulator, both single- and multi-channel up to the RF front end. The main difference is that ReGen functionality is implemented in the software, whereas an off-the-shelf RF simulator implements this functionality in FPGA, and therefore is capable of providing it in real time. The complete GNSS RF signals can be restored from the DIF signal generated by ReGen using off-the-shelf playback systems.

In the single-channel simulator, Doppler profile is calculated according to a given rule, for example as a constant or oscillating value. A navigation message can also be introduced differently to facilitate various tests.The single-channel simulator panel of the ReGen simulator allows us to specify a navigation message as a check code, a complete navigation message with all parameters, or a dummy navigation message with preamble only.

A navigation message is applied directly to a code in order to avoid possible misalignment between code chips and navigation message bits. A Doppler shift should be coherently introduced in both the code and carrier. As already described, the Doppler effect shifts code chips. More sophisticated simulator design accounts for a change of the chips' size as well. This feature may be useful for tests related to long coherent integration of the signal.

The main difference in the analog and digital simulators is in how they handle the Doppler shift. An analog simulator calculates Doppler shift explicitly and applies it to carrier and code. A digital simulator can also use the same approach, but it can also just calculate a code-based range and carrier-based range between a satellite and a receiver at each epoch. The Doppler effect is accounted for in exactly the same way as it appears in nature, just as a result of geometrical considerations.

This difference in the approach to the generation of the Doppler effect is reflected in the simulator's ability to simulate high dynamics. A digital simulator design makes it much easier to account for what would appear as higher order components in (3.26) and also for a change in the code chip length.

Now we should look at how the signal is generated in a digital simulator in comparison with real satellites, using ReGen as an example. In the real system the signal is generated at the time of transmission (TOT). In a simulator, the signal is generated at a time of reception (TOR). We have already looked at the difference as it is related to observables generation in the previous section. Now we look at signal generation based on these observables. We deduct the time required for the signal propagation from TOR and we get the *time at which the satellite transmits the signal in the user receiver timeframe*. This time, however, is different from the TOT, which we need in order to reconstruct the signal. The TOT is in the satellite timeframe. Or we can say that the TOT is in the GNSS time, as recognised by the satellite, because of its clock error. By deducting range we transfer TOR to the *time at which the satellite transmits the signal in the user receiver timeframe*. By deducting pseudorange we transfer TOR to TOT in the GNSS system timeframe.

We also need to apply satellite clock error in order to get TOT. The user timeframe is different from the GNSS system timeframe by the receiver clock error. We can omit this error for the signal simulation. In fact we should omit this error if we simulate an RF signal, because this error will be in the user receiver.

As we mentioned above, we do not have to add Doppler to the signal explicitly here. It will appear naturally through range change. In fact, it gives a nice explanation of the Doppler effect. It is just a delay of each sample of sine wave, which is changing due to a changing distance between a source and a receiver. To see that it is happening, we can look at the signal generation for a geostationary satellite. The phase change in the generated signal is due only to signal frequency (we assume that receiver and satellite clock drift is absent).

Signal generation can be done in a different way. We can add a sine wave generated at the user receiver position. In that case a Doppler effect would be added to signal frequency explicitly. That is what we will do, when we come to a receiver part, in order to generate a replica signal. We could in theory generate the complete signal like that, recalculating it for each sample at the TOT through TOR range. However, it is not practical for the following reasons:

1. We would need to do it for L1, not for IF, in order to have a correct Doppler, i.e. signal delay due to satellite/user dynamic. Therefore we would at first need to calculate a phase for L1 and then up-convert it to IF.
2. In practice, on a sample level the quality of the signal will not be sufficient if the signal is simulated from the ranges. The noise errors in range calculation may completely destroy the signal. The effect is rather different from noise errors, which are generated on top of a properly generated signal.
3. We need to generate IF, so after down-conversion the wavelength becomes much larger, but the error budget is still mostly defined by initial L1 frequency. Hence, the carrier accuracy is still intact after down-conversion.

The signal generation part requires scenario data to be generated with a sampling interval of 5–10 milliseconds for a low dynamics user. Within 5–10 milliseconds the data should be interpolated. During the signal generation process, Doppler frequency is rounded off to an integer. Therefore, the generated signal would always be at a frequency less than required. In order to compensate for this, the difference between real Doppler and its round-off representation should be calculated and compensated for.

3.5 RF signal generation

3.5.1 RF signal generation in simulator

A DIF signal generated by ReGen is in all respects similar to the DIF signal simulated by high-end off-the-shelf simulators (Figure 3.15). The difference is that the digital part is located in a PC. The rest of the high-end off-the-shelf simulator is represented by an RF front end. The RF part for a software simulator playback device (Figure 3.17) is similar to a digital simulator. The front end in both cases includes the following main parts:

1) Digital-to-analog converter (DAC),
2) Up-converter,
3) Attenuator.

At first the DIF signal data should be interpolated to the DAC frequency. An RF front end of a simulator should be designed in such way that the resulting RF signal meets GPS specification requirements. One demanding requirement in particular, especially for digital type simulators, is ***spurious free dynamic range (SFDR)***. SFDR can be defined as the measure of the ratio in amplitude between the signal and the largest harmonically or non-harmonically related spike. Essentially, it indicates a usable dynamic range. GPS SFDR is defined as 40 dB [1]. The spurious frequencies in the DIF are caused by the digital nature of the carrier generated by the NCO. The RF part introduces another source

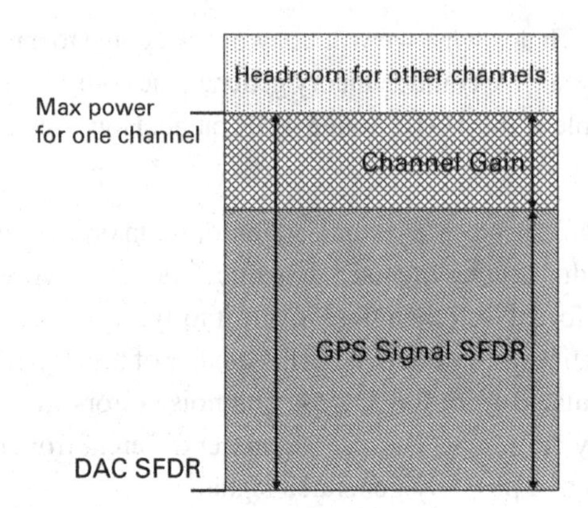

Figure 3.23 Calculation of SFDR requirements for multi-channel simulator.

of spurious frequencies in the DAC. A DAC SFDR is defined by its specification. A multi-channel simulator should provide SFDR on its channels along with the ability for fine control of the channel powers. The available headroom for channel gain is defined by the difference between required GPS SFDR and DAC SFDR. This headroom defines power and control resolution between the channels (see Figure 3.23).

The RF front-end functionality is also provided by a separate off-the-shelf ***record and play system (RPS)***. RPS allows the user to convert a previously recorded or generated DIF signal to an RF signal and retransmit it. The DIF signal generated by ReGen can be converted to an RF signal using RPS and consequently it can be acquired and tracked by conventional GNSS receivers.

3.5.2 RF signal generation in satellite transmitter

There are two main differences between signal generation in a satellite transmitter and a simulator, which we need to consider. The first is related to satellite clocks. They are connected to the system time and should be much more accurate than a simulator master clock. Although they are much more accurate, satellite transmitter clocks still have different drift parameters, though for the most part this is analytically compensated by the parameters of the drift model transmitted in the navigation message. All satellite signals in a simulator are generated using the same master clock, so its drift is not so important. For some applications, such as high sensitivity research, it may be required that the master clock drift be much lower than the drift of the receiver under test.

The second issue is related to relativistic effects [17]. For example, for GPS the ICD specifies that the satellite clock frequency be offset in order to compensate for relativistic effects. As a result a receiver sees a satellite clock generating frequency of 10.23 MHz, whereas its real frequency is 10.22999999543 MHz [1]. As far as the user on the Earth's surface is concerned these effects should generally be taken into account only for time transfer applications.

3.5.3 Pseudolites

A single-channel simulator can also be used as a pseudolite. Pseudolites, short for pseudo-satellite, were initially introduced for satellite simulation purposes and then for the purposes of augmenting GPS during aircraft approach and landing. Details of pseudolite design and applications are given in [18] and [19]. We can convert any RF simulator to a pseudolite simply by attaching an antenna to the RF output. This is, however, a do-not-do-this-at-home thing. It should not be done without special precautions to prevent an RF signal to escape from a designated and licensed (if required by law) lab. Otherwise it can easily cause interference with non-participating receivers. One of the solutions is to use a metal dome to cover both transmitter and receiver antennas. Such a solution can also be used as a field test device for a pre-flight test of airborne GNSS antennas.

Satellite clocks are of very high quality and perfectly synchronized. The pseudolites' clocks are much less accurate. A key to successful integration of pseudolites is to resolve the clock offsets among the pseudolite transmitters and communicate clock error to the user equipment. One approach is to use a reference station which is located at a known distance from the pseudolite and therefore can estimate the pseudolite clock error. The other approach is to provide an external synchronization. External synchronization can be done with the help of GPS. The pseudolites in this case operate like a constellation of satellites. Field simulation of GNSS started with GPS pseudolites located in Yuma state, which gave its name to the YUMA almanac format. Pseudolites are also being used to simulate GALILEO satellites in a real environment [20] and QZSS satellites [21].

In order to be implemented as a pseudolite, a simulator should implement special functions to reduce possible interference from the pseudolite with a non-participant receiver. This interference is also referred to as the near-far problem. If a pseudolite is located near a receiver, then the power of its signal can be significantly higher than the power of a satellite signal originating 20 000 km away from the receiver. Because the signal from the pseudolite is transmitted on the same frequency as a GPS signal, the satellite signal can be saturated, and effectively jammed. A reduction of this interference can be achieved by pulsing and frequency offset. Pseudolite pulsing is defined by an RTCM (The Radio Technical Commission for Maritime Services) standard. (RTCM also provides a standard for a GPS corrections format.) Pulsing provides pulse-wise interruption of the transmitted signal. It decreases the power of a pseudolite signal at non-participant receivers, by spreading the signal. A participant receiver uses the same pulsing scheme to receive the complete signal energy. Frequency shift allows us to move a signal out of a narrowed PLL band. This allows us to limit interference from a pseudolite during the tracking process. Frequency shift also limits pseudolite effect on the acquisition process if it is large enough to exceed the search range of the non-participant receiver, but will be within range of the participant receiver. This is actually a compromise to moving pseudolites out of the GNSS band.

Possible pseudolite applications can be found in social infrastructure [22] and robotics [19],[23]. The main problem for pseudolite applications in indoor robotics is multipath. Pseudolites currently occupy the GPS spectrum mainly because this allows researchers to use GPS off-the-shelf receiver front end and simulator components. However, application of GNSS pseudolites outdoors should be limited only to special cases, such as field simulation of a new satellite system design.

GPS and GLONASS up to now, and hopefully also other GNSSs in the future, such as Galileo, QZSS, and Compass, provide extremely powerful tools for navigation and geophysical researches, many of which have vital importance for humanity. Most of the methods, however, are extremely sensitive to interference. Any usage of signals with frequencies in close proximity to GNSS bands, let alone within GNSS bands, by other systems and signals should be restricted, if not prohibited. Unregulated usage of other systems on these frequencies may jeopardize first of all highly sensitive geodetic and geophysical receivers, with a huge impact on various often tremendously important geophysical, geodetic, and engineering applications. If not causing direct interference, such usage can cause increase of the noise floor, and consequently degradation of carrier-phase-based observables, cycle slips, and even loss of lock. Moreover, it may also unintentionally interfere with high-sensitivity applications and emergency services, thus directly putting human lives in danger. Therefore any wireless applications, RF ID transmitters, pseudolites and the like should be moved to other frequency bands for the sake of one of the biggest of humanity's achievements, GNSS, which goes far beyond the initially sought navigation and positioning applications. The GNSS business should be regulated in order to ensure that various GNSSs will not interfere with each other and especially with already successfully functioning systems, and that they will also not add a hazard by inevitably contributing with time to the space garbage collection in Earth orbits.

3.6 Project: GPS signal simulation with bundled ReGen simulator version

Let us use the ReGen simulator to simulate a signal for a given time and location.* In Chapter 5 we use an RF recorder to record GPS signal. One can use the same location and time for the recording and simulation.

Download a RINEX navigation file or YUMA almanac from the Internet. (Note that the RINEX file should be converted before loading into ReGen (see manual).) It is recommended to download the YUMA GPS almanacs from the US Coast Guard Navigation Center of Excellence (http://www.navcen.uscg.gov/?pageName=gpsAlmanacs). Use the ReGen software to simulate a GPS signal. Figures 3.22, 3.24, and 3.25 show ReGen orbital parameters, position, and DIF generator panels, which it is necessary to configure for signal generation. Similar information, i.e. satellite ephemeris (Figure 3.22), receiver antenna coordinates, time (Figure 3.24), are required from a user by any simulator. The information related to the DIF specification is the specific to DIF signal simulator (Figure 3.25). It is defined by the receiver baseband processor, and in conventional receivers specifies parameters of the receiver RF front end.

* Note that some of the functionality described here may not be included in the light version of ReGen, and available only in standard and professional versions.

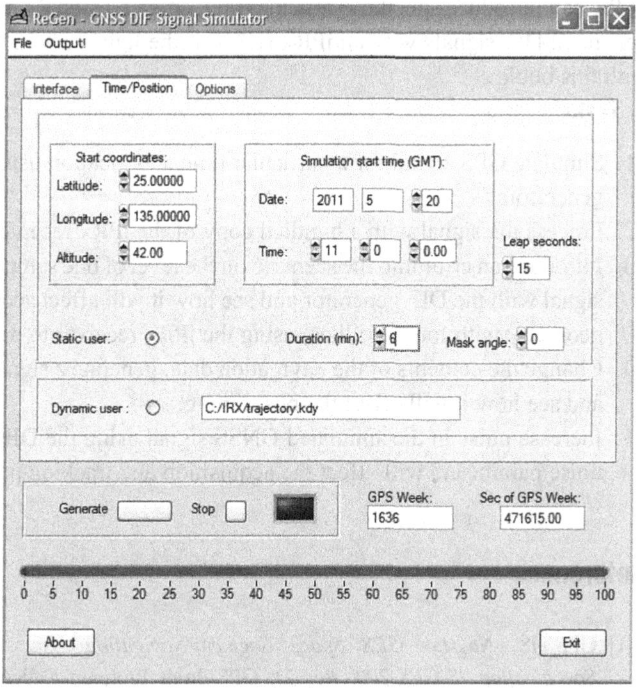

Figure 3.24 Receiver antenna coordinates and time of simulation input in ReGen DIF signal simulator.

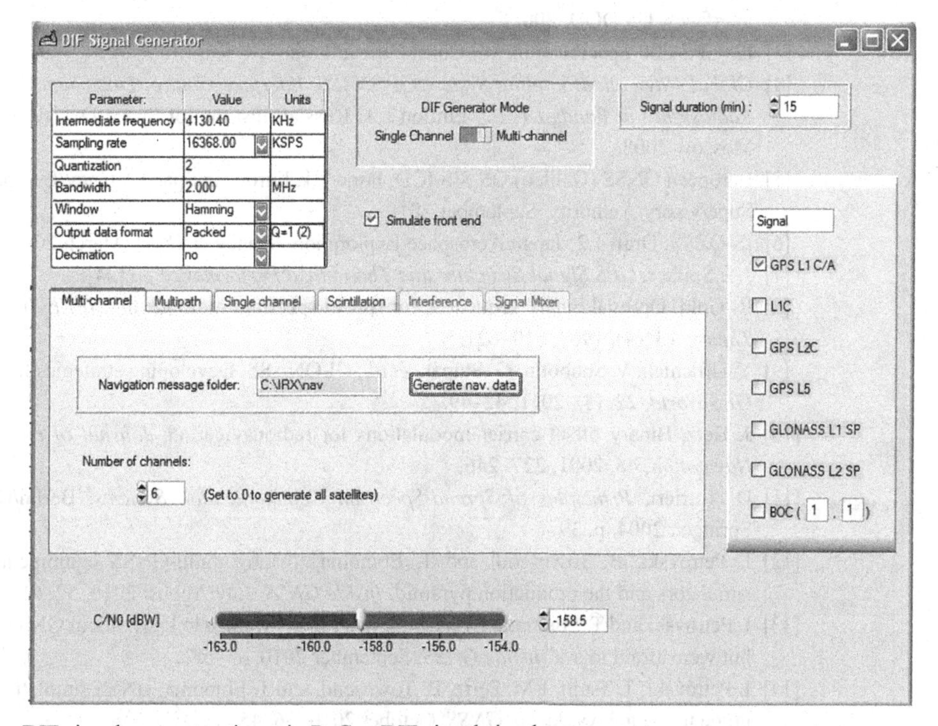

Figure 3.25 DIF signal parameters input in ReGen DIF signal simulator.

We can introduce specific errors into the generated signal. Further we can process the generated DIF signals with an iPRx receiver, the light version of which is also bundled with this book.

Tasks:

1. Simulate GPS signal for a particular time and location using scenario and DIF generators.
2. Process the signal with a bundled copy of the iPRx receiver.
3. Introduce an error into the scenario on the level of one sample size (15 m). Simulate a signal with the DIF generator and see how it will affect accuracy in the case of poor geometry with four satellites, using the iPRx receiver to process the signal.
4. Change the contents of the navigation data, generate a signal with the DIF generator and see how it will affect the positioning.
5. Increase noise in the simulated GNSS signal using the DIF generator. See how the noise parameters will affect the acquisition and tracking performance.

References

[1] GPS IS, *Navstar GPS Space Segment/Navigation User Interfaces, GPS Interface Specification IS-GPS-200, Rev D*, GPS Joint Program Office and ARINC Engineering Services, March 2006.

[2] ICD-GPS-705, Interface control document: Navstar GPS space segment/ navigation L5 user interfaces, US DOD, 2002.

[3] IS_GPS-800 Specification, US Coast Guard, 2008.

[4] *Global Navigation Satellite System GLONASS, Interface Control Document, Navigational Radiosignal in Bands L1, L2*, Edition 5.1, Russian Institute of Space Device Engineering, Moscow 2008.

[5] European GNSS (Galileo) OS SIS ICD, Issue 1.1, European Space Agency/European GNSS Supervisory Authority, September 2010.

[6] IS-QZSS, Draft 1.2, Japan Aerospace Exploration Agency (JAXA), March 2010.

[7] J. J. Spilker, *GPS Signal Structure and Theoretical Performance*, in [24].

[8] R. Gold, Optimal binary sequences for spread spectrum multiplexing, *IEEE Trans. Inform. Theory*, **13**, (4), 1967, 619–621.

[9] Y. Urlichich, V. Subbotin, G. Stupak, *et al.*, GLONASS. Developing strategies for the future, *GPS World*, **22**, (4), 2011, 42–49.

[10] J. Betz, Binary offset carrier modulations for radionavigation, *Journal of the Institute of Navigation*, **48**, 2001, 227–246.

[11] D. Torrieri, *Principles of Spread-Spectrum Communication Systems*, Berlin/Heidelberg, Springer, 2004, p. 39.

[12] I. Petrovski, B. Townsend, and T. Ebinuma, Testing multi-GNSS equipment, systems, simulators and the production pyramid, *Inside GNSS*, July/August 2010, 52–61.

[13] I. Petrovski and T. Ebinuma, Everything you always wanted to know about GNSS simulators but were afraid to ask, *Inside GNSS*, September 2010, 48–58.

[14] I. Petrovski, T. Tsujii, J-M. Perre, B. Townsend, and T. Ebinuma, GNSS simulation: A user's guide to the galaxy, *Inside GNSS*, October 2010, 36–45.

[15] L. Biagiotti and C. Melchiorri, *Trajectory Planning for Automatic Machines and Robots*, Berlin/Heidelberg, Springer, 2008.
[16] National Research Council, Committee on the Future of the GPS, *The Global Positioning System: A Shared National Asset*, Washington, DC, National Academy Press, 1995.
[17] N. Ashby and J. J. Spilker, *Introduction to Relativistic Effects on the Global Positioning System*, in [24].
[18] S. Cobb, *Theory and Design of Pseudolites*, Dissertation, Stanford University, 1997.
[19] S. Sugano, Y. Sakamoto, K. Fujii, *et al.*, It's a robot life, *GPS World*, **18**, (9), 2007, 48–55.
[20] G. Heinrichs, E. Löhnert, E. Wittmann, and R. Kaniuth, Opening the GATE, Germany's GALILEO test and development environment, *Inside GNSS*, May/June 2007, 45–52.
[21] T. Tsujii, H. Tomita, Y. Okuno, *et al.*, Development of a BOC/CA pseudo QZS and multipath analysis using an airborne platform, Proceedings of the Institute of Navigation National Technical Meeting 2007, San Diego, California, USA, January 22–24, 2007, pp 446–451.
[22] I. Petrovski, *et al.*, Pedestrian ITS in Japan, *GPS World*, **14**, (3), 2003, 33–37.
[23] I. Petrovski, *et al.*, Indoor code and carrier phase positioning with pseudolites and multiple GPS repeaters, Proceedings of the Institute of Navigation ION GPS/GNSS 2003, Portland, Oregon USA, September 2003.
[24] *Global Positioning System: Theory, and Applications*, Vol. I, B. W. Parkinson and J. J. Spilker (editors), Washington, DC: American Institute of Aeronautics and Astronautics Inc., 1996.

Exercises

Exercise 3.1. Compare a generated signal spectrum with a spectrum of recorded live satellite signals. A recorded live satellite signal is available from the book website. Note that the signals are recorded in packed format with four 2-bit samples placed in one byte. You can use Streamer software to unpack the signal.

Exercise 3.2. Use almanac files available in the book webpage or from the Internet RINEX and YUMA sites to introduce orbits into the ReGen simulator and see what parameters are available in each case in the ReGen orbit parameters panel.

4 Signal propagation through the atmosphere

In the previous chapter we described how a GNSS signal is generated. In this chapter we consider how a GNSS signal propagates through the atmosphere. Figure 4.1 shows how this chapter is related to the book's contents.

How a radio signal propagates through the Earth's atmosphere to a great degree depends on the signal carrier frequency. We consider here a relatively narrow frequency range, which is allocated to various GNSSs. Figure 4.2 shows linear and logarithmic scales for signals with various wavelengths. It is possible to see the GNSS frequency allocation in relation to light, radio, and audio signals only on a logarithmic scale. Radiowaves, GNSS, and visible light are located relatively close to each other on a wavelength scale. This proximity between GNSS signals and light allows us to use most of the mathematical tools we use for optics for GNSS signal propagation.

A GNSS signal is affected by the atmosphere. It causes ray bending, signal delays, and frequency, amplitude and phase fluctuations. In this chapter we consider mostly systematic effects of the atmosphere on GNSS signals. In this respect most of the theory that has been developed to describe the behavior of light in the part of the electromagnetic spectrum visible to humans can be applied to GNSS signals.

The ionosphere is a layer of ionized electrons surrounding Earth at altitudes between 300 and 500 km. The ionosphere can reflect radio waves at particular frequencies, and therefore it allows radio signals to be received from a transmitter located over the horizon from a receiver. Though it may sound a rather dull subject for a reader unfamiliar with it, the ionosphere serves as constant inspiration for science fiction, conspiracy theories, and technological innovations alike.

An interesting thing about the ionosphere is that it actually contradicts our current understanding of the laws of nature. We have learned that nothing in this universe can exceed the speed of light. However, the ionosphere defies this law. A GNSS signal in general consists of a carrier modulated by a spreading code. When the signal goes through the ionosphere the code is delayed. The carrier, as we learn in this chapter, is on the contrary advanced. If the original signal propagates from a satellite transmitter with the speed of light, then in the ionosphere, the carrier propagates with a speed higher than the speed of light. There is an argument put by physicists that the carrier does not bear any information and therefore this exceeding of the speed of light does not count. However, an engineer can find information in this carrier, as we see in the next chapters. We can actually use the carrier for positioning, under some conditions even without code. For example, we can use only a carrier for positioning when we have an initial position

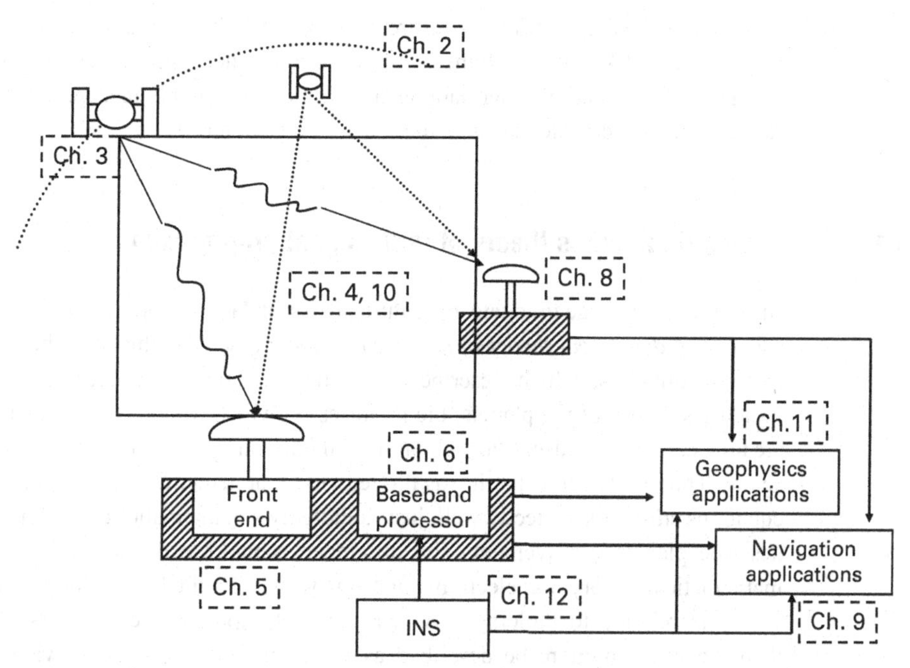

Figure 4.1 Subject of Chapter 4.

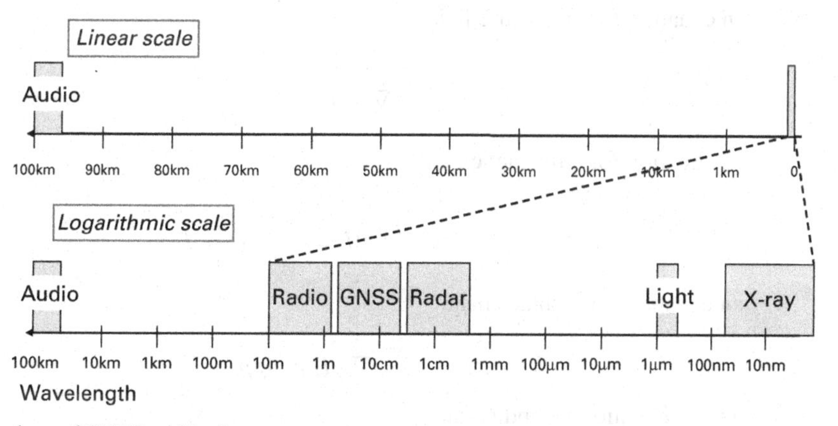

Figure 4.2 The place of GNSS within the wave spectrum.

estimate with a certain accuracy. Therefore, we may conclude that the ionosphere allows us to transmit the information with a speed greater than the speed of light.

In this chapter we consider relations between the ionosphere and a GNSS in two respects. Firstly, the ionosphere affects GNSS signal propagation, and if these effects are not accounted for the measurements become corrupted by errors. Therefore, GNSS signal propagation through the ionosphere should be modeled, and these models should be used when we retrieve code and carrier observables.

Secondly, we use GNSS for measurements of ionospheric parameters. In this case, we use receivers at known positions and put the ionospheric parameters in a state vector along with or instead of other parameters such as for example receiver coordinates and clocks. We consider such tasks in more detail in Chapter 11.

4.1 Geometrical optics theory of radio signal propagation

In 1865 a Cambridge graduate and future professor James Clerk Maxwell published his paper on *A dynamical theory of the electromagnetic field.* In this work he presented his famous equations, which describe an electromagnetic field. Somewhat similarly to Ptolemy's theory of the planets, the premises on which Maxwell built his mathematics are now considered false. Maxwell built his theory on the assumption of the existence of aether. Though it seems that it was proved later that aether does not exist, the Maxwell equations still work perfectly well, just as Ptolemy's mathematics can still perfectly well describe planetary movement. The difference is that we do not have an alternative mathematical theory for electromagnetism, which would be similar to Copernicus's theory in relation to Ptolemy's. As we saw in Chapter 2, the premises of Ptolemy's theory cannot anymore be described as incorrect from the point of view of modern science. Similarly, we still may come back to the aether concept, for example in relation to modern string theory.

Maxwell's equations can be written as follows:
an equation for an electric field,

$$\vec{\nabla} \times \vec{E} = -\frac{1}{c}\frac{\partial \vec{B}}{\partial t}, \tag{4.1}$$

an equation for a magnetic field,

$$\vec{\nabla} \times \vec{H} = \frac{1}{c}\frac{\partial \vec{D}}{\partial t} + \frac{4\pi}{c}\vec{J}, \tag{4.2}$$

an equation for displacement,

$$\vec{\nabla} \cdot \vec{D} = 4\pi\rho_e, \tag{4.3}$$

and an equation for induction,

$$\vec{\nabla} \cdot \vec{B} = 0. \tag{4.4}$$

In these equations $\vec{J}$ is the current density, ρ_e is the net charge density. We should be able in principle to analyze GNSS signal propagation through the atmosphere based on these equations. In a medium, Maxwell's equations should be complemented by constitutive equations, which describe the propagation media. Magnetic field is related to induction field through the magnetic permeability of the medium,

$$\vec{B} = \mu\vec{H}. \tag{4.5}$$

Displacement is related to electric field through the dielectric constant of the medium,

$$\vec{D} = \varepsilon(r,t)\vec{E}. \tag{4.6}$$

Magnetic permeability and dielectric constant define the speed of electromagnetic wave propagation, or ***phase speed***, in the medium,

$$c_p = 1/\sqrt{\varepsilon\mu}. \tag{4.7}$$

In our applications, magnetic permeability can be assumed to be a constant,

$$\mu = const \approx 1. \tag{4.8}$$

Dielectric constant on the contrary changes significantly with the medium. For the Earth's atmosphere it depends on coordinates and time and in general can be described through average and small stochastic components [1],

$$\varepsilon(r,t) = \varepsilon_0(r) + \Delta\varepsilon(r,t). \tag{4.9}$$

The stochastic component $\Delta\varepsilon\ (r,t)$ has spatial and temporal fluctuations. It is accounted for when we consider signal scintillations in Chapter 10.

Maxwell showed that his equations imply wave motion. One can deduce the general ***scalar wave equation*** from Maxwell's equations in the following form [2]:

$$\nabla^2 E(r) + k^2[1 + \Delta\varepsilon(r,t)]E(r) = -4\pi ikJ(r), \tag{4.10}$$

where $k = 2\pi\sqrt{\varepsilon\mu}\,f$ is the electromagnetic wavenumber. Wavenumber is the spatial analog of angular frequency. This equation describes radio signal propagation in any direction. Apparently from (4.7) the electromagnetic wavenumber can also be expressed as $k = 2\pi/\lambda$, where f is frequency and λ is the wavelength of the radio signal under consideration.

If it were possible to solve Equation (4.10) in general, than we would be able to describe a radio signal at any point, once and for all solving all problems related to its propagation through the atmosphere. If there is no source of electromagnetic field in the area which we need to describe, a condition which can be formalized as

$$J(r) = 0, \tag{4.11}$$

then the right-hand part of Equation (4.10) can be assumed to be 0. If we can also neglect a stochastic component,

$$\Delta\varepsilon \sim 0, \tag{4.12}$$

then Equation (4.10) can be simplified to

$$\nabla^2 E(r) + k^2 E(r) = 0, \tag{4.13}$$

which yields to a solution

$$E = E_0 \cos(\omega t \pm kx), \tag{4.14}$$

where E_0 is amplitude of the electric field, and x is the axis in which signal is propagating. This equation describes two waves propagating in opposite directions, because we did not specify where the source of the field is located. If we fix coordinates, then we can see how the signal changes at this point with time. It is a harmonic wave along the time axis,

$$E = E_0 \cos(\omega t). \tag{4.15}$$

If we fix an instant of time, then we can see that the signal's amplitude is distributed as a harmonic along the coordinate axis,

$$E = E_0 \cos kx. \tag{4.16}$$

Further complications of a mathematical description based on the wave Equation (4.10) come from the fact that a GNSS signal is modulated by a spread code. In navigational tasks one is mostly interested in propagation of a code signal which, depending on the medium, may propagate with a speed different from that of the carrier wave.

4.2 GNSS ray bending in the Earth's atmosphere

In general, a radio signal propagates in a medium with a speed different from its speed in vacuum. The ***refractive index*** shows how signal propagation speed in a medium is related to signal propagation speed in a vacuum, which we often refer to as the speed of light in vacuum. Refractive index can be defined as follows:

$$n(\omega, r) = \frac{c_0}{c_p}, \tag{4.17}$$

where c_p is signal propagation speed in a medium, and c_0 is the speed of light in vacuum.

Refractive index n allows us to look at the signal propagation from a geometrical point of view. It indicates how the signal ray path changes its angle of arrival when a signal crosses a border from a vacuum to a medium. In general, refractive index has a complex value. We are interested in the real part, which is related to the delay and ray bending. The imaginary part is related to signal absorption.

From (4.7), (4.8), and (4.17) the refractive index is related to the dielectric constant as follows:

$$\varepsilon = n^2. \tag{4.18}$$

The refractive index and corresponding dielectric constant may depend on signal frequency. If dielectric constant and corresponding electromagnetic wavenumber depend on signal frequency, then a medium is called ***dispersive***. A rainbow can be seen as an example of light propagation through a dispersive medium, where light waves with different frequencies bend differently when passing through raindrops into the atmosphere. A crystal gives another example of signal propagation through a dispersive medium (Figure 4.3).

This analogy with light waves is very important, because the main radio signal propagation theory is based on geometrical optics, where the same principles of light

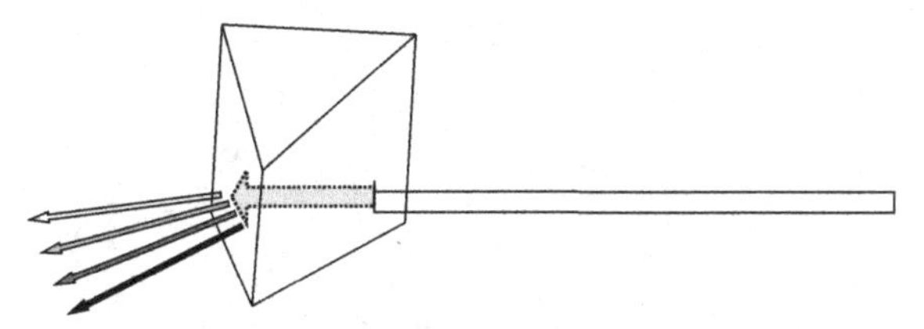

Figure 4.3 Signal propagation through a dispersive medium.

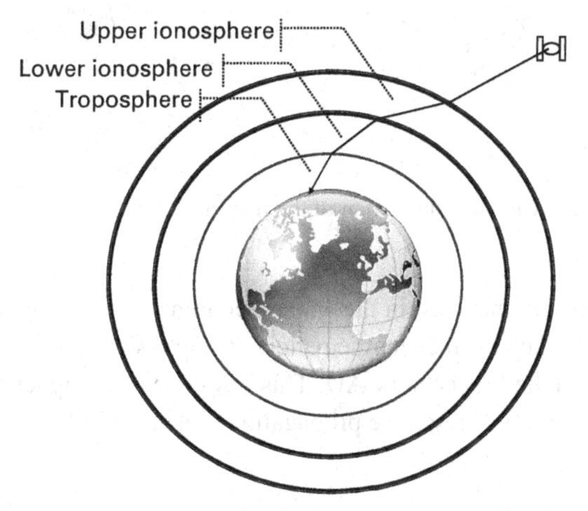

Figure 4.4 GNSS ray bending in atmosphere.

diffraction and refraction are applied to radio signals. Actually it was Maxwell who identified light as electromagnetic oscillation. Accordingly, if dielectric constant and electromagnetic wavenumber do not depend on signal frequency, then a medium is called ***non-dispersive***.

For GNSS applications we need to consider two layers of atmosphere. The troposphere, which is the lower atmosphere, is a non-dispersive medium. The ionosphere, which is the upper layer, is a dispersive medium. A GNSS signal beam is bent to the side of the medium with the higher refractive index. In the troposphere the beam bends towards the Earth, then it is less bent in the lower ionosphere and it is bent towards space in the upper ionosphere (see Figure 4.4).

The mechanism of refractive bending is described by Fermat's principle of least time. Fermat's principle states that a ray follows a trajectory that minimizes phase integrals,

$$\oint ds\sqrt{\varepsilon(s)} = \min. \tag{4.19}$$

This can be illustrated by the following analogy given by Richard Feynman in his *Lectures on Physics* [3]. We can compare the path of the ray to the path which a lifeguard on a shore would choose to get quicker to a drowning person in the water. In this case the lifeguard optimizes his path in order to take advantage of the fact that his speed on the

Figure 4.5 Richard Feynman's illustration of the refractive index.

ground is higher than in the water (see Figure 4.5). Originally the principle of least time was stated by a Greek mathematician and physicist, Heron of Alexandria, who worked in Egypt in the first century AD. This principle also implies that an electromagnetic signal can somehow foresee the propagation path ahead of time.

4.3 GNSS signal phase and group velocity in the ionosphere

A GNSS signal exhibits code delay and phase advance when coming through the Earth's ionosphere. It means that when we calculate range to a satellite from code and carrier phase measurements, we should account for this delay. As we saw in the previous chapter, pseudorange observables to each satellite can be expressed as follows:

$$\rho_i = r_i + d_I^i + \delta t_r \cdot c, \quad i = 1, \ldots n, \tag{4.20}$$

where r_i is the distance to the ith satellite, δt_r is receiver clock error and d_I^i is code delay due to the ionosphere.

Carrier phase measurements, corrected for ionospheric delay, calculated from a number of carrier waves can be expressed similarly,

$$\phi_i = \lambda_j N_i + \delta\phi_i - d_I^i + \delta t_r \cdot c, \tag{4.21}$$

where λ_j is a wavelength on L_j frequency, ϕ_i is carrier phase measurement, N_i is number of whole waves between a receiver and ith satellite, d_I^i has a minus sign because the carrier advances due to the ionosphere.

The code delay and phase advance of the GNSS signal occur in the ionosphere because the ionosphere is a dispersive medium, in which refractive index depends on signal frequency (4.17). This fact has tremendous implications. In order to explain this effect we consider an amplitude modulated carrier wave E_1. Such a signal, as well as any GNSS signal, can be presented as a number of Fourier series harmonics. If a medium is non-dispersive, then all harmonics propagate with the same speed and waveform shape is not disturbed. The speed of waveform propagation is equal to the speed with which harmonics are propagating. Phase velocity is defined (4.17) by

$$c_p = \frac{c_0}{n}, \tag{4.22}$$

where c_0 is the speed of light in vacuum.

In a dispersive medium a refractive index is defined by (4.17) and dielectric constant correspondingly depends on signal frequency. Refractive index can be derived in different ways. It can be shown that this dependency is as follows [1]:

$$\varepsilon = \varepsilon_0 \left(1 - \frac{e^2 N}{\varepsilon_0 m_e \omega^2}\right), \tag{4.23}$$

where N is total number of electrons per volume, e is electron charge, m is electron mass, ω is probing frequency. Similarly, phase velocity is defined through (4.22) and (4.18) as follows:

$$c_p = \frac{c}{\sqrt{1 - \frac{e^2 N}{\varepsilon_0 m_e \omega^2}}} \; {}^{(*)}. \tag{4.24}$$

Therefore, if the medium is dispersive then each harmonic propagates with its own speed. In this case our code phase is defined by those Fourier harmonics which have maximum amplitude and carry most of the wave energy. This defines the waveform envelope.

The group velocity is defined as speed of the waveform envelope (we consider all propagation equations in scalar form without loss of generality),

$$c_g = \frac{dx}{dt}. \tag{4.25}$$

For GNSS the bandwidth of the wavepacket is narrow in comparison to the carrier and the shape of the envelope waveform is retained, though it arrives with a delay.

Using the approach given in [4], we can look at the harmonic signal modulated by another harmonic. This can be done without a loss of generality, because any modulating signal including a square wave sequence can be represented by a large enough Fourier series of harmonics,

$$x = A \sin(2\pi f_m t) \sin(2\pi f_c t). \tag{4.26}$$

This can be transformed to

$$x = \frac{1}{2} A \cos[(\omega_c - \omega_m)t] - \frac{1}{2} A \cos[(\omega_c + \omega_m)t]. \tag{4.27}$$

(*) Note that there is an erratum in the reference, which defines it as a group velocity.

If the signal propagates in a dispersive medium, then the propagation time for each component will be different and the equation should be rewritten:

$$x = \frac{1}{2}A\cos[(\omega_c - \omega_m)(t + t_{\mathrm{L}})] - \frac{1}{2}A\cos[(\omega_c + \omega_m)(t + t_{\mathrm{H}})], \tag{4.28}$$

where t_{L} is the ionospheric delay for the lower frequency signal and t_{H} is the ionospheric delay for the higher frequency signal.

In a dispersive medium signal velocity depends on signal frequency with an inverse square law. Thus, the delay in a propagation in dispersive medium can be expressed for a signal with a lower frequency as

$$t_{\mathrm{L}} = \frac{K}{(\omega_c - \omega_m)^2}, \tag{4.29}$$

and for a signal with a higher frequency as

$$t_{\mathrm{H}} = \frac{K}{(\omega_c + \omega_m)^2}, \tag{4.30}$$

where K is a proportionality constant. After substitution into (4.28) and assuming

$$\omega_c^2 \gg \omega_m^2 \tag{4.31}$$

it yields the following:

$$x = A \cdot \sin\left[\omega_m\left(t - \frac{K}{\omega_c^2}\right)\right] \cdot \sin\left[\omega_c\left(t + \frac{K}{\omega_c^2}\right)\right]. \tag{4.32}$$

Here we see that the code is delayed and the carrier advanced by the same amount. We can show with a GNSS receiver that code delay and phase advance are indeed equal in value.

For a ***monochromatic wave***, which is defined as

$$\omega t - kx = \text{const}, \tag{4.33}$$

the group velocity can be expressed as

$$c_g = \frac{dx}{dt} = \left(\frac{dk}{d\omega}\right)^{-1} \tag{4.34}$$

and phase velocity as

$$c_p = \frac{x}{t} = \frac{k}{\omega}. \tag{4.35}$$

Expressing k through n from (4.22) and (4.35), and substituting into (4.34), we can write.

$$c_g = c_0\left(n + \omega\frac{dn}{d\omega}\right)^{-1}. \tag{4.36}$$

In a non-dispersive medium the refractive index n does not depend on frequency and therefore $\frac{dn}{d\omega} = 0$. This leads to $c_g = c_p = c_0$ for wave propagation in a non-dispersive medium, which can also be expressed as

$$c_g \cdot c_p = c_0^2. \tag{4.37}$$

At the ***cutoff frequency*** the refractive index becomes zero. The phase velocity is increasing, and group velocity is becoming zero, because the second component in (4.36) becomes very large. In this case the wave will not propagate further.

In the case of the ionosphere, the refractive index satisfies the condition $n^2 < 1$. Therefore, from (4.17), phase velocity is higher than the speed of light. Moreover, the product of group and phase velocities is still constant and equation (4.37) is correct. A decrease of the group velocity is compensated by an increase in phase velocity. The GNSS signal code will arrive at a user antenna with a delay, while the carrier arrives earlier than it should have if it had traveled with the speed of light. Physicists normally explain this apparent inconsistency with the special theory of relativity by noting that the carrier wave does not carry any information, and therefore special relativity theory still holds. However, as we will in the next chapters, carrier waves actually carry information. We can derive positioning information using carrier waves. Therefore a phase advance in the ionosphere may in fact violate the special theory of relativity.

The way phase travels can be expressed through the refractive index,

$$l = \int c_0 dt = \int \frac{c_0}{c_p} dx = \int n(x, \omega) dx. \tag{4.38}$$

The delay can be calculated as a difference between this distance and ideal propagation when $n = 1$ as follows:

$$d = \int (1 - n(x, \omega)) dx. \tag{4.39}$$

4.4 GNSS models of propagation in the ionosphere

4.4.1 Model formalization through total electron contents

A structure of the Earth's ionosphere is schematically presented in Figure 4.6. The names of the layers were introduced by E. V. Appleton, another Cambridge graduate and recipient of Nobel Prize for Physics, who actually made a huge contribution not only to study of the atmosphere, but also to the development of radio. He started with letter D in order to have letters available to name layers both below and above. But it was found later that there were no other layers to be discovered below layer D.

Solar radiation and cosmic rays ionize the upper parts of the Earth's atmosphere and create free electrons and positively charged ions. Ions are the reason for atmospheric electricity, without ions we would have no thunderstorms, no lightning. The amount of solar radiation which reaches the Earth's atmosphere depends on time of day and time of

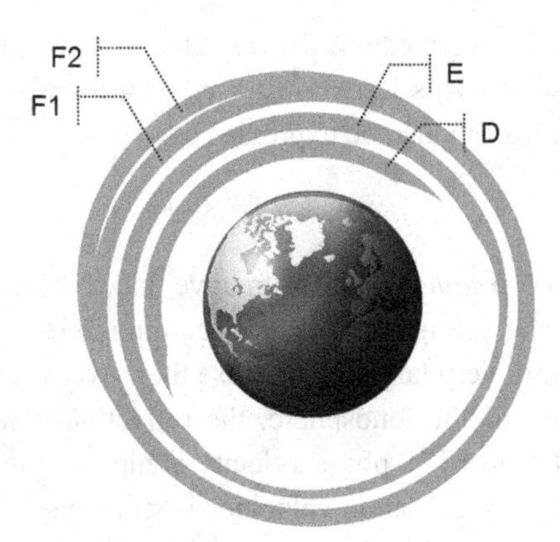

Figure 4.6 The Earth's ionosphere structure (not to scale) (after [5]), showing F, E, and D layer development in relation to the Sun's position relative to the Earth.

year. The number of sunspots is correlated with the amount of solar radiation and accordingly affects the level of ionization. The intensity of solar radiation varies through an 11 year cycle.

The dynamics of diurnal (daily), seasonal, and solar cycle variations in the layers of the atmosphere can be seen with the ReGen simulator (see Chapter 10 for more details).

We have shown in the previous section that group and phase velocity of a GNSS signal depends on refractive index. In a dispersive medium such as the ionosphere, refractive index depends on signal frequency. This dependency is caused by plasma ionization and in turn is a function of the number of free electrons along the ray path.

The total delay depends on refractive index along the path. Refractive index depends in turn on number of electrons. We can express the refractive index as a function of signal frequency (f) as follows [6],[7]:

$$n = 1 - \frac{K_x}{2} N_e \left(\frac{1}{f}\right)^2 \pm \frac{K_x K_y}{2} N_e H_0 \cos\theta \left(\frac{1}{f}\right)^3 - \frac{K_x^2}{8} N_e^2 \left(\frac{1}{f}\right)^4, \tag{4.40}$$

where N_e is the electron density, expressed in units of electrons per cubic meter, H_0 is the magnetic field strength, θ is the angle between the signal propagation direction and the vector of the Earth's magnetic field. Coefficients are expressed as follows:

$$K_x = \frac{e^2}{4\pi^2 \varepsilon_0 m_e}, \tag{4.41}$$

$$K_y = \frac{\mu_0 e}{2\pi m_e}, \tag{4.42}$$

where e is the electron charge, ε_0 is the dielectric constant in a vacuum, m_e is the mass of an electron, μ_0 is the magnetic permeability in a vacuum.

From (4.38) we can calculate delay as a nominal distance and an additional component which is a function of electron density. We are interested in total number of electrons integrated along the ray path. Therefore we introduce slant ***total electron content (TEC)***, which is defined as electron density integrated along the ray path,

$$\mathrm{TEC} = \int N_e(s)ds. \tag{4.43}$$

It is expressed in TEC units (TECU), with one TECU equal to 10^{16} electrons per a cylindrical volume with one square meter cross-section aligned along the line-of-sight.

Following [8] we can ignore the third and fourth terms in (4.40), which are a few orders smaller than the second term, and rewrite the equation as follows:

$$n = 1 - \frac{K_x}{2} N_e \left(\frac{1}{f}\right)^2. \tag{4.44}$$

Integrating refractive index (4.44) along the ***line-of-sight (LOS)*** ray path, we can rewrite (4.39) for the delay in the ionosphere as follows:

$$d_I = \frac{K_x}{2} \cdot \mathrm{TEC} \cdot \left(\frac{1}{f}\right)^2, \tag{4.45}$$

where

$$\frac{K_x}{2} \approx 40.3 \cdot 10^{16} \left[\frac{\mathrm{m}}{\mathrm{TECU} \cdot \mathrm{sec}^2}\right]. \tag{4.46}$$

From Equation (4.45) we can calculate a delay caused by TEC for all GNSS frequencies. For example, for GPS L1=1.57542 GHz we get approximately 0.162 [m/TECU].

The slant TEC is calculated along the ray path from a satellite to a receiver and is therefore a function of satellite and user position. Being unique to each user, slant TEC cannot be used for mapping the ionosphere. Therefore we use a vertical TEC (VTEC), which is TEC along the local vertical. Maps of VTEC can be supplied to users and each user should recalculate VTEC to LOS slant TEC for each satellite. The ionospheric maps are provided by IGS through the Internet in IONEX (IONosphere EXchange format).

A reduced TEC map, broadcast by GPS satellites, is considered in the next section.

In order to recalculate VTEC to slant TEC, we use a single layer model (see Figure 4.7). In this model we assume that all electrons given by VTEC are concentrated in the single layer located at the specific altitude. An altitude of 350 km is chosen for the GPS broadcast model. CODE uses 450 km for ionosphere analysis.

A mapping function needs to be constructed in order to transfer VTEC to slant TEC values. These mapping functions are also required in order to construct VTEC maps from ground network observations,

$$\mathrm{TEC} = M_{\mathrm{TEC}}(z) \cdot \mathrm{VTEC}. \tag{4.47}$$

The point at which a LOS ray penetrates the layer is called a ***pierce point***. The satellite zenith distance at the pierce point can be expressed as

$$\sin z' = \frac{R}{R+H} \sin z, \tag{4.48}$$

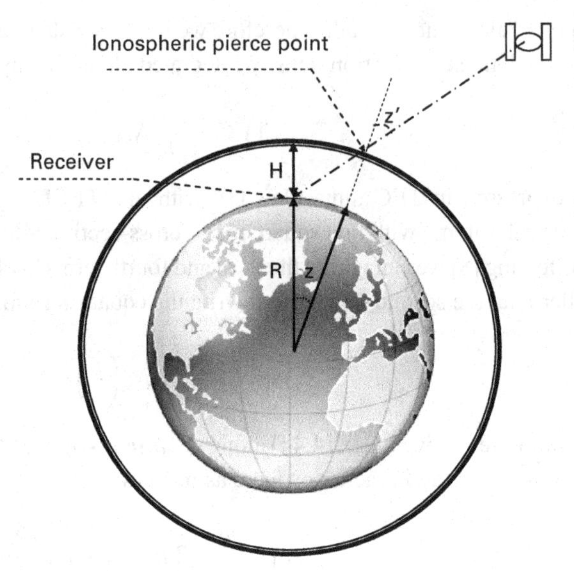

Figure 4.7 Single layer ionosphere model.

where R is the Earth's radius, H is the layer altitude, z is the satellite zenith distance at the receiver location.

From geometrical considerations, the single-layer mapping function can be defined as

$$M_{\mathrm{TEC}}(z) = \frac{1}{\cos z'} = \frac{1}{\sqrt{1 - \sin^2 z'}}. \tag{4.49}$$

4.4.2 GPS broadcast ionospheric model

GPS has adopted a Klobuchar ionospheric model. The Klobuchar model is a single-layer model, which takes only first members from a harmonic series (4.40) and defines a TEC map as a cosine-shaped bulge rotating synchronously with the Sun.

The mapping function for a GPS broadcast ionosphere model [9] implements a single layer at 350 km altitude and is expressed as

$$M_{\mathrm{TEC}}(z) = 1 + 2\left(\frac{z+6}{96}\right)^3, \tag{4.50}$$

where z is satellite zenith distance t from the receiver in degrees.

GPS satellites broadcast eight parameters for a single-layer model. An example of such parameters as they should be specified in the ReGen simulator is presented in Figure 4.8a, and this model is calculated and visualized in a ReGen ionospheric panel in Figure 4.8b and c.

The broadcast model should be used by a single-frequency user to correct for ionospheric delay using the algorithm described in GPS ICD [10]. It is estimated that this model corrects for more than 50% of single-user RMS error due to the ionosphere.

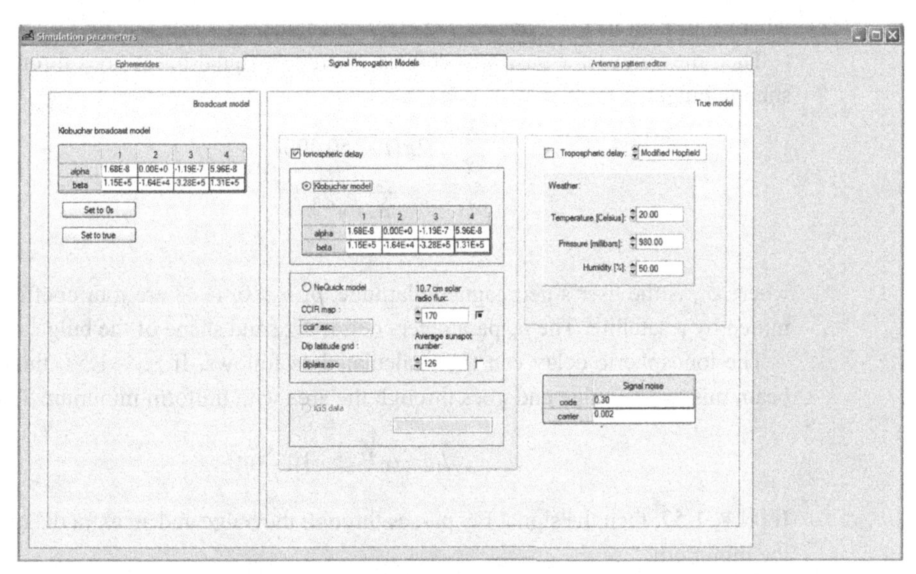

(a)

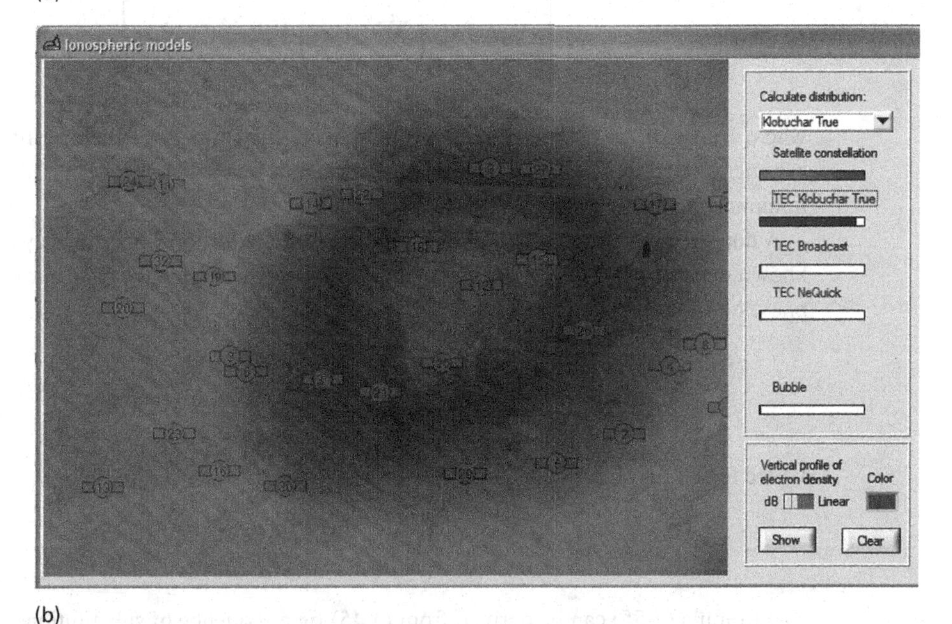

(b)

Figure 4.8 Klobuchar model parameters in ReGen GUI panel (a), and TEC distribution according to Klobuchar model in ReGen GUI panel (b).

We present the algorithm following [9], and encourage readers to refer to the ICD algorithms for any development. An ionospheric correction for a single user is calculated as follows.

Slant factor is calculated as a function of satellite elevation,

$$F = 1 + 16(0.53 - E)^3, \tag{4.51}$$

where F is slant factor, E is satellite elevation angle.

Then the value of x basically defines whether a signal beam goes through an ionospheric bulge:

$$x = \frac{2\pi(t - 50\,400)}{\sum_{n=0}^{3} \beta_n \phi_m^n} [\text{rad}], \tag{4.52}$$

where φ_m is the user's geomagnetic latitude, $\beta_n, n = 0, \ldots 3$ are four coefficients transmitted by a satellite. The β_n parameters define size and shape of the bulge.

The ionospheric delay can then calculated as follows. If $|x| > 1.57$, then the signal beam misses the bulge and goes through the area with uniform minimum TEC,

$$d_I = c_0 F \cdot 5 \cdot 10^{-9} [\text{m}] \tag{4.53}$$

If $|x| \leq 1.57$, then the signal ray passes through the bulge and an extra delay is added to the minimum:

$$d_I = c_0 F \left[5 \cdot 10^{-9} + \left(\sum_{n=0}^{3} \alpha_n \phi_m^n \right) \left(1 - \frac{x^2}{2} + \frac{x^4}{24} \right) \right] [\text{m}], \tag{4.54}$$

where $\alpha_n, n = 0, \ldots 3$ are four other coefficients transmitted by the satellite. The α_n parameters define the shape of the bulge and how much extra delay is added to the minimum.

A dual frequency user should be able to compensate for ionospheric delays using the known dependency of the GNSS signal delay on carrier frequency. The corrected code phase observables should be calculated as follows [10]:

$$\rho = \frac{\rho_j - \gamma_{i,j} \cdot \rho_i}{1 - \gamma_{i,j}}, \tag{4.55}$$

where for L1 and L2 P-code users

$$\gamma_{1,2} = \left(\frac{f_1}{f_2} \right)^2 = \left(\frac{1575.42}{1227.6} \right)^2. \tag{4.56}$$

The formula (4.55) can be derived from (4.45) by a sequence of substitutions.

A similar algorithm should be applied to other civil dual frequency signals. The algorithms for various combinations of L1-C/A, L2C, and L5-I/L5-Q users are similar to (4.55) and also include broadcast inter-signal correction terms (ISC) and group delay T_{GD} for each satellite,

$$\rho = \frac{(\rho_j - \gamma_{i,j} \cdot \rho_i) + c_0(\text{ISC}_j - \gamma_{i,j} \cdot \text{ISC}_i)}{1 - \gamma_{i,j}} - c_0 T_{\text{GD}}. \tag{4.57}$$

Accordingly,

$$\gamma_{1,5} = \left(\frac{f_1}{f_5}\right)^2 = \left(\frac{1575.42}{1176.45}\right)^2 \tag{4.58}$$

and

$$\gamma_{2,5} = \left(\frac{f_2}{f_5}\right)^2 = \left(\frac{1227.6}{1176.45}\right)^2. \tag{4.59}$$

4.4.3 Ionospheric error compensation in GLONASS receivers

A GLONASS navigation message does not provide a single-frequency user with information on ionospheric parameters. GLONASS receivers can compensate for ionospheric errors using (4.55) or the following analytical model [11]:

$$d_{\mathrm{I}} = \frac{d_{\mathrm{I}}^0}{\sqrt{1 - \left(\frac{R_{\mathrm{E}} \cos E}{R_{\mathrm{E}} + h_{\mathrm{LAYER}}}\right)^2}}, \tag{4.60}$$

where d_{I}^0 is minimum delay due to the ionosphere, R_{E} is the Earth's radius, E is satellite elevation angle, h_{LAYER} is single-layer altitude, which should be on the level corresponding to the electron density profile maximum.

The minimum delay can be calculated, for example, based on the Klobuchar model from (4.53). The slant factor should be adjusted accordingly.

Most GLONASS receivers also support GPS. In single-frequency GPS/GLONASS receivers, ionospheric delay can be compensated for using a GPS broadcast Klobuchar model for GLONASS satellites. The correction algorithms should be adjusted for GLONASS frequencies.

GLONASS transmit open access signals on L1 and L2 frequencies, therefore dual frequency users can find an ionospheric error on one frequency using the difference between measured code phases on two frequencies. Using (4.45) and substituting,

$$\Delta\rho = d_{\mathrm{I}}^{f_1} - d_{\mathrm{I}}^{f_2} = k(K_x, \mathrm{TEC})\left(\frac{1}{f_1^2} - \frac{1}{f_2^2}\right) = d_{\mathrm{I}}^{f_1}\left(1 - \frac{f_1^2}{f_2^2}\right), \tag{4.61}$$

$$d_{\mathrm{I}}^{f_1} = \Delta\rho\left(1 - \frac{f_1^2}{f_2^2}\right)^{-1}, \tag{4.62}$$

where $\Delta\rho$ is the difference in code phase measurements on the two frequencies. This difference includes errors resulting from unaccounted differences in code phase measurements on both frequencies. This error includes inter-frequency hardware biases on satellite and receiver, noises, ionospheric errors of higher orders, and so on.

For GLONASS, $\frac{f_1}{f_2} = \frac{9}{7}$ and the corresponding (4.62) ionospheric correction for a GLONASS L1 is given as follows:

$$d_{\mathrm{I}}^{f_1} = 1.531 \cdot \Delta\rho. \tag{4.63}$$

4.4.4 Ionospheric error compensation in GALILEO receivers – NeQuick model

The previously considered broadcast Klobuchar and IGS models were single-layer models and did not require a knowledge of vertical electron density profile. Therefore such models can be implemented using only ground-based observations, in this case a GPS global network. Recent advances in ionosphere mapping using occultation techniques and LEO satellites allow us to get information on vertical electron density profile and so to introduce multi-layer models.

The GALILEO system is planning to adopt a NeQuick model, which is a multi-layer model, sometimes called a profiler. It was first introduced in [12] and then continuously developed. The difference between single-layer models, such as Klobuchar and IGS global models, and a multi-layer model such as NeQuick is shown in Figure 4.9.

If we compare a TEC distribution given by a Klobuchar model with IGS and NeQuick models, we can see that the latter shows more harmonics in a TEC distribution map. Figures 4.8b and 4.10 demonstrate TEC distribution generated with a bundled ReGen signal generator for Klobuchar and NeQuick distributions.

NeQuick uses empirical data collected from ionograms, which are a result of ionosphere probing with various frequencies. Based on ionograms, NeQuick builds an electron density profile. Examples of such profiles constructed with ReGen for specific times and places are shown in Chapter 10.

Input for a NeQuick program is position and time and either solar flux or sunspot number. Sunspot number and solar flux are related by the following:

$$R_{12}=\sqrt{167273. + (F10.7 - 63.7) * 1123.6}-408.99, \tag{4.64}$$

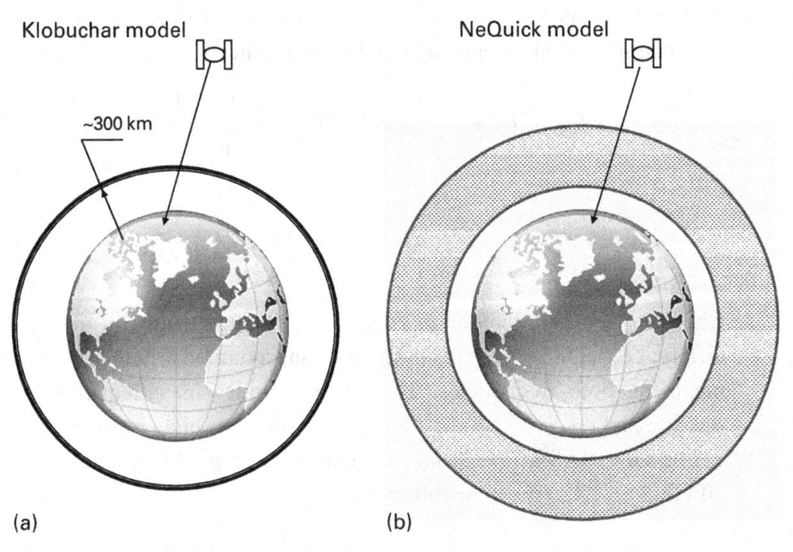

Figure 4.9 Single-layer (a), and multi-layer (b), ionosphere model (not to scale).

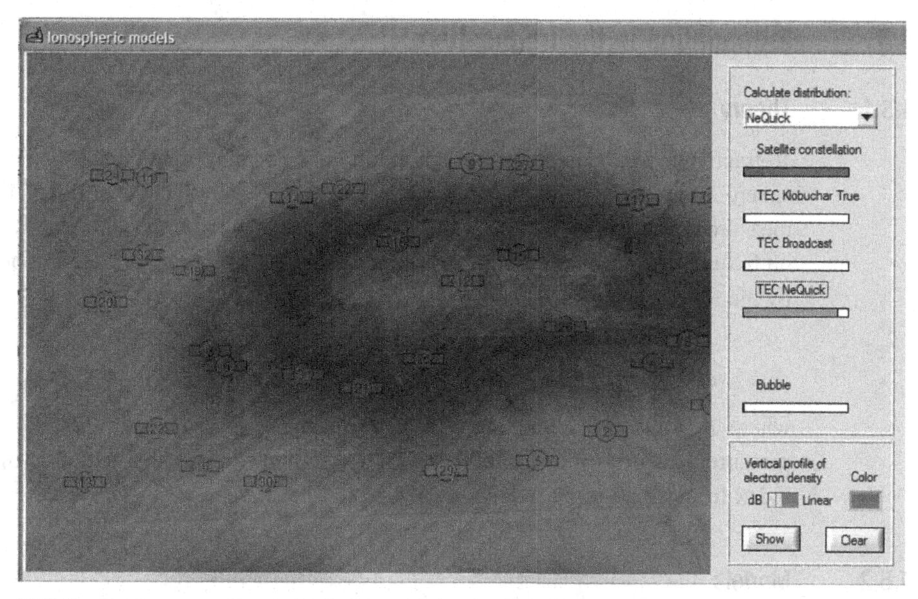

Figure 4.10 TEC distribution according to NeQuick model on ReGen GUI panel.

where R_{12} is a sunspot number, F10.7 is solar radio flux per unit frequency at a wavelength of 10.7 cm. This frequency is near the peak of the observed solar radiation.

NeQuick allows us to calculate electron concentration at any point in the ionosphere. We can integrate electron density along the line of sight between a receiver and a satellite. The resulting slant TEC can be directly translated to the ionospheric error.

In the case of implementation in GALILEO, it is proposed [13] that three parameters be calculated using the global reference station network and broadcast to a user on the following day.

4.4.5 Signal attenuation in the ionosphere

GNSS signals and radio waves in general are attenuated when they propagate through a medium. The attenuation is caused by collisions between electrons and other particles [1]. Refractive index is a complex variable, and so far we have considered only its real part. Refractive index radio wave attenuation is defined by

$$E = E_0 e^{-\frac{\omega}{c_0}\int \chi dx} = E_0 e^{-\Gamma}, \tag{4.65}$$

$$\Gamma = 20\lg\frac{E_0}{E} = 20\frac{\omega}{c_0}\int \chi dx \lg e, \tag{4.66}$$

$$\alpha = 20\frac{\omega}{c_0}\chi \lg e = 4.6\cdot 10^{-2}\frac{N_e\nu}{\omega^2+\nu^2}. \tag{4.67}$$

4.5 Propagation in the troposphere

4.5.1 Theory

Refractive index, temperature, and humidity have been routinely measured in the troposphere using balloon flights, airborne sampling, and reflective probing. From these measurements, it was found that the refractive index in the troposphere does not depend on frequency, and for the frequency range between 1 MHz and 30 GHz can be described by the following equation [1]:

$$n = 1 + \frac{77.6}{T}\left(p_{\mathrm{H}} + 4810 \cdot \frac{p_{\mathrm{W}}}{T}\right)10^{-6}, \tag{4.68}$$

where T is absolute temperature, p_{H} is atmospheric pressure in millibars, p_{W} is partial pressure of water vapor in millibars. As we showed before, this variation of refractive index from 1 will cause a delay according to (4.39).

4.5.2 Models

As can be seen from Equation (4.68), tropospheric delay has two components, a hydrostatic delay, which depends on dry pressure, and a wet delay, which is a function of water vapor. Hydrostatic delay can be modeled and determined much more precisely than wet delay.

Similarly to ionospheric models, tropospheric models describe slant delay in the troposphere as a product of zenith-angle-dependent mapping function and zenith delay [14]. Mapping functions map the actual path delay as a function of elevation angle and meteorological conditions to the zenith delay:

$$d_{\mathrm{T}} = M_{\mathrm{H}}(z)d_{\mathrm{H}} + M_{\mathrm{W}}(z)d_{\mathrm{W}}, \tag{4.69}$$

where $M_{\mathrm{H}}(z)$ is a hydrostatic mapping function , d_{H} is zenith hydrostatic delay in meters , $M_{\mathrm{W}}(z)$ is a wet mapping function, d_{W} is zenith wet delay in meters, and z is zenith angle in radians.

The various models are different in the way the zenith delays and mapping functions are calculated. The dry or hydrostatic component is responsible for 90% of tropospheric delay and can be well modeled. It demonstrates slow temporal variations on the level of one centimeter per six hours. The wet component has much higher variability, and it is hard to model due to dependency on highly variable water vapor pressure. The wet component can achieve up to 40 cm. The accuracy of the hydrostatic zenith delay (first term) is at the millimeter level while the wet delay (second and third terms) are calculated with an error of a few centimeters. Therefore, if the atmospheric pressure measurement is available, the hydrostatic delay is fixed and the wet delay is estimated. Moreover, the wet delay is sometimes divided into two terms, which are dependent on the azimuth of the satellite, in order to take into account the asymmetry of meteorological conditions.

A good approximation of tropospheric delay for navigation applications is given by the following mapping function. This simple approximation is sufficient for low accuracy or high elevation angles. However, at lower elevation angles, the nonuniform and finite width spherical shell model requires a more complicated function:

$$M_{\mathrm{H}}(z) \triangleq M_{\mathrm{W}}(z) \triangleq \frac{1}{\cos z}. \tag{4.70}$$

There are many more precise alternatives suitable for geodetic applications. For navigation applications another mapping function is given by Black and Eisner,

$$M_{\mathrm{H}}(z) \triangleq M_{\mathrm{W}}(z) \triangleq \frac{1.001}{\sqrt{0.002001 + \sin^2 \alpha}}, \tag{4.71}$$

where α is satellite elevation angle.

The Niel model [15] uses a three-term continued fraction as a mapping function,

$$M(\alpha) = \frac{\dfrac{1}{1 + \dfrac{a}{1 + \dfrac{b}{1+c}}}}{\dfrac{1}{\sin(\alpha) + \dfrac{a}{\sin(\alpha) + \dfrac{b}{\sin(\alpha) + c}}}}, \tag{4.72}$$

where a,b,c are the coefficients for hydrostatic and wet mapping functions. To correct the mapping function according to the height, the following form is adopted:

$$\Delta M(\alpha) = \frac{dM(\alpha)}{dh} H, \tag{4.73}$$

where H is the height of the site above sea level, and

$$\frac{dM(\alpha)}{dh} = \frac{1}{\sin(\alpha)} - f(\alpha, a_{ht}, b_{ht}, c_{ht}), \tag{4.74}$$

where $f(\alpha, a_{ht}, b_{ht}, c_{ht})$ is a similar three-term continued fraction with parameters determined by least squares fits to the height corrections at the nine elevation angles.

The widely used Saastamoinen model [16] is given by

$$d_{\mathrm{T}} = \frac{0.002277}{\cos z}\left[p_{\mathrm{H}} + \left(\frac{1255}{\mathcal{T}} + 0.05\right)p_{\mathrm{W}} - \tan^2 z\right]. \tag{4.75}$$

This model implicitly contains both components and mapping functions.

Input values for temperature $\mathcal{T}$, atmospheric pressure p_{H}, and humidity are derived from a standard atmosphere model as functions of receiver height:

$$p_{\mathrm{H}} = p_{\mathrm{H}_0}(1 - 0.0000226(h - h_0))^{5.225}[\text{millibar}], \tag{4.76}$$

$$\mathcal{T} = \mathcal{T}_0 - 0.0065(h - h_0)[\text{Celsius}], \tag{4.77}$$

$$\mathcal{H} = \mathcal{H}_0 \cdot e^{-0.0006396(h-h_0)}[\%]. \tag{4.78}$$

The partial pressure of water vapor can be obtained from humidity,

$$p_W = \mathcal{H} \cdot e^{\left(-37.2465+0.213166\mathcal{T}-0.000256908\mathcal{T}^2\right)}. \tag{4.79}$$

Temperature must be transformed from Celsius to Kelvin by

$$\mathcal{T}[\text{Kelvin}] = \mathcal{T}[\text{Celsius}] + 273.16. \tag{4.80}$$

The reference values are given by the standard atmosphere model for altitude $h = 0$,

$$\begin{aligned} p_H &= 1013.25 \text{ millibar}, \\ \mathcal{T} &= 18°\text{Celsius}, \\ \mathcal{H} &= 50\%. \end{aligned} \tag{4.81}$$

4.6 Modeling atmospheric errors in a receiver and in a simulator

In this chapter we use a ReGen signal simulator to generate ionospheric maps using Klobuchar and NeQuick models. Note that there are always two ionospheric models in GPS and future Galileo simulators. One ionospheric model is used for calculating ionospheric delays and phase advances in generated code and carrier phases. This model is denoted as a ***true model***. The other model is used to set up parameters in broadcast navigation messages. A simulator should allow us to introduce parameters for both true and broadcast models. Figure 4.8a shows such a panel in a ReGen signal simulator. True and broadcast models can be identical, in which case a simulated ionospheric error should be compensated for completely. Figure 4.11 shows the concept of introducing an error into the simulated signal. A true model can be introduced as a more detailed and complicated model as in Figure 4.11a, or just filled with a different set of parameters (Figure 4.11b). In the case of ionospheric error, a true model can be calculated based on IGS maps or NeQuick mode. It is important that a more complex model be used as a basis for calculating a simplified model, or they should be derived from measurements taken at the same time. For example, the Klobuchar model depicted in Figure 4.8b is calculated

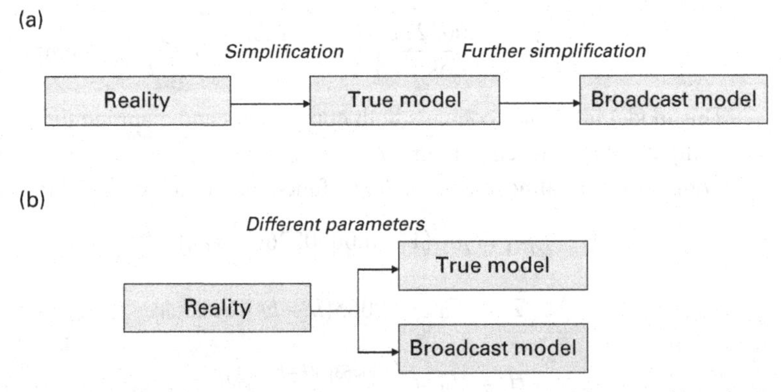

Figure 4.11 (a, b) Concept of introducing an inaccuracy in to a simulated signal with true and broadcast models.

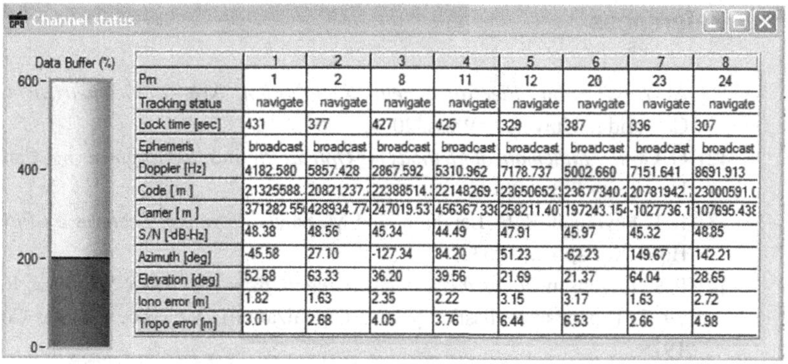

Channel status

Data Buffer (%)

600- 400- 200- 0-

	1	2	3	4	5	6	7	8
Prn	1	2	8	11	12	20	23	24
Tracking status	navigate	navigate	navigate	navigate	navigate	navigate	navigate	navigate
Lock time [sec]	431	377	427	425	329	387	336	307
Ephemeris	broadcast	broadcast	broadcast	broadcast	broadcast	broadcast	broadcast	broadcast
Doppler [Hz]	4182.580	5857.428	2867.592	5310.962	7178.737	5002.660	7151.641	8691.913
Code [m]	21325588.	20821237.	22388514.	22148269.	23650652.	23677340.	20781942.	23000591.
Carrier [m]	371282.55	428934.77	247019.53	456367.33	258211.40	197243.15	-1027736.1	107695.43
S/N [-dB-Hz]	48.38	48.56	45.34	44.49	47.91	45.97	45.32	48.85
Azimuth [deg]	-45.58	27.10	-127.34	84.20	51.23	-62.23	149.67	142.21
Elevation [deg]	52.58	63.33	36.20	39.56	21.69	21.37	64.04	28.65
Iono error [m]	1.82	1.63	2.35	2.22	3.15	3.17	1.63	2.72
Tropo error [m]	3.01	2.68	4.05	3.76	6.44	6.53	2.66	4.98

Figure 4.12

from parameters taken during a year with low solar activity, and the NeQuick model is taken from a year of high solar activity, so the amount of error is different. Such models can be used together in order, for example, to estimate a worst case scenario of incorrect ionospheric modeling for academic or research purposes.

Generally, however, if a NeQuick model is used as a true model for GPS simulation, or GPS and GALILEO are simulated in one session, the Klobuchar and NeQuick models should be properly correlated. This is relevant only for a limited number of demanding applications. To derive parameters of one model through given parameters of another model involves basically a simulation of TEC estimation algorithms (see Chapter 12). We can only derive Klobuchar model parameters from a NeQuick model. This is because a Klobuchar model is a single-layer model and therefore it does not have enough information to fill the vertical profiles of a NeQuick model.

4.7 Project

1. Generate a scenario with ReGen using the following models as a true ionospheric model:
   a. Klobuchar model,
   b. NeQuick model.
2. Generate a scenario with ReGen without an ionospheric error.
3. Create a RINEX file for all the above cases and see how code and carrier phase measurements are affected.
4. Use the scenario with applied true models to generate a GPS signal with a ReGen simulator with and without ionospheric error in a broadcast model.
5. Run an iPRx receiver and process simulated signals with various combinations of true and broadcast models.
6. See what errors are calculated for each satellite by the receiver and how the positioning is affected in each case. Figure 4.12 shows a screenshot of an iPRx status panel with ionospheric errors estimated for each satellite.
7. Run an iPRx receiver in carrier smoothing mode to see how it is affected by code-carrier divergence.

References

[1] A. D. Wheelon, *Electromagnetic Scintillation,* Vol. I. *Geometrical Optics*, Cambridge, Cambridge University Press, 2001.

[2] G. Gbur, *Mathematical Methods for Optical Physics and Engineering*, Cambridge, Cambridge University Press, 2011.

[3] R. P. Feynman, R. B. Leighton, and M. Sands, *Feynman Lectures on Physics*, Portland, OR, Book News, Inc., 1963.

[4] P. S. Jorgensen, *Ionospheric Measurements from NAVSTAR Satellites*, Report SAMSO-TR-79–29 (Space and Missile Systems Organization), Air Force Systems Command, December 1978.

[5] R. D. Hunsucker and J. K. Hargreaves, *The High-Latitude Ionosphere and its Effects on Radio Propagation*, Cambridge, Cambridge University Press, 2003.

[6] K. G. Budden, *The Propagation of Radio Waves*, Cambridge, Cambridge University Press, 1985.

[7] F. Brunner and M. Gu, An improved model for dual frequency ionospheric correction of GPS observations, *Manuscripta Geodaetica*, **16**, 1991, 205–214.

[8] S. Schaer, *Mapping and Predicting the Earth's Ionosphere Using the Global Positioning System*, Volume 59 of *Geodätisch-geophysikalische Arbeiten in der Schweiz*, Schweizerische Geodätische Kommission, Institut für Geodäsie und Photogrammetrie, Switzerland, Eidg. Technische Hochschule Zürich, Zürich, 1999.

[9] J. A. Klobuchar, Ionospheric time-delay algorithm for single-frequency GPS users, *IEEE Transactions on Aerospace and Electronic Systems*, **23**, 1987, 325–331.

[10] GPS IS, *Navstar GPS Space Segment/Navigation User Interfaces*, GPS Interface Specification IS-GPS-200, Rev D, GPS Joint Program Office, and ARINC Engineering Services, March, 2006.

[11] *GLONASS : Design and Operations Concepts*, A. Perov and V. Harisov (editors), 4th edition, Moscow, Radiotechnica, 2010 (in Russian).

[12] G. Di Giovanni and S. M. Radicella, An analytical model of the electron density profile in the ionosphere, *Adv. Space Res.*, **10** (11), 1990, 27–30.

[13] S. M. Radicella, The NeQuick model genesis, uses and evolution, *Annals of Geophysics*, **52**, (3/4), June/August 2009, 417–422.

[14] F. Kleijer, *Troposphere Modelling and Filtering for Precise GPS Leveling*, Volume 56 of *Publications on Geodesy*, Netherlands Geodetic Commission, Delft, The Netherlands, 2004.

[15] A. E. Niel, Global mapping functions for the atmosphere delay at radio wavelengths, *Journal of Geophysical Research*, **101**, (B2), February 10, 1996, 3227–3246.

[16] J. Saastamoinen, Contributions to the theory of atmospheric refraction, *Bull. Géodésique*, 1973, (105), 270–298, (106), 383–397, (107), 13–34.

5 Receiver RF front end

We have described how a GNSS signal is generated in Chapter 3. In Chapter 4 we examined how a GNSS signal propagates through the Earth's atmosphere. In this chapter we discuss how a radio frequency signal is converted by a GNSS receiver front end to a digital format for further processing. The scope of this chapter is shown schematically in Figure 5.1. The purpose of this chapter is to describe a design of a receiver front end and the operations of its major components, and to analyze how the design of these components affects a GNSS signal.

5.1 RF front end for software GNSS receiver

5.1.1 Generic GNSS receiver

A generic receiver flowchart is presented in Figure 5.2. The receiver in the flowchart does not include a navigation processor. A receiver acquires a ***radio frequency (RF)*** signal coming through atmosphere from a satellite. The signal goes to a front end. On the output of the front end we have a digitized intermediate frequency signal, which can then be processed either in digital hardware or in the software. In the front end the signal is filtered by a bandpass filter and then down-converted from RF to ***intermediate frequency (IF)***. (Figure 5.3 shows an example of a simplified frequency plan for the front end.) Next the IF signal goes through an ***analog to digital converter (ADC)***, and the resulting ***digitized IF (DIF)*** signal goes from the front end to a baseband processor. The baseband processor performs signal acquisition and tracking. The acquisition and tracking are achieved by means of correlating the incoming signal in the shape of DIF with a replica signal which is generated within the receiver. The baseband processor outputs raw measurements, in particular code phase, carrier phase, Doppler, and signal-to-noise ratio (SNR) measurements. Baseband processor operation is described in detail in the next chapter.

The measurements from a baseband processor can then be recorded for further processing or passed to a receiver navigation processor along with ephemeris information to compute antenna coordinates. If the receiver has to output coordinate estimates immediately, then the ephemeris information can come either from a navigation message or from an outside source, such as a cellular phone network, in the case of assisted GPS (AGPS). If the raw measurements are recorded for post-processing, one can use more precise ephemeris information from IGS or CODE, as described in Chapter 2. As we see

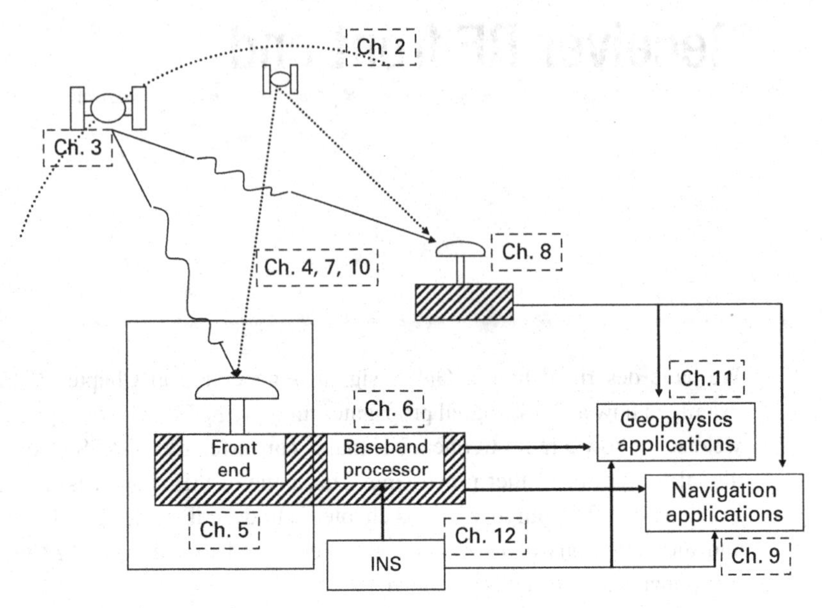

Figure 5.1 Subject of Chapter 5.

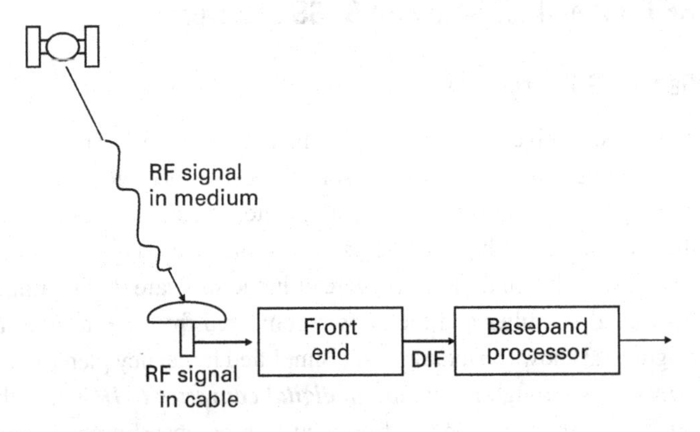

Figure 5.2 Software GNSS receiver design concept.

later in this chapter, one can record not only raw measurements, but also the entire digitized IF signal from the front-end output, and process it later.

5.1.2 Software GNSS receiver

A software GNSS receiver concept has been developed in recent years [1]–[6], which can be described by the same Figure 5.2. We intentionally exclude a navigation processor from this figure, because for many applications a navigation processor may be outside the receiver. The difference between conventional and software receivers is in how a baseband processor is implemented. A conventional

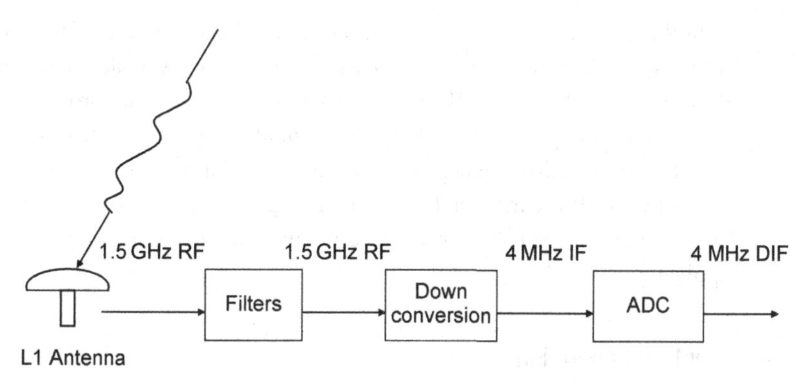

Figure 5.3 Simplified front-end frequency plan.

GNSS receiver normally has its baseband processor implemented in FPGA or ASIC. The navigation functions of the conventional receiver are normally located in an embedded processor. The software receiver implements a baseband processor in software, which allows it to collocate with a navigation processor and even to put all this functionality into a general processor along with an application part which uses raw measurements for a particular task.

The front end is the only hardware part of the software receiver, with all other parts allocated to a general processor. Therefore, a software receiver in general may become a solution of choice for many low-end applications where price reduction is essential. For high-end applications, an advantage of a software receiver is its flexibility. Baseband processor functions of the conventional receiver with ASIC implementation are hard-coded. If it is required to change something, the hardware should be redesigned. In the case of an FPGA processor, the changes can be done without change of the design if they are within FPGA resources, but FPGA programming still gives much less flexibility because it is hardware dependent. Changes in the software baseband processor can be much more easily implemented.

There are various implementations of the software receiver concept. For research and educational purposes we mostly concentrate on a software receiver implemented on a general purpose personal computer (PC) running under Windows OS. A free version of a real-time GNSS receiver is available for download from the book website. A front end for this free version of the software receiver is also readily available. It allows the receiver to work in real-time with live satellites. The book website also contains prerecorded data, which a reader can process with the receiver while working with the book examples and projects.

Other implementations of the software receiver are optimized in terms of computational and memory load and already used in numerous applications. One such application is a cellular phone. An advantage of using a software receiver in such applications is cutting the cost of the solution. The only hardware which is required for a software receiver is a front end. The baseband processor is moved to a general processor, where a navigation processor already resides.

In this chapter we concentrate on the front-end component of the receiver. The basic design of the front-end core is mostly the same for software and hardware receivers. The difference is in how the DIF signal is supplied to a baseband processor.

The range of applications which may benefit from a particular solution depends on the quality of the measurements, which in the case of a software receiver is only limited by the front end. If the front end can provide high quality measurements, then the resulting observables can be used for a whole range of navigational, geodetic, and geophysical applications.

5.1.3 Front-end operation

A front-end flowchart is presented in Figure 5.4 for hardware (a) and software (b) receivers. When we look at a front end, we consider two types of circuit dealing with two types of signal. The first type is analog RF signals and the second digital signals, which have a relatively low frequency defined by the front-end clocks. Today RF circuitry for GNSS signals is already well developed and provided as off-the-shelf modules or chips. When using an off-the-shelf module or chip, one needs to develop an interface between it and the receiver baseband processor. This interface is also basically an analog circuitry, which works with digitized and down-converted GNSS signals.

When an RF signal is received by the antenna it goes to the RF front-end module, where it is first amplified by a ***low noise amplifier (LNA)***. The LNA may also be realized separately from the front-end module. After the LNA the signal is filtered and down-converted to IF. Down-conversion is done by mixing the incoming signal and a harmonic wave in a mixer device. Then the IF signal is digitized and quantized by ADC. After that the DIF signal should be supplied to the baseband processor. In the case of a software receiver, we need to provide an interface between the front-end module and host computer. This interface consists of a logic device to convert the data stream to the required format, a memory buffer and one or another port interface devices. One can use USB, LAN, parallel port or FireWire interfaces to transfer data to a host computer.

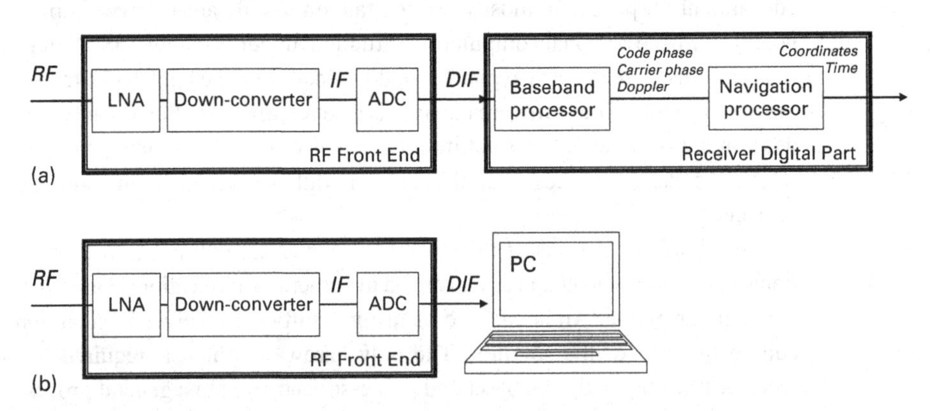

Figure 5.4 Front-end flowchart for hardware (a), and software (b), receiver.

5.1.4 Front-end bandwidth

Requirements for GPS, GALILEO, and GLONASS front ends are defined in particular by a signal's bandwidth. In Chapter 3 we looked at GPS L1 C/A code. As seen in Chapter 3, GPS L1 C/A code has bandwidth 2.046 MHz, defined by its chip rate. A narrow-band front-end limits the signal to the main lobe of the spectrum only. Other higher frequency components are removed from the signal as it passes through the front end. This process also removes out-of-band interference. However, due to an effective decrease in sampling rate it also decreases resolution of the signal processing algorithms in the baseband processor. Wide-band front ends include several side lobes as well. This can be useful for some applications, including multipath mitigation.

GPS L1C, L2C, and L5 front ends have their bandwidths defined by the corresponding signals. GPS L1C, L2C, and L5 have their bandwidths equal to 4.092 MHz, 2.046 MHz, and 24 MHz respectively. These values, as we have seen in Chapter 3, are defined by code design and chip rate. GLONASS L1 front-end bandwidth is defined not only by the GLONASS signal chip rate, but also by the frequency range, which contains L1 signals from all GLONASS satellites. Therefore a GLONASS L1 front end should have a minimum bandwidth of about 8 MHz.

Front ends can be roughly classified as narrow-band or wide-band front ends. A wide-band GPS front end is required for geodetic dual frequency receivers in order to be able to process L1 P-code or L5 signals. Navigation applications work with a narrow-band L1 front end with bandwidth limited to 2.046 MHz for GPS, 4.092 MHz for GALILEO and 8 MHz for GLONASS. Low-end GPS L1 receivers can even work with a front end which provides a narrower bandwidth, for example 1.8 MHz. Narrowing bandwidth for such applications allows a baseband processor to operate on a lower sample rate.

5.2 Antenna

An antenna connects front-end hardware to the physical world. A simple antenna is a quarter wavelength dipole, or piece of wire. Surprisingly enough, this piece of wire can provide a standard off-the-shelf receiver with enough signal power to make a positioning. Figure 5.5 shows a Garmin receiver successfully working with such a simple wire antenna. More elaborated designs, still constructed out of wire (made after [17]), provide even more signal power. Figure 5.6 shows a wire antenna operating with an iPRx software receiver. Impedance of the hardware elements should match each other, the interface to a host computer, and the physical world. Impedance mismatch results in loss of power, which depending on the hardware elements, will in turn result in more power consumption, less optimal performance and loss of sensitivity up to a total operational failure. The physical world, as far as a receiver is concerned, is represented by a radiowave signal propagating in free space. The impedance of the medium to the electromagnetic wave is calculated as the ratio of the electric field to the magnetic field,

Figure 5.5 Quarter-wavelength dipole antenna working with Garmin GPS receiver.

Figure 5.6 Homemade wire antenna working with iPRx receiver.

$$Z \equiv \frac{E}{H} = \sqrt{\frac{\mu}{\varepsilon}}. \tag{5.1}$$

In free space the radiowave impedance Z_0 is therefore about 377 Ω (ohm). The impendence in a medium is a function of the refractive index,

$$Z = \frac{Z_0}{n}. \tag{5.2}$$

A requirement for all RF hardware circuits is to have an impedance value equal to 50 Ω. Coaxial cables also usually have an impedance of 50 Ω. The 50 Ω standard was selected in the nineteen thirties as a compromise solution between 30 Ω (best power handling) and 77 Ω (lowest loss) for coaxial cables. The antenna must convert radiowave signal impedance to a value about 50 Ω, which can be easily matched to a cable leading to a front end.

5.2.1 Antenna gain pattern

An omni-directional antenna transmits signal equally distributed in all directions. Signal power is then equally distributed over a sphere. Power on a surface unit can be expressed as

$$P_0 = \frac{W_T}{4\pi r^2}, \tag{5.3}$$

where W_T is transmitting antenna power, r is distance from the transmitting antenna phase center to a point at which power is measured. Because of this distribution over a sphere's surface, the dependence of electromagnetic wave power on distance (similarly to gravitational force) follows an inverse square law.

An electric field induces a current in a receiving antenna. The current goes through the cable to the front end. ***Antenna aperture*** (or effective area S_A) is defined as the ratio of antenna produced power to the power of the received signal,

$$S_A = \frac{P}{W_R}, \tag{5.4}$$

where p is antenna produced power, W_R the power of the received signal. ***Antenna gain*** is defined by antenna aperture,

$$G_A = \frac{4\pi S_A}{\lambda^2}, \tag{5.5}$$

where G_A is antenna gain, expressed as a ratio, λ is radiowave wavelength. The antenna gain is therefore defined by how many squared wavelengths fit in the antenna effective area. If the signal frequency is changed, then the aperture should be changed as well in order to maintain the same gain.

Antenna aperture (effective area) can be estimated, as a rule of thumb, as a half of the physical antenna area,

$$S_A \approx 0.5 S_G, \tag{5.6}$$

where S_G is physical (or geometrical) antenna area.

Transmitting and receiving antennas are reciprocal, therefore, antenna pattern and aperture are the same for an antenna whether it is transmitting or receiving. Commonly-used light emitting diodes (LEDs) can be viewed as an example of receiving–transmitting reciprocity when they are used to measure atmospheric water vapor [8]. When current goes through a LED it emits a light with a specific wavelength. LEDs are reciprocal, and when they are exposed to light they produce a current, which depends on the light spectrum and the LEDs color. The spectral response of a LED is in general different from an emission spectrum, and therefore they require calibration. Water vapor absorbs light waves in the infrared; therefore one can derive information on water vapor by observing two LEDs on the ground, one of which is as close to the infrared part of the spectrum as possible and the other as far as possible.

A ***directional antenna*** provides a gain in accordance with its antenna pattern. Antenna gain pattern defines the signal power transmitted by an antenna as a function of direction. GNSS satellite transmitting antennas are directional with a narrow antenna pattern. Because the angle in which a satellite antenna is radiating is reduced, the directional gain goes up. For a hemispheric pattern all energy is concentrated in half of the original area. Therefore

$$P_0 = \frac{2 \cdot W_{\mathrm{T}}}{4\pi r^2}. \tag{5.7}$$

This gives an extra 3 dB gain. The satellite antennas also have sidelobes. These sidelobes become useful for spaceborne applications.

Receiver antennas usually have a hemispheric antenna pattern. Antenna gain is normally specified by its value in the direction of maximum gain in its pattern profile. Power generated in the receiving antenna can be defined as

$$P = \frac{W_{\mathrm{T}} S_{\mathrm{A}}}{4\pi r^2}. \tag{5.8}$$

A signal simulator should be able to simulate an antenna gain pattern for the receiver antenna. Figure 5.7 demonstrates an example of the antenna pattern editor in a ReGen simulator. For spaceborne applications a signal simulator should also be able to simulate transmitting antenna gain pattern, including sidelobes. Simulation of transmitting antenna gain pattern is not necessary for any ground-based and even aviation applications.

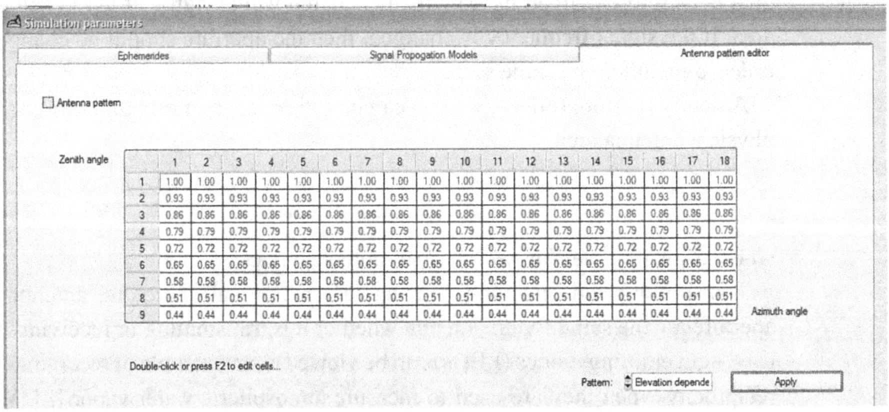

	1	2	3	4	5	6	7	8	9	10	11	12	13	14	15	16	17	18
1	1.00	1.00	1.00	1.00	1.00	1.00	1.00	1.00	1.00	1.00	1.00	1.00	1.00	1.00	1.00	1.00	1.00	1.00
2	0.93	0.93	0.93	0.93	0.93	0.93	0.93	0.93	0.93	0.93	0.93	0.93	0.93	0.93	0.93	0.93	0.93	0.93
3	0.86	0.86	0.86	0.86	0.86	0.86	0.86	0.86	0.86	0.86	0.86	0.86	0.86	0.86	0.86	0.86	0.86	0.86
4	0.79	0.79	0.79	0.79	0.79	0.79	0.79	0.79	0.79	0.79	0.79	0.79	0.79	0.79	0.79	0.79	0.79	0.79
5	0.72	0.72	0.72	0.72	0.72	0.72	0.72	0.72	0.72	0.72	0.72	0.72	0.72	0.72	0.72	0.72	0.72	0.72
6	0.65	0.65	0.65	0.65	0.65	0.65	0.65	0.65	0.65	0.65	0.65	0.65	0.65	0.65	0.65	0.65	0.65	0.65
7	0.58	0.58	0.58	0.58	0.58	0.58	0.58	0.58	0.58	0.58	0.58	0.58	0.58	0.58	0.58	0.58	0.58	0.58
8	0.51	0.51	0.51	0.51	0.51	0.51	0.51	0.51	0.51	0.51	0.51	0.51	0.51	0.51	0.51	0.51	0.51	0.51
9	0.44	0.44	0.44	0.44	0.44	0.44	0.44	0.44	0.44	0.44	0.44	0.44	0.44	0.44	0.44	0.44	0.44	0.44

Figure. 5.7 ReGen signal simulator antenna pattern editor panel (elevation dependence in the parameters is exaggerated).

5.2.2 Polarization

A GPS signal is ***right-hand circularly polarized (RHCP)***. An electric field component of an electromagnetic wave is oriented perpendicular to the wave's path. When the signal is linearly polarized it means that the electric field component oscillates in the same plane in the same direction consistently. In this case, field direction is aligned to the antenna. A vertical antenna will create a vertically, linearly polarized electromagnetic wave. When the signal is circularly polarized, the electric field component is rotating in the plane oriented perpendicular to the wave's path. A receiving antenna which is aligned with the polarization has maximum sensitivity, because the oscillations of the electromagnetic field, which induce the current in the antenna, are aligned with the antenna.

Natural light sources, such as the sun or candles, transmit non-polarized light. The electric field component is oriented randomly in the plane perpendicular to the wave's path. Non-polarized light becomes polarized by scattering in the atmosphere. That is why we see skies as blue. When the signal propagates through the atmosphere, it may encounter sudden changes in polarization caused by scattering, which for line-of-sight propagation is mostly negligible.

A GPS transmitting antenna is circular polarized because it eliminates the necessity to match the orientation of the receiving antenna to the orientation of the satellite's antenna. Right-hand circular polarization (RHCP) of a GPS signal means that looking from the source, the rotation is clockwise. To provide an optimal reception, a receiving antenna should have the same type of polarization as the transmitting antenna. We have described a multipath error in Chapter 3, which for many applications occupies the first place in an error budget. A reflected GNSS signal changes polarization, and therefore becomes attenuated by an antenna, so that such polarization allows a GNSS signal to be less affected by multipath. A second reflection, however, restores signal polarization, and a twice reflected signal is not attenuated by the antenna.

5.2.3 Antenna design

Antenna designs for GNSS vary from thin microstrip patch antennas to large multipath mitigating helical coils. Different applications have different antenna requirements. When choosing an antenna for a particular application, one should look at antenna gain pattern and multipath mitigation characteristics. For airborne applications one should also consider antenna aerodynamic performance and mounting area (see Figure 5.8). For geodetic applications, stability of antenna phase center is important. In navigation applications we may also need geodetic grade antenna for reference stations, which supply GNSS corrections to users (Figure 5.9).

Power produced by an antenna can be expressed through current induced in the antenna and ***radiation resistance*** as follows:

$$P = \frac{1}{2} R_{\mathrm{RAY}} I^2. \tag{5.9}$$

The radiation resistance can be expressed for a simple dipole antenna as follows [9]:

Figure 5.8 Aircraft mounted antenna.

$$R_{\mathrm{RAY}} = \frac{2\pi}{3}\sqrt{\frac{\mu_0}{\varepsilon_0}}\left(\frac{l}{\lambda}\right)^2, \tag{5.10}$$

where l is length of dipole. From the above equations we can see that small antennas operate less effectively. Antennas with sizes of the order of the wavelength have better characteristics. If an antenna has a length of $\lambda/2$, then a stationary current can be established in the antenna. When the length of the antenna is $\lambda/4$, then it acts as a half-wave antenna, as it generates a symmetrical image in a conductor plane.

A simple patch antenna is presented in Figure 5.10. A patch antenna consists of a very thin metal patch on a dielectric substrate, which is set on top of preferably a large metal ground plane. The height of the dielectric layer is usually chosen to be

$$h \leq 0.02\lambda_d, \tag{5.11}$$

where λ_d is a signal wavelength [10] in the dielectric.

If we remove the ground plane, then the patch antenna operates as a simple dipole antenna. The ground plane allows us to double antenna gain in the same way as we described above for any directional antenna. The length of the patch (L) optimally should be equal to half of the wavelength. The width (W) is less important. It can be chosen from the signal wavelength and dielectric parameters as follows [10]:

$$W = \frac{\lambda_0}{2}\left(\frac{\varepsilon_r + 1}{2}\right)^{-1/2}, \tag{5.12}$$

where λ_0 is the signal wavelength, ε_r is the dielectric constant of the dielectric substrate.

Figure 5.9 Reference station antenna on top of JAXA research facilities building (on the right with radome , on the left w/o radome).

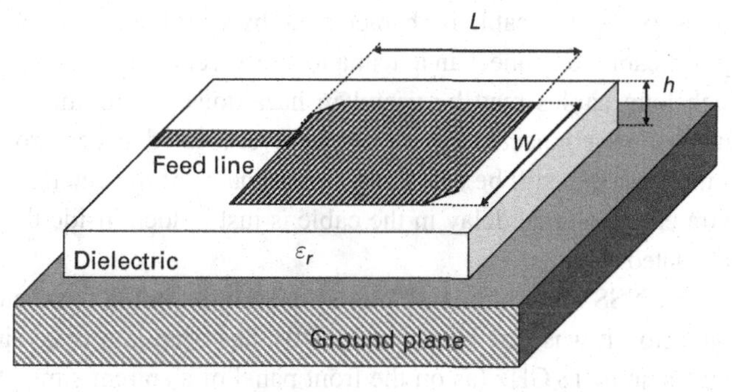

Figure 5.10 Patch antenna.

The patch length (L) for the specific dielectric can be calculated as follows [10]:

$$L = \frac{\lambda_d}{2} - 2\Delta l = \frac{\lambda_0}{2\sqrt{\varepsilon_{\text{eff}}}} - 2\Delta l, \tag{5.13}$$

where ε_{eff} is effective dielectric constant and Δl is an edge extension correction term, given by:

$$\varepsilon_{\text{eff}} = \frac{\varepsilon_r + 1}{2} + \frac{\varepsilon_r - 1}{2}\left(1 + \frac{12h}{w}\right)^{-1/2}, \tag{5.14}$$

and

$$\Delta l = 0.412h\left(\frac{\varepsilon_{\text{eff}} + 0.3}{\varepsilon_{\text{eff}} - 0.258}\right)\frac{\frac{W}{h} + 0.264}{\frac{W}{h} + 0.8}. \tag{5.15}$$

The frequency of the induced wave will be higher than in free space because of the dielectric. If a ceramic with high dielectric constant is used for loading, the size of a patch antenna can be significantly reduced to fit, for example, into cellular phones [11]. Further miniaturization can be achieved by using quarter-wave antennas. In comparison with a dipole antenna, the quarter-wave antenna has its ground plane replacing the half-wave dipole null potential. In order to produce and receive a circularly polarized signal, an antenna either has two feeds or has a rectangular shape instead of square and has one or two of its corners clipped.

An antenna is designed to work at a specific frequency. Therefore it can be modeled also as a pass-band filter. A GPS antenna usually has a bandwidth of about 2% of the signal center frequency. Therefore bandwidths for L1, L2, and L5 antennas are about 31.5 MHz, 24.6 MHz, and 23.5 MHz respectively [12].

5.2.4 Cables and cable connectors

There are several connectors which we use to connect an antenna to a receiver. Cables are characterized by a power loss, which depends on cable design and material and is also a function of cable length. There are coaxial and optical RF cables for GNSS applications. For a coaxial cable a length of more than 50 meters is not recommended. More than a few meters of coaxial cable may require use of an external amplifier to compensate for power losses. An optical cable is characterized by very low power loss. It is possible to use an optical cable to connect an antenna to a receiver over a distance from 3 to 10 kilometers. Such long cables may be useful in hazardous environments, where distantly located antennas are connected to a remote receiver located in a control center [13]. The receiver in that case can still be calculating coordinates of the antenna phase center without any extra errors. Signal delay in the cable is just hidden inside the receiver clock error and calculated with it.

For GNSS applications we mostly use the following connectors: (a) the Navy N- type connector, it was first introduced 1940s, has been improved since, and can be used for signals up to 18 GHz (as on the front panel of a spirent simulator (Figure 3.11)); (b) an

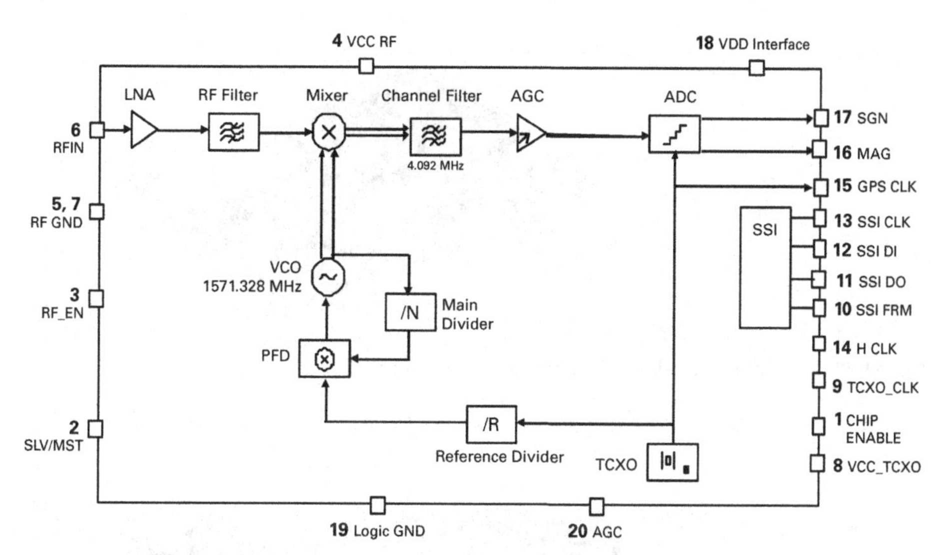

Figure 5.11 Front-end module design based on GRM8652 module (courtesy of Rakon Ltd.).

SMA connector (seen on a PCB (Figure 5.13)) and a bayonet Navy connector (BNC) can be used for a somewhat smaller range up to 2 GHz; (c) a threaded version (TNC) allows us to work with signals up to 12 GHz.

5.3 Front-end design

In the next section we look at the front end in detail. We discuss a GPS L1 front end, illustrating it with a GRM8652 module from Rakon. A schematic of such a front end is given in Figure 5.11. The main components are filters, LNA, a down-converter, the main component of which is a mixer, automatic gain control (AGC), ADC and a high quality TCXO clock. The SSI block is responsible for serial communication with a software receiver. Figure 5.12 shows a 3D drawing of the module. Figure 5.13 shows the module as a component of an iP-Solutions front end.

5.4 Front-end clock

5.4.1 Front-end clock and receiver clock

A front-end clock facilitates conversion of an RF signal to a DIF signal. Through the quality of DIF signal the clock's parameters affect the baseband processor. The clock parameters, however, do not affect the navigation processor directly. The signal replicas generated in the receiver, both for carrier and spread code, are not affected by this clock drift. The carrier and spread code of the incoming signal are on the contrary affected by the clock drift. This difference affects acquisition and tracking. It may result in decreased signal acquisition capabilities and lower accuracy of the tracking loops.

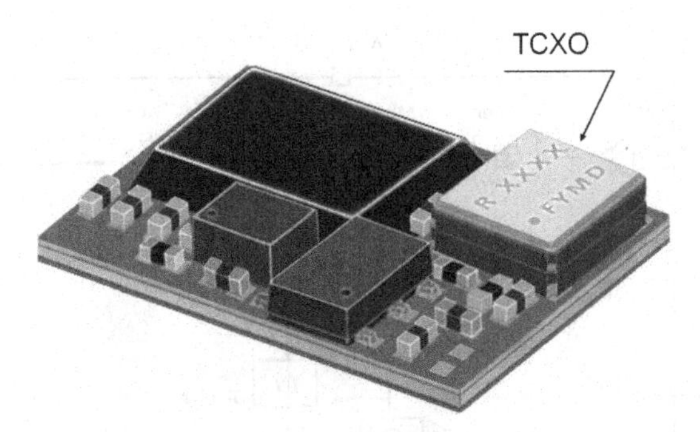

Figure 5.12 GRM8652 module, 3D drawing (courtesy of Rakon Ltd.).

Figure 5.13 TCXO as a part of the GRM8652 module (inset) on FE PCB (courtesy of Rakon Ltd. and iP-Solutions).

Receiver clock error, which we discussed in Chapter 1, comes from the receiver's internal timekeeping. In software receivers this time is initially set to the time in the navigation processor. For receivers working with PCs, this time reference comes from the host PC. If the receiver operates in real-time mode, then the receiver clock is set to the PC clock. If the receiver operates in post-processing mode, then the receiver clock is set to the time of the beginning of data recording. This initial setup is, however, not necessary. It is used only in order to assist in acquisition and positioning.

After the receiver has acquired the signal (either in real-time or in post-processing mode), the initial time is set to the time mark provided in the navigation message. After that, the time is kept basically by dead reckoning applied to the acquired signal code sequence. Therefore the front-end clock may affect only the signal quality, but not the timekeeping, because timekeeping essentially comes from a GNSS satellite.

As we have mentioned, the quality of the front-end clock affects performance of a baseband processor. A front-end clock can usually be represented by a phase-locked loop (PLL) and an oscillator. A PLL is used in the front-end clock circuitry for two main purposes. The first is to generate frequencies others than the one generated by an oscillator. The other is to clean up noise from the noise frequency by removing short-term phase variations. Off-the-shelf GNSS modules provide control over a PLL, making it possible to tune the PLL to specific user requirements, such as oscillator frequency and corresponding sampling rate.

The most simple oscillator type is a voltage-controlled crystal oscillator (VCXO). A VCXO has a stability in the range of ±20 ppm. A simple model explaining VCXO features can be made using a simple amplifier schematic [14]. A zero phase response of an open-loop amplifier is in close proximity to its oscillation frequency when it operates as an oscillator in a closed loop. Application of a varicap diode with variable capacitance allows us to create a simple VCXO from an amplifier. Most commonly used clocks in GNSS receivers are temperature compensated crystal oscillators (TCXO). An excellent quality Rakon TCXO in our example provides a user with up to 0.5 ppm stability and also provides low power consumption. A TCXO clock is shown as part of the front-end module in Figure 5.12 and the inset in Figure 5.13, marked by a letter R.

For some applications an oven controlled crystal oscillator (OCXO) is required. In an OCXO, an oscillator is contained inside a temperature controlled enclosure, which maintains the crystal at a constant temperature. It thus provides a superior stability. An iPRx front end with an OCXO is depicted in Figure 5.14 with top cover removed.

5.4.2 OCXO versus TCXO

Let us look at OCXO parameters, taking as an example an OCXO embedded in an iPRx front end as shown in Figure 5.14. This OCXO has 5×10^{-10} aging per day, 3×10^{-9} stability. Figures 5.15a and b show frequency variations of clock noise for TCXO (a) compared with OCXO (b). Figures 5.15c and d show power spectrum density for TCXO (c) and OCXO (d). These plots were obtained by processing the DIF signal in static mode.

When we looked at GNSS signal spectrums in Chapter 3, we looked at an estimate of the mean-square (power) spectrum. It was calculated using the Welch method and measured in power units. Power spectral density (PSD) shows how the power of the signal is distributed among its Fourier frequencies. Therefore it is measured in units of power/frequency. The PSD of clock noise of a TCXO is much larger than an OCXO. In particular, the frequency at which the slope of PSD becomes flat is very important, because it indicates the required bandwidth for the tracking phase-locked loop (PLL) (we look at tracking loops in the next chapter). Obviously, the noise bandwidth of a PLL can be narrowed for an OCXO, and the phase error is reduced as a result. From the frequency variation plots it is clearly seen that the frequency of the TCXO is drifting.

Figure 5.14 iPRx front end with OCXO clock.

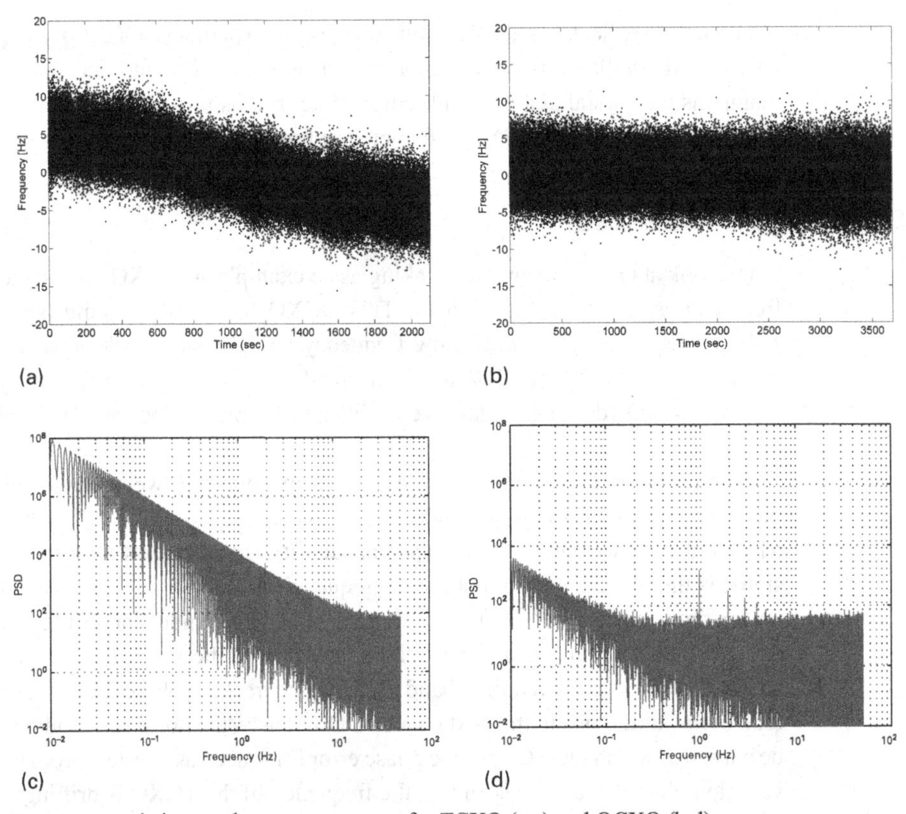

Figure 5.15 Frequency variations and power spectrums for TCXO (a,c) and OCXO (b,d).

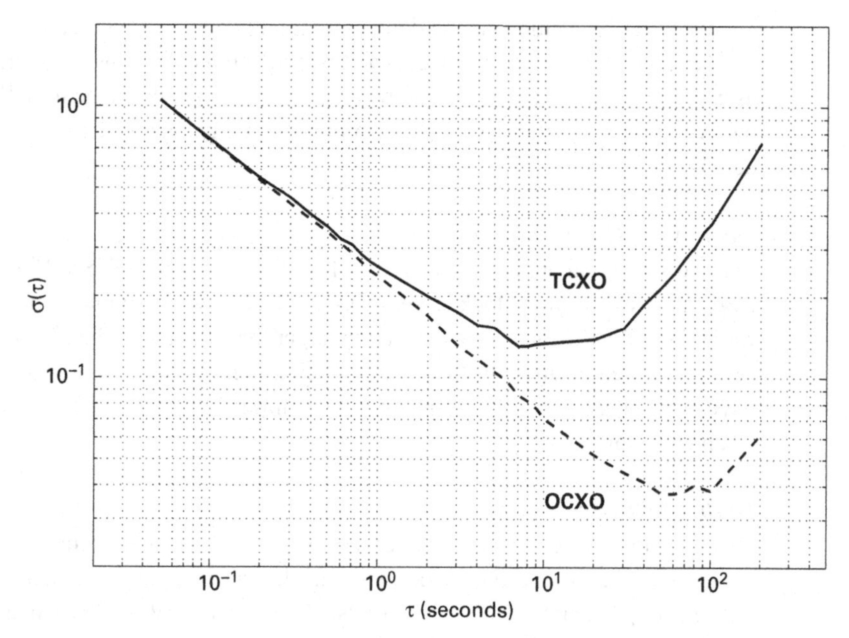

Figure 5.16 Allan deviation for OCXO and TCXO.

Short-term clock accuracy is described by Allan deviation. The Allan deviation $\sigma_A(\tau)$ is defined as half of the time average of the squared differences between successive frequency readings over an interval τ,

$$\begin{aligned} \sigma_A^2(\tau) &= \frac{1}{2} E\left[(y_{n+1} - y_n)^2\right], \\ y_n &= \frac{\delta f(n)}{f_0}, \end{aligned} \tag{5.16}$$

where f_0 is the nominal frequency, δf is the random frequency fluctuation with a null mean value, E is a mathematical expectation operator. Figure 5.16 shows Allan deviation for an OCXO and a TCXO implemented in an iPRx front end. The superior stability of OCXO to that of TCXO is clearly seen. However, apart from cost being prohibitive for many applications, OCXO consumes considerably more power than TCXO. Therefore the choice of clock is made based on application requirements.

High-stability front-end clocks are important for many applications. We look here at three examples.

Inertial navigation and satellite navigation tightly integrated systems

In such systems information about a vehicle's, in particular an aircraft's, movement comes from an inertial navigation system (INS). This information is implemented in a receiver baseband processor to remove vehicle dynamics from the acquisition and tracking. The satellite dynamics can also be accounted for, because an almanac can provide satellite orbit information accurate enough for such purposes. If all dynamics is removed, the acquisition can be done more rapidly and tracking becomes more reliable and accurate. A front-end clock drift affects acquisition and tracking in the same way as user and satellite dynamics. Therefore, in order to get full advantage from INS and GNSS receiver tight integration, we need to implement a high quality clock. See Chapter 6 for details.

Geophysical applications
Many geophysical applications are based on analysis of the GNSS RF signal. In many cases these applications look at very subtle effects, such as signal scintillations. In this case these effects should be significantly larger than background noise coming from the receiver implementation. In particular, for scintillation measurements the receiver should have low phase noise. This parameter is limited by front-end clock quality. See Chapter 10 for details.

High sensitivity
As we see in the next chapter, high sensitivity in a GNSS receiver requires coherent signal integration over a long time. The longer a receiver integrates a signal, the weaker the signal it can acquire. If the antenna is static, then time of coherent integration by the receiver is limited only by the clock drift.

5.4.3 Clock role in simulators and record and playback systems

When a receiver is working with a simulator, the simulator clock operates instead of a satellite clock. The simulator should have a clock with parameters better than the receiver under test. Drift of a simulator clock does not affect accuracy of receiver positioning in the same way as drift of a satellite clock. Satellite clocks, though very accurate, drift independently of each other, so their drift is directly translated into receiver positioning error. Simulated satellite clocks are all drifting with the same simulator clock drift, so this drift is almost all compensated for together with a common receiver clock drift. We say "almost" because the measurements taken by a receiver from each satellite are related to slightly different time instants within a signal code period, for example one millisecond for a GPS L1 C/A. Different clock errors for various satellites are simulated according to Figure 3.25.

Another limitation on simulator clock drift is related to the receiver acquisition process. The simulator clock error is added to the receiver clock error. As such it may affect a receiver acquisition process and tracking in the same way as we have seen before. Also, simulator clock quality directly translates to simulated signal quality. Generally off-the-shelf simulators are equipped with high quality OCXO clocks.

The clock issue in simulation becomes much more important when we consider ***record and playback systems (RPS)***. RPS uses a receiver front end to record a DIF signal. After that, RPS can play back the recorded signal by restoring it through a simulator front end. The simulator front end converts the DIF signal to an analog RF signal by a sequence of operations inverse in relation to the receiver front-end operations. The signal goes through a ***digital to analog converter (DAC)***, and then to an up-converter. As a result the signal comes to the receiver under test after it is has been affected by at least one more

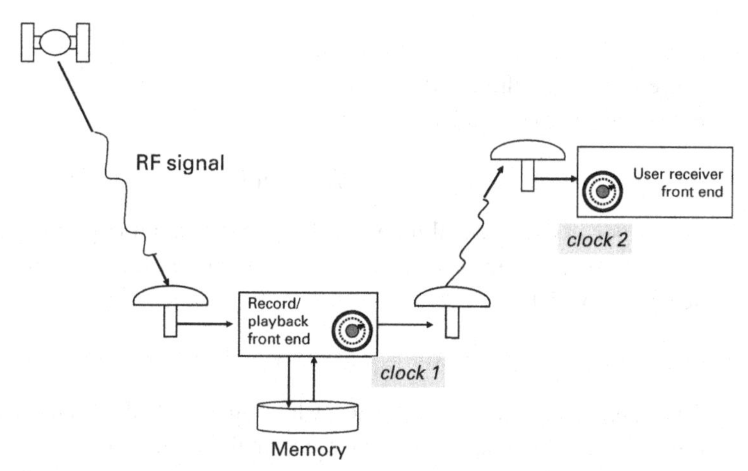

Figure 5.17 Double effect of RPS clock.

clock (see Figure 5.17). If an RPS uses different front ends to record and play back the signal, then two extra clocks will affect the signal quality. In this case it is important to set up requirements for an RPS clock or clocks in accordance with the specification of the receiver under the test. An implementation of OCXO would be a requirement for some systems.

5.5 Down-conversion

The down-conversion process shifts a spectrum of a signal along the frequency axis. A signal from a particular satellite is not centered at the signal center frequency. As described in Chapter 3, a LOS projection of satellite velocity can be up to 800 meters per second. The received signal frequency is increased by the Doppler effect caused by this motion if the receiver and satellite are converging, and decreased if they are moving apart,

$$f_R = f_T - f_T \frac{v_{LOS}}{c}, \tag{5.17}$$

where f_R is received signal frequency, f_T is transmitted signal frequency, c is speed of light, and v_{LOS} is relative velocity between the satellite and the receiver along the line-of-sight. The Doppler shift is less than 6 kHz for a low dynamic vehicle. It is important that the value of Doppler shift is not changed in the down-conversion process.

A device known as a mixer has a received RF signal and a low-frequency signal (LO) in its input. A signal in the mixer output is a sum of two harmonics, one with frequency equal to the difference between and another with frequency equal to sum of the input signal frequencies. After filtering out the upper signal we have only one harmonic with the following frequency:

$$f_{\mathrm{IF}} = f_{\mathrm{R}} - f_{\mathrm{LO}}. \tag{5.18}$$

Correspondingly, the IF of the received signal is a sum of the IF of the transmitted frequency and the Doppler frequency,

$$f_{\mathrm{IF}} = f_{\mathrm{IFT}} + f_{\mathrm{D}}. \tag{5.19}$$

Let us look at this in detail following [11]. In the circuit implementation the product on the mixer output is represented by some complicated waveform with a main frequency described by (5.18),

$$x_{\mathrm{IF}} = x_{\mathrm{RF}} \cdot x_{\mathrm{LO}} = \sin(\omega_{\mathrm{RF}} t) \cdot x_{\mathrm{LO}}. \tag{5.20}$$

The mixer can be realized on diodes, which are switched on and off depending on wave polarity. The LO signal should be large enough to control the diodes, so that they switch depending on the sign of the LO wave. When the diode is off, the RF is not passed. As a result the signal on the mixer output can be seen as a product of an incoming harmonic signal and a square wave with chip rate equal to the doubled LO frequency. The square wave can be expressed by a Fourier series,

$$x_{\mathrm{LO}} = \frac{4}{\pi}\left(\sin(\omega_{\mathrm{LO}} t) - \frac{1}{3}\sin(3\omega_{\mathrm{LO}} t) + \frac{1}{5}\sin(5\omega_{\mathrm{LO}} t) - \ldots\right). \tag{5.21}$$

The other frequencies are filtered out by filters and the output signal has an envelope with IF frequency defined by (5.18),

$$x_{\mathrm{IF}} = \frac{2}{\pi}(\sin(\omega_{\mathrm{RF}} + \omega_{\mathrm{LO}})t + \sin(\omega_{\mathrm{RF}} - \omega_{\mathrm{LO}})t). \tag{5.22}$$

5.6 Analog-to-digital conversion

5.6.1 Defining a sampling frequency

At the final step an IF signal should be digitized. Digitization includes two processes – signal sampling and quantization. Sampling of the band-limited analog signal in the receiver can be viewed as a multiplication of the incoming IF signal (after it is band-pass filtered) by a periodic train of unit impulses. The spectrum representation in the signal can be expressed as a Fourier transform of signal multiplication. The spectrum of the digitized DIF signal can be found through a Fourier transform of the DIF signal as follows:

$$X_d(f) = \sum_{n=-\infty}^{n=\infty} x_d(n) \mathrm{e}^{-\mathrm{j}2\pi f n}. \tag{5.23}$$

The resulting signal is a convolution of the IF signal spectrum and impulse train spectrum and expressed as follows:

$$X_{\mathrm{DIF}}(f) = F\left[x(t)\sum_{n=-\infty}^{n=\infty}\delta(t-nT)\right] = X_{\mathrm{IF}}(f)\otimes\left[\sum_{m=-\infty}^{m=\infty}\delta(f-mf_{\mathrm{S}})\right] \quad (5.24)$$

where X_{DIF} is the DIF signal spectrum on the ADC output, $x(t)$ is the analog IF signal on the ADC input, X_{IF} is the analog IF signal spectrum, T is sampling period, f_{S} is sampling frequency, δ is a delta function. The resulting DIF signal has a spectrum consisting of repeated images of the spectrum of the analog IF signal. If the sampling frequency is smaller than the IF signal bandwidth, then the spectrum lobes of the DIF images overlap or ***alias***. The IF signal is digitized without loss of information only if this overlap does not occur. Then the signal can be restored from its spectrum by inverse Fourier transform as follows:

$$x(n) = \int_{-1/2}^{1/2} X_d(f)e^{j2\pi fn}\,df. \quad (5.25)$$

The requirements to prevent signal alias define a minimum sampling frequency, called the ***Nyquist frequency***, as follows:

$$f_{\mathrm{N}} = 2\cdot B, \quad (5.26)$$

where B is the analog signal bandwidth. As we can see from (5.26), Nyquist frequency can be defined by the signal bandwidth rather than the IF signal highest frequency, which is a sum of the central IF frequency and half of the signal bandwidth. This definition is clear from the fact that the signal can be freely transformed along the frequency axis without distortion, as we saw in Section 5.4. Therefore, for the purpose of finding the Nyquist frequency without loosing generality, we can consider an IF signal with zero central frequency. This Nyquist frequency sets the conditions at which we can restore the signal without loosing the information.

Readers of this book have access to a set of recorded DIF signals and a DIF signal simulator. In order to process these signals in the receiver baseband processor we need to specify two values:

1) signal sampling,
2) intermediate frequency.

The recorded signals have a different sampling rate and IF for records made with TCXO and OCXO front ends. The front end with OCXO uses sampling 16.3676 MS (mega-samples) and IF = 4.1304 MHz. For academic versions of the simulator, ReGen sampling rate is fixed at 16.368 MS and IF to 4.1304 MHz.

It is necessary to understand that samples per second and cycles per second (Hz) are different. Figure 5.18 presents an explanation of the difference between those two. The samples can be expressed as a Fourier series. In the case of simulated data our sample signals have odd harmonics at 8.184 MHz, 24.552 MHz, 40.92 MHz, and so on.

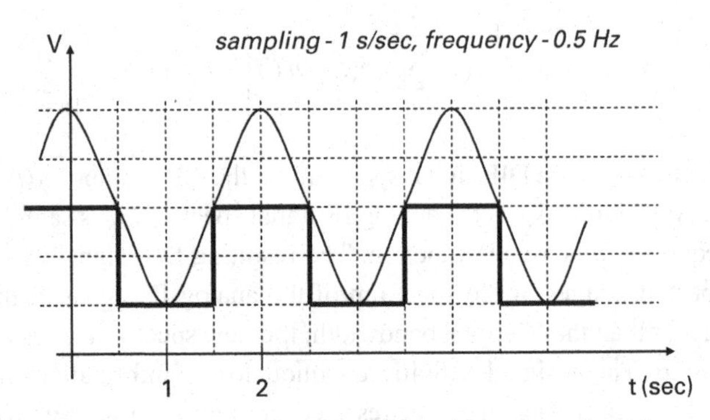

Figure 5.18 Difference between sampling (in samples per second) and frequency (in Hz).

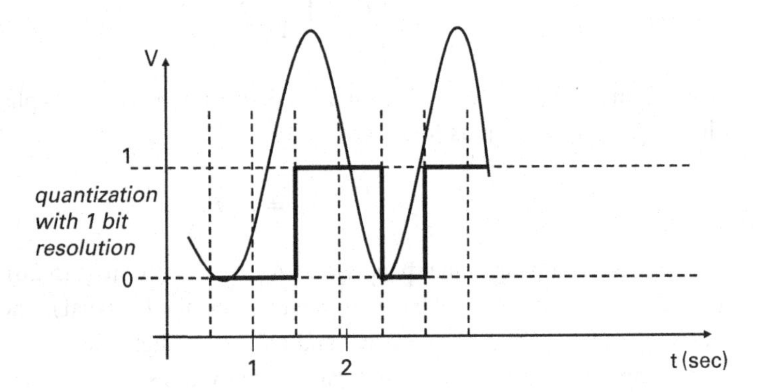

Figure 5.19 Signal quantization with 1-bit resolution.

5.6.2 Quantization

Each sampled value can be represented by an N-bit word. The word can be in one of 2^N states. Therefore an analog IF signal can be represented by 2^N levels of the DIF signal. Most commercial receivers have 1- or 2-bit quantization. In particular 1-bit quantization means that the analog signal is represented by two levels (see Figure 5.19). In terms of hardware implementation it means that one pin with two voltage states (high and low) is enough for the front-end output.

Most high-end simulators have at least 14-bit DACs. It allows them to simulate more complex waveforms, interference, and multipath more rigorously. Most RPSs have their quantization level limited by the receiver front end, which is used to record data. Figure 5.20 shows a comparison between the quantization process of the live satellite signal, a simulated and a recorded signal. If a satellite signal is recorded with RPS with a bit resolution lower than a user receiver, then user receiver bit resolution in that test will be limited by RPS. A DIF signal simulator allows us in general to simulate a signal with high bit resolution; however, special attention should be paid to quantization noise.

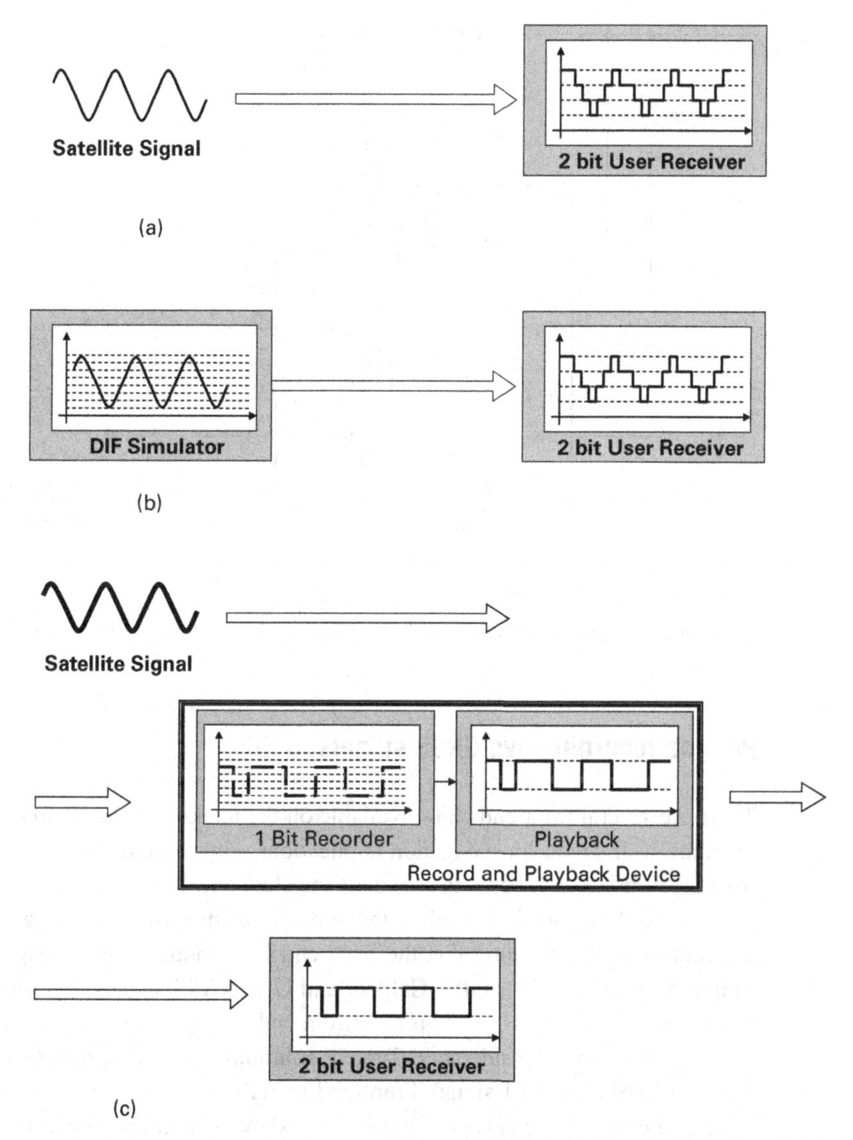

Figure 5.20 A comparison between quantization process of a live satellite signal (a), simulated (b), and recorded (c), signal.

Readers can see more on comparison of RPSs and simulators in terms of quantization and sampling in [15] and [16].

Though a simulator may often be superior to an RPS system in terms of quantization levels, the RPS can be markedly superior to a simulator when it plays back a simulated DIF signal. A ReGen simulator can simulate data with 32-bit quantization, though it is a time-consuming process. It makes such systems the only solution for tasks such as, for example, simulating interference from a large number of sources, because the number of channels can be unlimited.

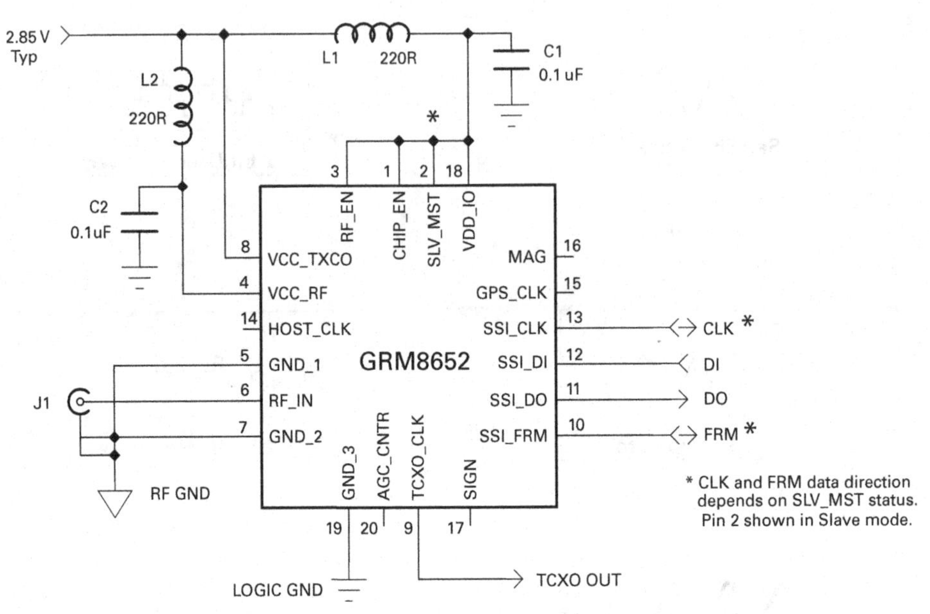

Figure 5.21 RF front-end module interface, GRM8652 Software Application (courtesy of Rakon Ltd.).

5.7 Project: recording live GNSS signals to PC

There are several front-end chips available on the market for the L1 frequency. Most of them are well suited for navigation applications. Unless something special is required from the front end, it is better to use an off-the-shelf one. It is even possible to use L1 front ends for GNSS signals on other frequencies by down-converting and filtering the incoming signal, providing that the front-end bandwidth fits to the signal bandwidth. In this book we consider GPS, Galileo, and GLONASS signals. We use a Rakon [17] module and MAXIM [18] chip as core front ends to provide a DIF signal.

In this section we consider an E-Type iP-solutions front end with a Rakon module as a core to record a GPS L1 signal. Front-end module design corresponds to one we have discussed earlier in this chapter. Figure 5.21 shows the module interface, which corresponds to the schematics in Figure 5.11.

The front end has built-in LNA and can work with passive and active antenna. It features on-chip AGC (automatic gain control), which allows us to change the dynamic range of the front end on the fly. It frees the baseband processor from the overhead of gain adjustment. An onboard clock is used to provide sampling frequency and LO frequency for down-conversion. Some parameters of the front end can be programmed through the SPI, which can also be programmed to provide streaming output. An E-Type front end implements the hardware configuration using sign and magnitude pins.

The software receiver which we use in this book works on a personal computer, therefore its front end uses a few more components [19]. Those components are required to transfer the signal from the front-end core to a host computer through a USB

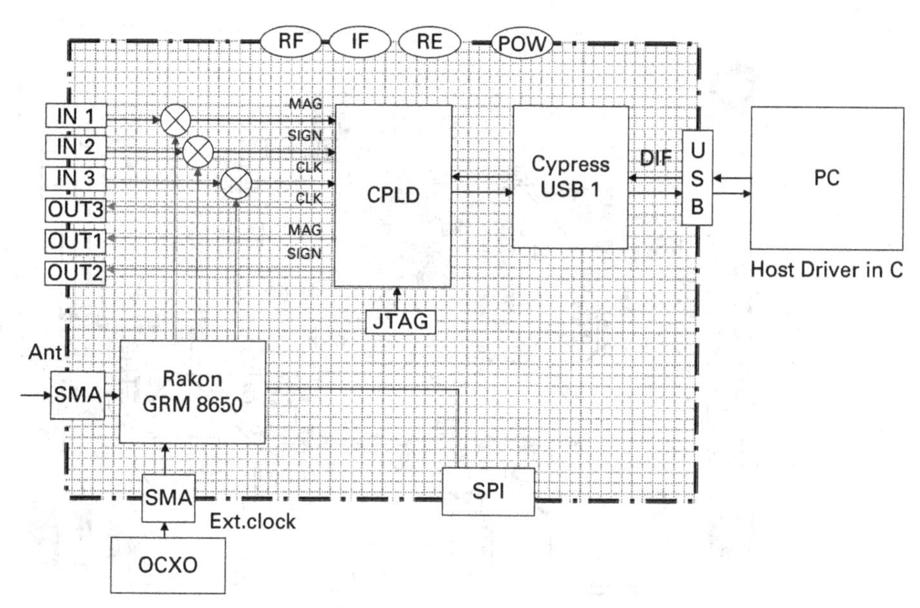

Figure 5.22 USB front-end concept.

(Figure 5.22). The concept is implemented in front ends which also allow playback of a DIF signal. The playback may be useful when testing or developing a real-time receiver. The implementation in the front-end printed circuit board (PCB) is shown in Figure 5.23, and the assembled front end is shown in Figure 5.24.

The PCB features the following main components:

1) USB interface,
2) Memory buffer,
3) Logic device for creating a data stream.

The logic device can be a CPLD (complex programmable logic device) or FPGA (field programmable gate array). The memory buffer can be realized as a separate device or be within the logic device.

USB interface functions are handled on the PC side by CyPress USB drivers. On the PCB side the USB interface functionality is provided by programming a microcontroller. An EPROM device contains a program for the microcontroller. The memory buffering is provided by a CPLD which also packs streaming data into the chosen format.

Recording an RF signal normally serves the following purposes:

1) All further work or study can be done behind a desk once the field work is finished.
2) It cuts cost by not repeating expensive tests, such as those which involve flight tests, borrowed expensive equipment, and so on.
3) It records signals which may contain unique geophysical information, such as signals recorded at certain places during periods of high scintillation.

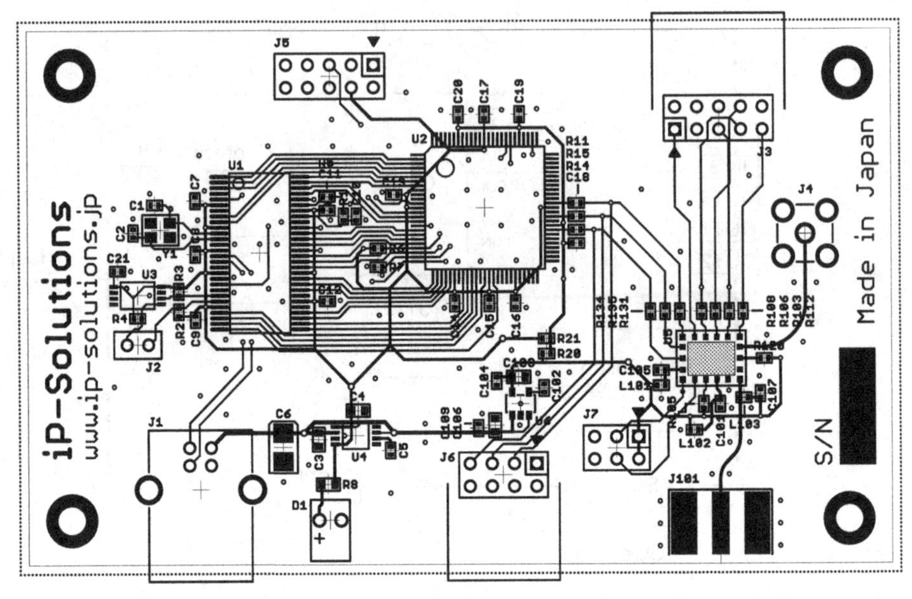

Figure 5.23 USB front-end PCB layout.

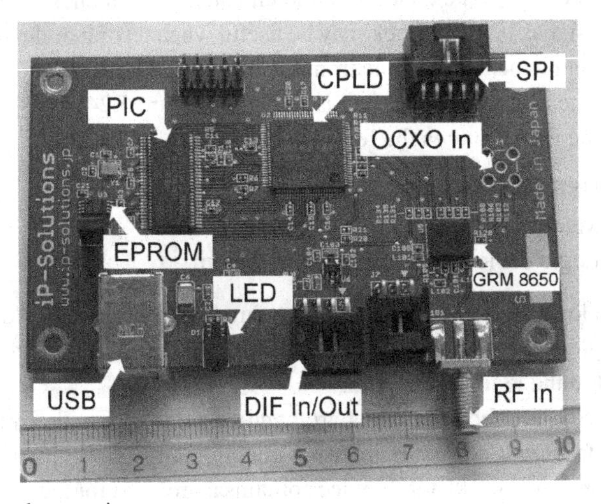

Figure 5.24 USB front-end implementation.

A USB GNSS front end is available from iP-Solutions. If you remove the enclosure you may see a board similar to the one depicted in Figure 5.24. It is supplied with free recording software (see Figure 5.25). The software interface allows a user to set up a recording time and shows recording progress with a clock dial. It also shows if there are any errors in the recorded file, which can come from a user loading a computer CPU with

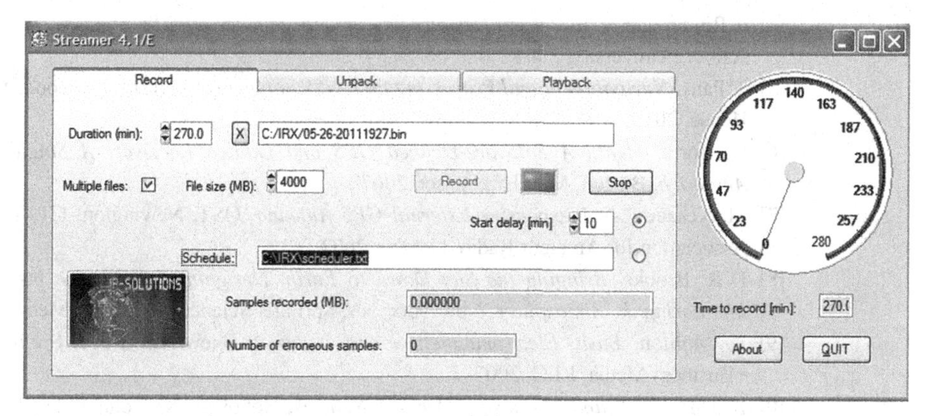

Figure 5.25 Streamer software recording a few hours of GPS signal.

other tasks. In particular, antivirus or LAN software may interfere with RF recording on non-real-time operating systems, such as Windows.

When the recording is finished you have a complete GPS/GALILEO L1 signal recorded for use with your applications. In the same way, using a different front-end model (iP-Solutions Type-M front end with Maxim core), one can record GLONASS signals.

If you do not have a front end, you can use the recorded signal files which accompany this book. Several data sets with signals recorded in field tests are available for download from the book website. A signal can also be simulated with ReGen software. The academic version of the ReGen software accompanies this book.

In all cases the recorded and simulated signals are such that they can be viewed on the output of a receiver front end, with parameters specified either by the recorder front end or by the simulator.

The same front end can be used to stream data to user application software instead of recording it to the memory. An academic version of an iPRx receiver, which accompanies this book, can be used as an example of such an application. The iPRx receiver can work with recorded and streaming signals. Using a front-end API, one can make the iPRx front end work in real time with one's application, such as a software receiver.

References

[1] J. Tsui, *Fundamentals of Global Positioning System Receivers: A Software Approach*, New York, NY, John Wiley & Sons, 2000.

[2] D. Akos, *A Software Radio Approach to Global Navigation Satellite System Receiver Design*, Athens, OH, Ohio University, 1997.

[3] A. Dempster and C. Rizos, *Implications of a "system of systems" receiver, Surveying & Spatial Sciences Institute Biennial International Conference*, Adelaide, Australia, 28 September–2 October 2009.

[4] P. Rinder and N. Bertelsen, *Design of a Single Frequency GPS Software Receiver*, Aalborg, Aalborg University, 2004.

[5] T. Pany, *Navigation Signal Processing for GNSS Software Receivers*, Norwood, MA, Artech House, 2010.

[6] K. Borre, *et al.*, *A Software-Defined GPS and Galileo Receiver: A Single Frequency Approach*, Boston, MA, Birkhäuser, 2007.

[7] M. Kesauer, *An Inexpensive External GPS Antenna*, QST, Newington, CT, The National Association for Amateur Radio, October 2002.

[8] D. R. Brooks, *Bringing the Sun Down to Earth. Designing Inexpensive Instruments for Monitoring the Atmosphere*, New York, NY, Springer Science + Business Media B.V., 2008.

[9] A. Moliton, *Basic Electromagnetism and Materials*, New York, NY, Springer Science +Business Media, LLC, 2007.

[10] K. Chang, *RF and Microwave Wireless Systems*, Hoboken, NJ, John Wiley & Sons, Inc., 2000.

[11] A. Scott and R. Frobenius, *RF Measurements for Cellular Phones and Wireless Data Systems*, Hoboken, NY, John Wiley & Sons, Inc., 2008.

[12] E. D Kaplan and C. J. Hegarty (editors), *Understanding GPS, Principles and Applications*, second edition, Boston, MA, Artech House, 2006.

[13] I. Petrovski, *et al.*, *LAMOS-BOHSAI™: LAndslide Monitoring System Based On High-speed Sequential Analysis for Inclination*, ION GPS'2000, USA, Salt Lake City, September 2000.

[14] R. Lacoste, *Robert Lacoste's the Darker Side. Practical Applications for Electronic Design Concepts*, Burlington, MA, Elsevier Inc., 2010.

[15] I. Petrovski and T. Ebinuma, GNSS simulators, Part 2: Everything you wanted to know . . . but were afraid to ask, *Inside GNSS*, September 2010, 48–58.

[16] I. Petrovski, T. Tsujii, J-M. Perre, B. Townsend, and T. Ebinuma, GNSS simulation: A user's guide to the galaxy, *Inside GNSS*, October, 2010, 36–45.

[17] Datasheet: *Rakon GRM8650 High Sensitivity RF Front-End Module for GNSS Systems*, Rakon Ltd., Auckland, New Zealand, 2008.

[18] Datasheet: *MAX2769 Universal GNSS Receiver*, Maxim Integrated Products, Sunnyvale, CA, 2007.

[19] R. G Lyons, *Understanding Digital Signal Processing*, 3rd edition, NJ, USA, Prentice Hall, 2011.

Exercises

Exercise 5.1. A satellite antenna footprint should cover the Earth's surface. Calculate the gain resulting from reducing a satellite transmitter antenna pattern to one necessary to cover only the Earth's surface.

Exercise 5.2. Calculate if there is a difference in antenna gain for the same antenna for L1 and L2 GPS signals.

6 Real-time baseband processor on a PC

In this chapter we describe operation of a GNSS receiver baseband processor. The place which this chapter occupies in the book is schematically presented in Figure 6.1. It is one of two major components of a GNSS receiver. In the previous chapter we described the other component of the receiver – the RF front end, which takes an RF signal from an antenna, amplifies, down-converts, filters, and digitizes it. A digitized intermediate frequency signal (DIF), also described in the previous chapter, is taken from the output of the front end to the input of the baseband processor. The baseband processor processes the DIF signal and outputs all information carried by the GNSS signal, pseudoranges or code phase observations, carrier phase observations, Doppler, signal to noise ratio, navigation message, and so on.

6.1 Do we need all the receiver or just a baseband processor?

If we compare the structure of a generic receiver as it is usually presented, and as illustrated in Figure 6.2, with the flowchart of this book presentation in Figure 6.1, we can see that a navigation processor is omitted from the receiver design. This has been done on purpose. In numerous applications today the navigation processor embedded in the receiver is not actually used, though all conventional receivers have one.

Let us look at a few examples. Geodetic receivers collect raw data, most often in RINEX format. Only this RINEX format is used by geodetic software for precise coordinate estimation. Also, only raw data are used in other applications which require high accuracy, whether they are real-time or post-processing, such as deformation monitoring, surveying, satellite ephemeris estimation, or atmospheric parameters estimation.

If, on the other hand, we look at navigation applications, we see the same tendency. Modern aircraft navigation complexes use coupled GNSS and inertial navigation systems (INS). These systems can be integrated on different levels. We look at different types of such integration in detail in the next chapter. For now we can just state that it is less advantageous to integrate these systems in a positioning domain, i.e. to use user coordinates calculated by the receiver, instead of raw data. The simplest way to demonstrate this is to look at a situation when only one or two satellites are available, which may happen for example during an aircraft maneuver. In such a case a GNSS receiver generally cannot make a positioning, and therefore the information from GNSS cannot

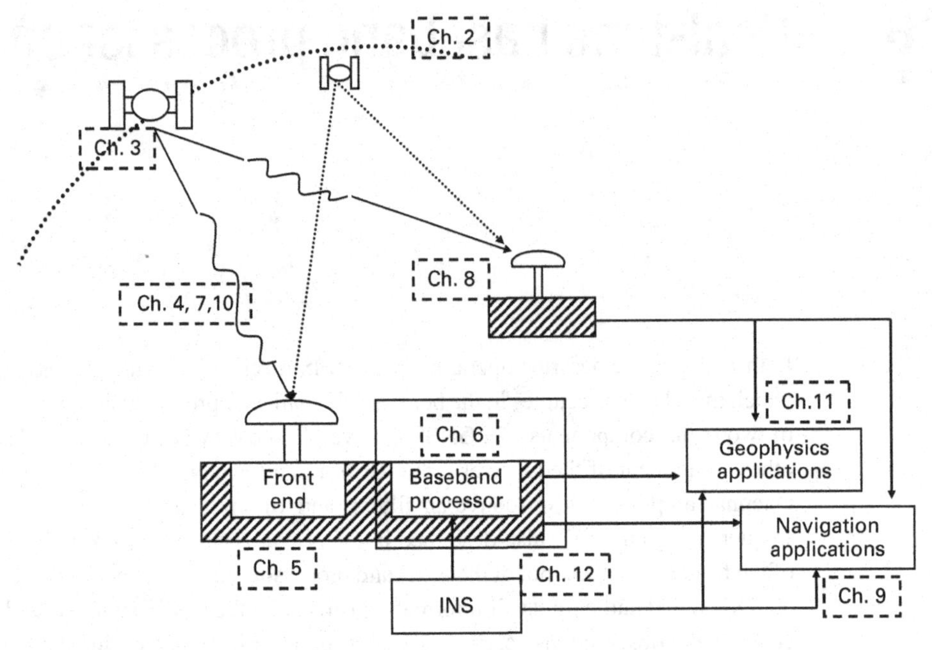

Figure 6.1 Subject of Chapter 6.

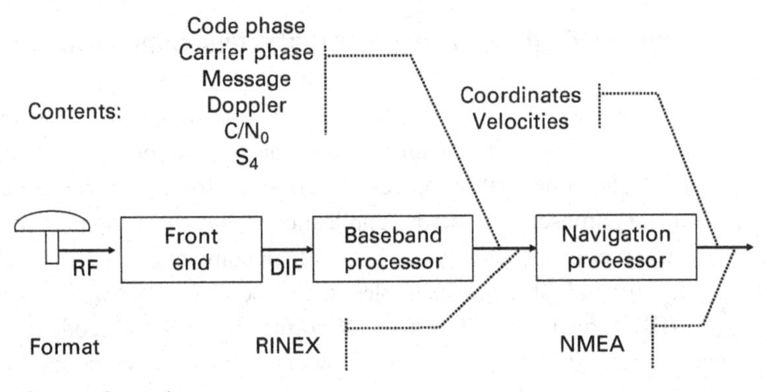

Figure 6.2 Structure of a generic receiver.

be used at all. On the level of raw data, however, all the information can be utilized along with data from an INS.

On the other end of the range of navigation applications, we can look at mobile devices. They are designed to minimize power consumption, size, and weight. Therefore, the functions of a navigation processor are moved to the device general processor. As a result device manufacturers would need only a GNSS receiver without a navigation processor from their suppliers. This tendency goes further, and a baseband processor, traditionally implemented in FPGA or ASIC, can also be implemented in the software in the general processor. This tendency is reflected in the fact that manufacturers have started to call some devices a GNSS receiver, though they are only an RF front end, without navigation and baseband processors. So the device which they sell is clearly not a receiver but just a front end, because it is missing essential components.

In comparison with mobile applications, the presence or absence of a navigation processor in geodetic and high-end navigation receivers does not make a big difference for a user from a price, weight, or size perspective. However, it is much more convenient to have it embedded and it may also still be required for some applications. For the purpose of our book, however, we can finish our description of the receiver as such with the baseband processor. All functions related to processing of GNSS raw data are considered as separate from the receiver tasks, which can be implemented either in the processor embedded into the receiver, or in the device general processor, or in a vehicle CPU, or in an outside computer, or anywhere else.

We can see a trend in GNSS similar to high-end cameras, where a user, instead of a picture, can now record a raw data snapshot with a lot of extra data, which may be useful for further processing. In GNSS now, instead of logging raw data, it is possible to collect DIF, which allows us not only to calculate raw data, but also saves information on the radio signal, such as scintillation, which (as we show in the next chapters) may be utilized later in conjunction with raw data to derive much more information, for example about atmospheric or sun conditions. This type of GNSS data collection may appear more and more appealing with further technological advances in storage media and processing powers.

We have considered such a recording device in the previous chapter. Basically, any modern geodetic receiver can be converted to record its DIF data, instead of sending them to a baseband processor. The quality of DIF signal depends on the front-end quality. The recorded DIF signal allows access not only to GNSS observations, but to a signal itself.

6.2 Operation in general

In this section we consider generic operations of a baseband receiver for particular implementation in a software receiver. A schematic of the baseband processor is given in Figure 6.3. The blocks in the figure show the main operations of the baseband processor with jobs distributed over a number of threads for real-time implementation. Block outputs show the data, which are calculated in each process.

A radio signal is applied to an input of the baseband processor in digitized form. We have described a GNSS signal from one satellite in Chapter 3 as follows (3.21):

$$A = A_0 \sin((\omega_0 + \omega_D)t + \varphi) \cdot D \cdot B, \tag{6.1}$$

where ω_0 is signal central angular frequency, ω_D is Doppler angular frequency, D is a spread code and B is a navigation message. Here we modify this equation in order to describe a signal from multiple satellites. Code and carrier phase measurements exist only as relative measurements, so they can be taken in relation to a receiver generated signal replica. We need to remember, though, that the receiver timescale is not fixed so

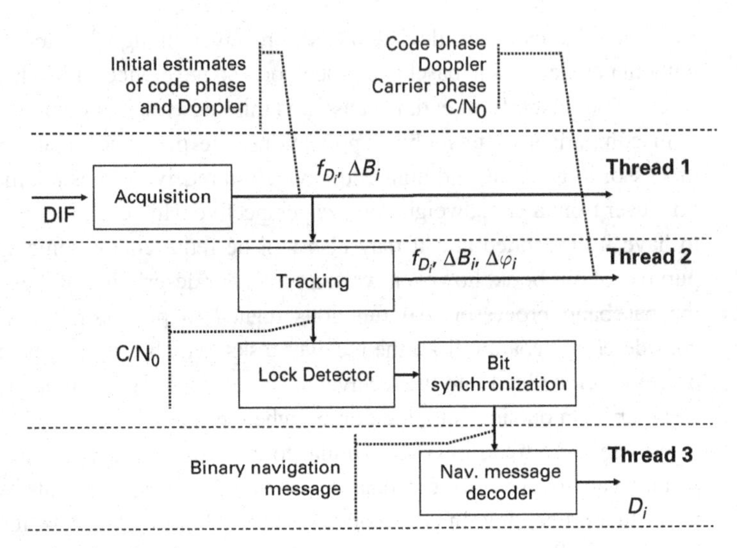

Figure 6.3 Software receiver baseband processor.

this would not give any extra information. The signal from multiple satellites can be presented then as follows:

$$\sum_{i}^{N} A_i = \sum_{i}^{N} \{A_{0_i} \sin((\omega_0 + \omega_{D_i})t + \varphi_i) \cdot D_i \cdot B_i\} \tag{6.2}$$

where N is number of satellites.

A purpose of a baseband processor is to find the Doppler shift $f_{D_i} = \omega_{D_i}/2\pi$ of the signal from nominal frequency $f_0 = \omega_0/2\pi$, code phase $\Delta B_i = B_i - B_R$, carrier phase $\Delta\varphi_i = \varphi_i - \varphi_R$, encoded navigation message D_i, and estimated amplitude of the signal A_{0_i} for each satellite. A receiver-generated replica gives B_R and φ_R in the receiver time-frame. There is also some additional information which can be derived from the signal, for example statistics of carrier and code fluctuations with time.

A purpose of acquisition is to start the process by finding approximate values for Doppler shift f_{D_i} and code phase ΔB_i in (6.2). These approximate values are then transferred to tracking. In the code tracking loop the values ΔB_i and f_{D_i} are refined. Tracking may also include a carrier tracking loop, where carrier phase $\Delta\varphi_i$ is defined.The final operation is to record a navigation message D_i through an operation called bit synchronization.

A baseband processor can be reduced to provide only acquisition functionality for some applications. These applications apply snapshot positioning, when only short snapshots of a GNSS signal are available. In this case ephemeris information normally derived from the contents of a navigation message should be supplied from outside. A positioning algorithm should also resolve ambiguity in code measurements, because a spreading code repeats itself. For example the GPS L1 C/A code repeats itself approximately every 300 km. With longer codes it becomes a lesser problem. The methods of acquisition and tracking for GPS receivers in general and software GPS receivers are well developed now [1]–[6]. We describe them using the real-time iPRx software receiver bundled with this book.

6.3 Acquisition

6.3.1 Search area for acquisition

The purpose of acquisition is to find an encoded spreading code signal in the incoming DIF signal. Acquisition starts by looking for a particular satellite signal,

$$A_i = A_{0_i} \sin((\omega_0 + \omega_{D_i})t + \varphi_i) \cdot B_i. \tag{6.3}$$

In general, an acquisition will ignore the existence of a navigation message D_i, one bit of which for GPS L1 contains 20 spread code sequences.

The acquisition process is presented in Figure 6.4. At the first step acquisition removes carrier phase from the signal by multiplying it by carrier replica. We know the nominal value of the signal carrier f_0 as it is transmitted from all satellites. The Doppler shift f_{D_i} is mostly defined by satellite movement, which is roughly about 4 km/sec and could be up to 800 m/sec along the line-of-sight (LOS) to the receiver, and is in a range of ± 6 kHz for a static or low dynamics user and up to ± 10 kHz for high dynamics. This range is the search range for carrier frequency of the incoming signal. As we described in Chapter 4, some Doppler changes can be introduced during signal propagation through the Earth's atmosphere. A drift of the receiver clock also manifests itself as a Doppler shift of the incoming signal. Therefore, we need to make a search within the possible Doppler range for the correct frequency of incoming signal.

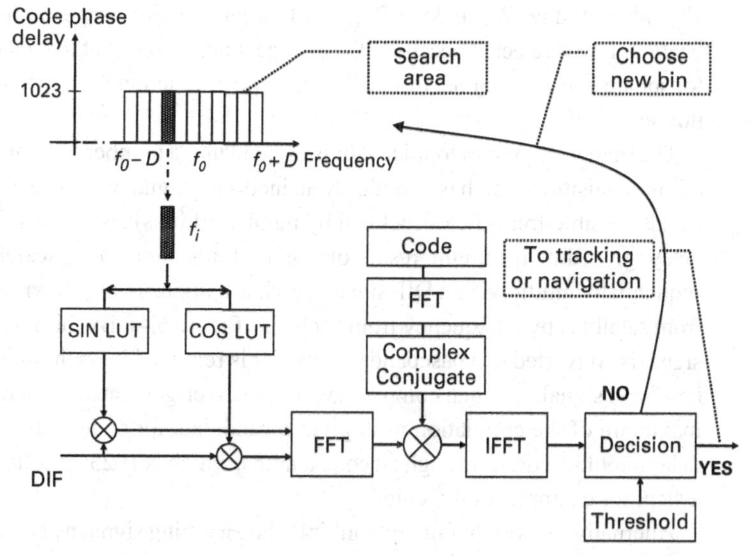

Figure 6.4 Acquisition in software receiver.

Then we check if the remaining signal contains satellite PRN code. For that purpose we generate a replica of satellite spread code in a receiver and check if there is a correlation between the replica and incoming signal. To do that, we look at the ***convolution*** between incoming signal and replica.

We do not know how far the satellite is from the receiver antenna, so we need to search through all possible shifts between incoming signal and replica code. Therefore, the search is conducted in two-dimensional space, with code delay and frequency offset to the nominal frequency due to the Doppler shift. This search can be done by a brute force method, just going through all possibilities sequentially. This would be too long for a software receiver, so there are a number of methods to optimize the speed of this search.

The methods of acquisition for software receivers are slightly different from hardware receivers, because they are in particular adapted to operate in sequence on a computer, whereas hardware receivers usually implement this search in parallel either on an FPGA or ASIC platform. This is one item which makes a software receiver much slower than hardware receivers. Even with a multi-core multi-processor computer a software receiver cannot possibly beat a hardware receiver with a multi-correlator, which may have up to tens of thousands of hardware correlators working in parallel. Nevertheless, a modern software receiver, such as the iPRx bundled with this book, can acquire the number of satellites necessary for a position fix well within a second. The time required depends on computer power and signal strength. The flow chart of the algorithm in Figure 6.4 shows the parallel code phase search algorithm, which use circular correlation. It involves fast Fourier transform (FFT), which allows fewer iterations and optimizes the algorithm further [7],[4].

6.3.2 Circular correlation algorithm

Fast Fourier transform plays a predominant role in software receiver acquisition algorithms today. When Jean Baptiste Fourier initially proposed his method for publication, it was rejected because of Lagrange's objections that this method does not work for square wave functions. Ironically, this type of function is exactly what we consider in this section.

The search range for Doppler shift is divided into a number of frequency bins. The idea of the acquisition search is to multiply an incoming signal with a locally generated replica for all possible frequencies (defined by number of bins) and delays. The correct bin and delay give the maximum result of the multiplication. The search is conducted by sequentially multiplying a DIF signal which is a digitized and down-converted RF signal from satellites by a frequency from each bin (Figure 6.4). By this multiplication, the DIF signal is converted to a baseband (i.e. carrier is removed from the signal). Converted to a baseband signal, it is then compared with a receiver-generated replica of a spread code. A maximum of the correlation peak gives a candidate for code phase. The search in code delay should be done through all code lengths, which is 1023 chips for GPS L1 C/A, with a step, for example, of 0.5 chip.

Alternatively, we can initially multiply the incoming signal by spread code and then by frequency. In this case we can significantly speed up an algorithm by FFT. The result of the multiplication of an incoming signal by a spread code with each possible delay can be transferred to a frequency domain in one step and analyzed there. If the replica spread code has the same delay as one in the incoming signal, then multiplication will strip the incoming signal off the spread code. The spectrum of the signal resulting from multiplication will show Doppler shift as the peak. FFT allows it to make a transfer from time domain to frequency domain in a very computationally efficient way.

However, we can make an even quicker algorithm if we go back to the initial method (Figure 6.4) and strip the incoming signal of carrier first and then search for the code delay in the frequency domain. That is because the first part would require us to go sequentially through all the frequency bins, which can be chosen from 18 to 72 in the case of an iPRx receiver, instead of the 1023 possible code delays. So we make an FFT to transfer both replica and incoming baseband signal into the frequency domain and compare them there. Those signals are spread codes, and therefore they are exactly those signals for which Fourier's work was not accepted for 15 years. After finding the correlation in the frequency domain, an inverse FFT gives a correlation versus delay in the time domain.

The overall algorithm can be represented as follows:

$$\mathrm{IFFT}(\mathrm{FFT}(B_i) \times \mathrm{FFT}(R)) = B_i \otimes R, \tag{6.4}$$

where FFT(IFFT) define direct (inverse) Fourier transform operation, B_i is the incoming baseband signal and R is a spread code replica, generated in the receiver. The left part of the equation gives our exact circular correlation algorithm as it is depicted in Figure 6.4. We denote Fourier transform operation in this equation as FFT only in order to relate it to the figure. We note that FFT is only an algorithmic realization of a Fourier transform operation. The right part of the equation defines a convolution between two signals, which means sample by sample shifting and multiplication of two signals in the time domain (see Figure 6.5). The result vector should have a distinctive peak at the shift equal to the code phase. This type of algorithm is called circular, because it will look through all possible delays between an incoming baseband signal and receiver code replica, by shifting them as in a circular buffer.

The spreading code properties define the shape of autocorrelation function (see Chapter 3). The shift between replica spread code and incoming baseband signal can be conducted in steps of a half chip. Figures 6.6 and 6.7 show autocorrelation functions for a BPSK Gold code (GPS L1 C/A signal) and BOC(1,1) (L1C) code respectively. Figure 6.8 shows a typical correlation output for a search area for GPS C/A BPSK code for an iPRx receiver for eight channels.

In order to use FFT, we need the number of signal samples to be equal to 2^n. Therefore, in order to enjoy the advantages of FFT, we need to pad our samples with zeros. It will, however, slightly decrease the correlation performance. Otherwise, FFT performance will be the same as standard ***digital Fourier transform (DFT)***. There are several acquisition algorithms [4],[6], but a circular correlation, FFT-based algorithm is currently the optimal choice for a software receiver.

When the output of the circular correlation algorithm is generated, we need to estimate whether the satellite signal is present. The decision is usually made based on the set

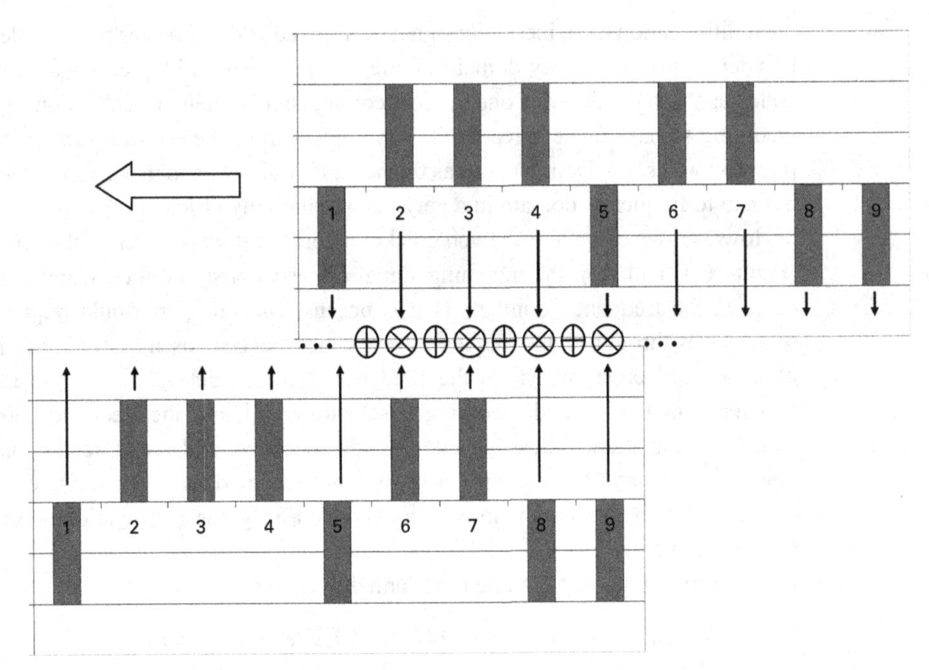

Figure 6.5 Representation of convolution process.

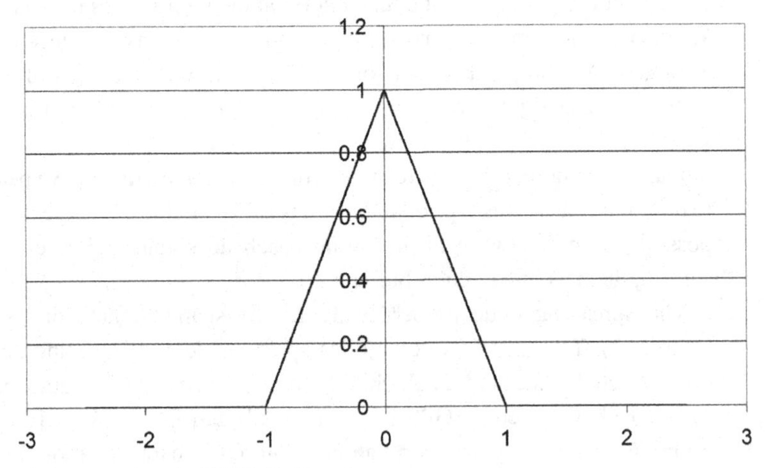

Figure 6.6 Autocorrelation function for BPSK Gold code.

thresholds and probabilities for Type I and Type II errors, which are errors of wrong detection and missed detection. The threshold cannot be set once and for all. It depends first of all on signal strength, and therefore may vary with different antennas and in different environments. In order to make this process automatic, we need to estimate a noise floor in each session. This can be done from the results of acquisition of a satellite, which is not presented in a signal. Alternatively, it can be done through acquisition of a signal with a non-existing PRN [4], or with a large frequency offset.

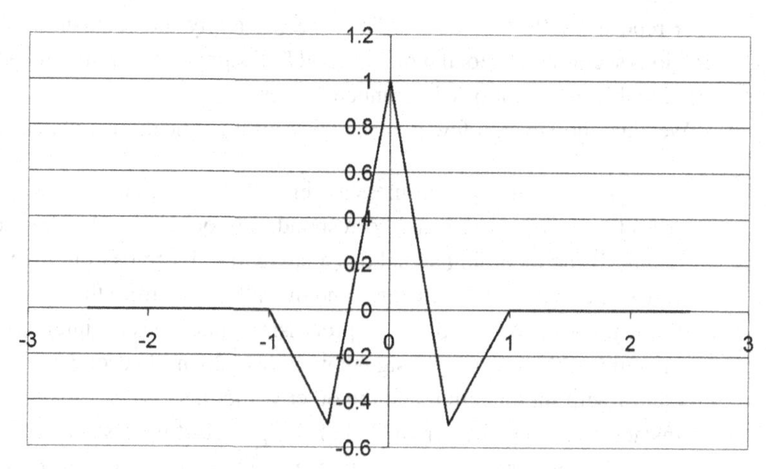

Figure 6.7 Autocorrelation function for BOC(1,1) code.

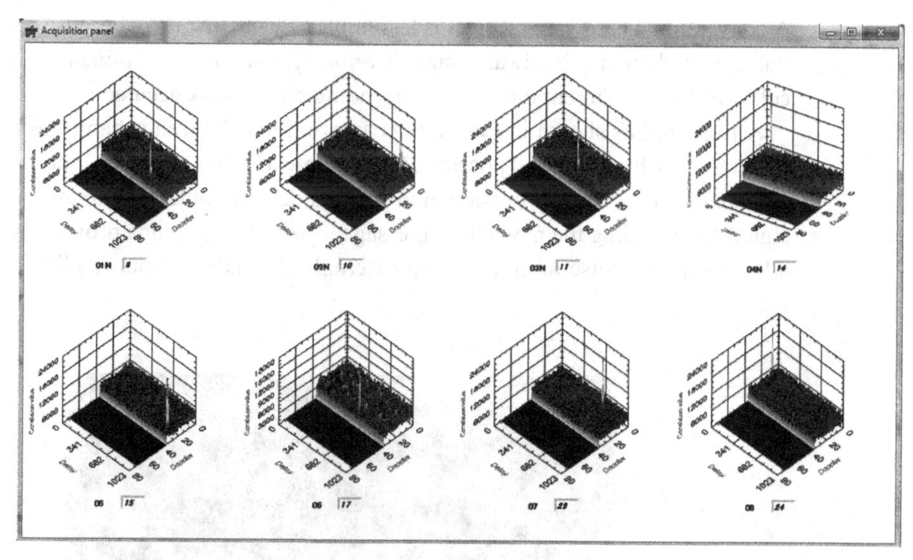

Figure 6.8 iPRx acquisition panel.

6.3.3 Coherent versus incoherent integration

Let us suppose that we have to find a GPS signal in ***n*** milliseconds of DIF signal. We can do it in two ways. We can process chunks of signal in code length sequences (one millisecond for GPS L1 C/A code), then sum up the results of ***n*** acquisitions. This is called ***incoherent*** integration. Alternatively we can construct an ***n*** millisecond replica out of $\boldsymbol{n}$ repeated GPS C/A codes. Then we can apply the acquisition algorithm from the previous section to the total $\boldsymbol{n}$ millisecond DIF signal. In this case of ***coherent integration*** the signal-to-noise ratio will be much higher.

We have, however, a few issues, which limit applications of coherent integration:

1. A single bit of navigation message in GPS C/A signal has a length of 20 complete spread code sequences, and correspondingly occupies 20 milliseconds of the signal length. If the incoming signal has a navigation bit transition, then the polarity of the codes in that bit will be changed and overall results from the coherent integration will be degraded. For the price of processing time we can integrate, for example, an incoming 20 millisecond signal with two 20 millisecond replicas, one of which comes with bit transition and the other without.
2. Instability of the receiver clock and Doppler shift from satellite and receiver antenna movement will affect the appearance of the spread code in the incoming signal in comparison with an ideal replica. It is possible to construct a replica that will compensate for these changes, but the price in terms of processing load may be prohibitive for any implementation near real-time. The various algorithms of such a compensation are considered in [8] in detail.

A conventional hardware receiver is less limited by the first item, because correlators are implemented in the hardware, and therefore it can use an optimal combination of coherent and incoherent integration depending on its clock accuracy.

A front end as shown in Figure 6.9 can provide enhanced capabilities for coherent integration, which makes it possible, for example to receive a signal indoors. One of the antennas can be located outside, and one inside. The outside antenna can provide a signal for decoding the navigation message. The indoor antenna provides a signal with a low signal-to-noise ratio. By using external information about satellite Doppler and

Figure 6.9 Multi-antenna front end for high sensitivity research.

synchronized navigation bit sequence decoded from an outside antenna, one can integrate the indoor signal almost indefinitely, as long as front-end clock stability will allow. In real-life applications, a decoded navigation message can come as outside corrections from a reference station.

6.3.4 Frequency resolution

Baseband processors for low-end receivers and especially receivers for cellular phones may implement only an acquisition in the baseband processor. They implement snapshot positioning using only a snapshot of a GNSS signal. In this case, a baseband processor needs to output only code delays. The accuracy of code delays regarding acquisition is basically limited by DIF sampling rate. We cannot overlap an incoming baseband signal with a receiver code replica with an accuracy better than the sample. For example, for 16.367 M samples per second, the distance between samples is about 18.3 meters, which limits range measurements. Added propagation and receiver errors (see Table 3.3) will give us an estimate of user range error. The product of range error multiplication by DOP (Chapter 1) gives us user positioning error. We can see now why the specification of E911 service gives 50 meters, when we normally get much better performance from a GNSS receiver. It is because usually a receiver implements tracking loops, which allow it to improve accuracy, which in that case will be limited by its chip rate.

If baseband processors implement tracking loops, they are normally initialized using code phases (or code delays) and Doppler frequency estimates, which they receive from the acquisition. In order for a tracking loop to pull-in, the frequency from the acquisition should be defined with a certain accuracy. The frequency resolution possible for any signal depends on its length in samples [9]. We can introduce a digital angular frequency defined in radians per sample as follows:

$$\tilde{\omega} = \frac{2\pi f}{f_S}, \tag{6.5}$$

where f_S is sampling frequency. Then the frequency resolution we can get from the K samples can be defined as follows:

$$\Delta\tilde{\omega} = \frac{2\pi}{K}. \tag{6.6}$$

Similarly defining Δf through $\Delta\tilde{\omega}$ we can express it through the length of the processed signal chunk:

$$\Delta f = \frac{f_S}{K} = \frac{1}{KT_S} = \frac{1}{\Delta T}, \tag{6.7}$$

where T_S is sampling interval, ΔT is the duration of the processed signal length in seconds. This puts requirements on the length of the signal, which usually requires much longer than a few milliseconds to find a code phase.

6.4 Tracking in a software baseband processor

At the acquisition stage, a receiver finds approximate values for signal frequency and code phase, which translates into code delay. Then these values are transferred to tracking loops. Tracking loops refine these values to be more accurate and keep them updated. There are two tracking loops, one for carrier tracking, and another for code tracking. Both tracking loops implement a phase lock loop concept, which has been developed by the same Sir Edward Appleton [10] we have already mentioned in Chapter 4. The details of code and carrier tracking loop theory for GPS applications are given in [1]–[6], and general information on code and carrier tracking theory can be found in [11] and [12], so here we give an overview of the particular implementation which is necessary in order to control and understand the software receiver bundled with the book.

Both tracking loops should operate in parallel, as they are using each other's information in order to wipe out the carrier from the baseband signal and spread code from the carrier. Such a structure of tracking loops is shown in Figure 6.10. A code tracking loop estimates code delay and applies it to remove code from the DIF signal. This codeless DIF signal is then tracked by the carrier tracking loop. The carrier tracking loop in its turn estimates carrier phase and is applied to the original DIF signal in order to remove carrier from it. The signal converted to baseband is tracked by the code tracking loop.

A carrier tracking loop is designed as a ***phase-locked loop (PLL)***. In a PLL, a phase comparator compares phases of a locally generated carrier with the carrier of the incoming signal and adjusts phase by applying feedback in such a manner that the phase difference is kept to a minimum. A code tracking loop, also called a ***delay lock loop (DLL)***, is also designed to keep the phase difference between two codes small. This

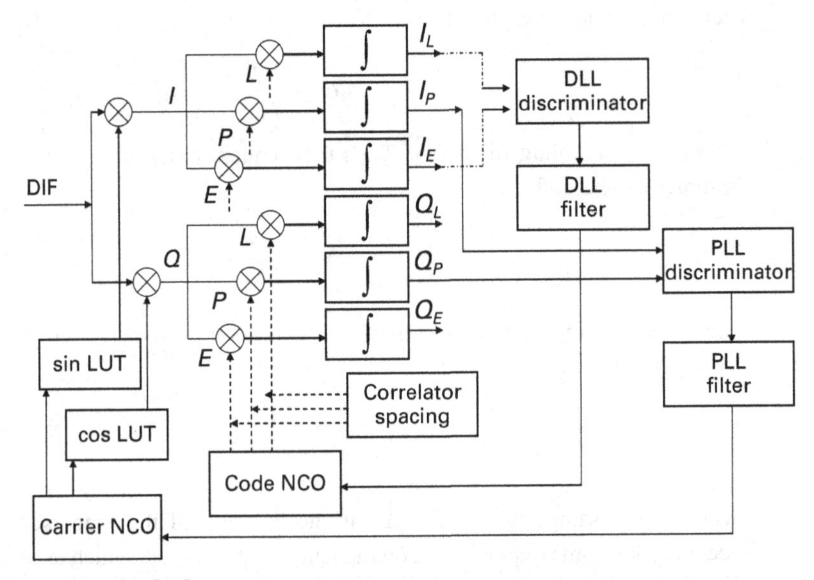

Figure 6.10 Tracking loops after [13].

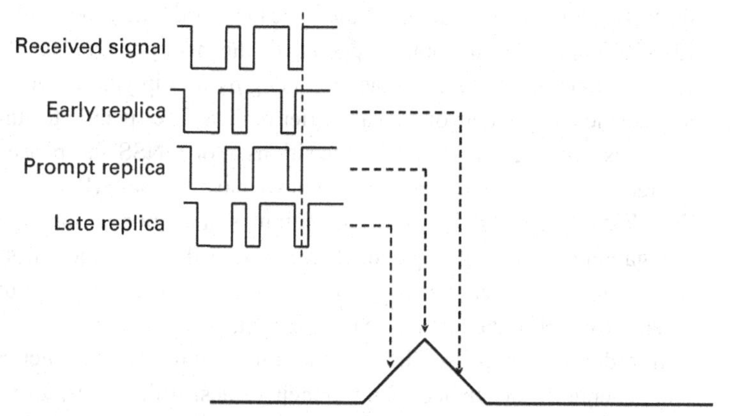

Figure 6.11 Early, prompt, and late replicas.

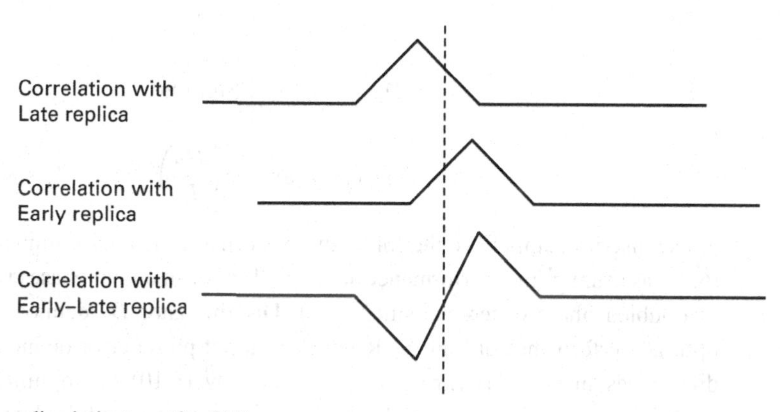

Figure 6.12 Forming discriminators after [14].

is done by keeping cross-correlation between a local replica and incoming baseband signal at a maximum. In the simplest case, a receiver generates three local replicas shifted by a half chip. The value of the correlation between the incoming baseband signal and replicas is defined by an autocorrelation function. An autocorrelation function for GPS C/A code is shown in Figure 6.11 together with three code replicas. The figure also shows points on a Gold code autocorrelation function, which should ideally correspond to late, early, and prompt replicas. When the values of correlation of these replicas with an incoming signal correspond to those shown in the figure, the prompt replica coincides with the signal. Figure 6.12 shows an early-minus-late function, which is basically a simple early-minus-late DLL discriminator, as shown in Figure 6.10. We use a discriminator because it allows us just to keep our early-minus-late observable near zero and hence the incoming code in phase with the prompt replica.

The implementation of these two loops then differs in terms of mechanism of comparison and feedback implementation. These mechanisms are defined by code and phase discriminators. In our implementation, we have combined code and carrier tracking loops [3]. Carrier and code are both stripped from the incoming DIF signal. The carrier tracking loop is implemented as a Costas loop [12], by multiplying the incoming signal by sine and cosine components of a local carrier replica. The specific feature of a Costas loop is that it is not affected by phase transitions. For GNSS the phase transition is due to navigation bits. After the multiplication we have in-phase (I) and quadrature (Q) arms in the Costas loop. An I-arm would output signal modulation. It is represented by a navigation message bit sequence. If we remove the navigation message , then the I-arm output from a carrier tracking loop should be 0. The I and Q outputs are integrated, usually for a period equal to the code length.

In order to keep phase difference to a minimum, we construct carrier discriminators and generate feedback according to their values. A discriminator is characterized by its value as a function of phase difference. The phase difference should be within certain limits so that the discriminator can be used to maintain a lock on the signal. The most common discriminator for a carrier tracking loop can be defined as follows [3]:

$$D_{\mathrm{PLL1}} = Q_{\mathrm{P}} \times I_{\mathrm{P}}, \tag{6.8}$$

$$D_{\mathrm{PLL2}} = Q_{\mathrm{P}} \times \mathrm{Sign}(I_{\mathrm{P}}), \tag{6.9}$$

$$D_{\mathrm{PLL3}} = \arctan\left(\frac{Q_{\mathrm{P}}}{I_{\mathrm{P}}}\right). \tag{6.10}$$

These discriminators have the following performance [3]. Discriminator D_1 defined by (6.8) has near optimal performance at low SNR with an output phase error defined as sine of doubled phase difference $\sin(2 \cdot \Delta\varphi)$. Discriminator D_2 defined by (6.9) has near optimal performance at high SNR with an output phase error defined as sine of phase differences in $\Delta\varphi$. Discriminator D_3 defined by (6.10) has optimal performance. It requires more computations and usually is implemented with lookup tables. Its output phase error is defined as phase difference $\Delta\varphi$. These discriminators are shown in Figure 6.13.

To provide a code tracking functionality the in-phase (I) and quadrature (Q) arms are multiplied locally by replicas of code created in the processor. These replicas are generated as prompt, early, and late versions. The number of such replicas implemented directly is defined by the number of samples within a chip length. The results of multiplications of these replicas and incoming baseband signal depend on the shape of the spread code autocorrelation function (Figure 6.11).

A prompt replica is generated initially in accordance with code phase estimation from the acquisition step. Early and late replicas are generated with some shift in code phase, called ***correlator spacing***. This phase shift can be set in the range from ½ of the chip length to much smaller values. Smaller correlator spacings between the replicas make the tracking loop more accurate, but can get out of lock more easily due to user dynamics. The value of the correlator spacing is basically limited by sampling rate. Further decreasing of the correlator spacing may not improve the performance. This is because a real baseband signal has a trapezoidal shape rather than rectangular [1]. If correlator

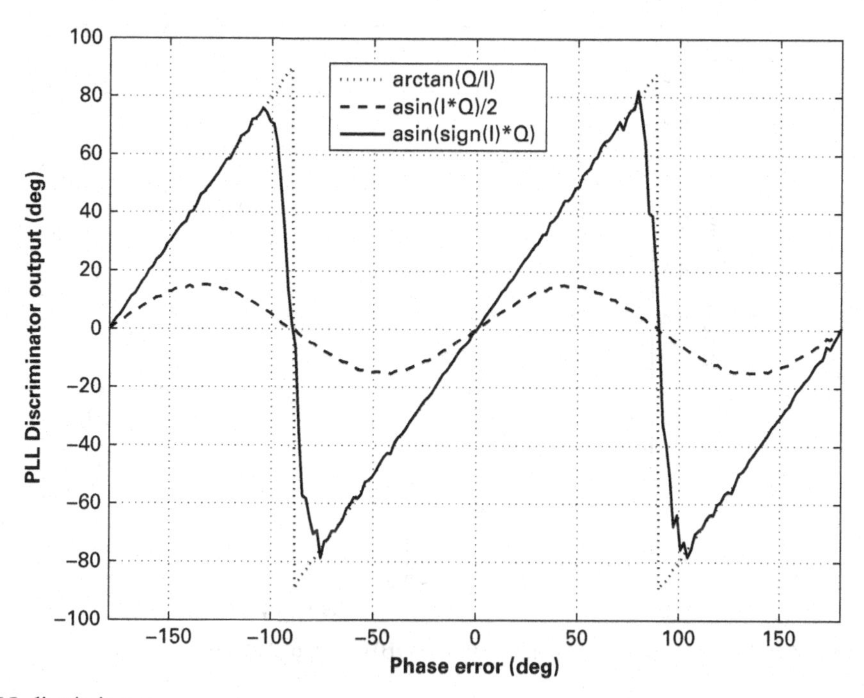

Figure 6.13 PLL discriminators.

spacing is getting smaller the loop performance will improve while it is less than a ***rise time***.

As we have seen in the previous chapter, the front-end bandwidth may affect baseband processor performance. Some higher frequency components are removed from the signal on its path through the front end. Due to the effective decrease in sampling rate it also decreases resolution of the signal processing algorithms in the baseband processor. We can see that narrow bandwidth rounds the corners of the correlation peak. Figure 6.14 shows the autocorrelation function GPS L1 C/A code for the front-end bandwidth equal to 20, 6, and 2 MHz.

The results from all multiplications go to a code discriminator, which provides a feedback loop on code delay value. The purpose of a discriminator is to generate feedback in such a way that a locally generated prompt replica will have minimum shift with an incoming signal. A discriminator can be calculated in various ways, and we consider a few discriminators here [3],[6].

The early-minus-late discriminator requires minimum computer resources, because Q_i values are not computed,

$$D_{\mathrm{DLL1}} = I_{\mathrm{E}} - I_{\mathrm{L}}. \tag{6.11}$$

This is a ***coherent discriminator***. This means that it requires coherent carrier tracking in order to operate. The discriminators below are non-coherent.

The early-minus-late power discriminator is defined as follows:

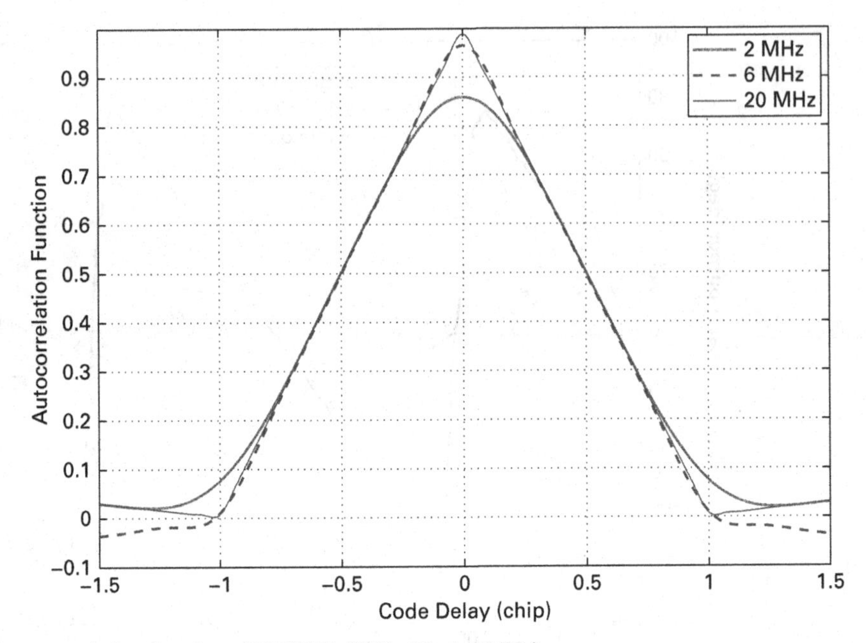

Figure 6.14 Auto-correlation function of BPSK(1) (BW = 20, 6, 2 MHz).

$$D_{\mathrm{DLL2}} = 1/2\left(I_{\mathrm{E}}^2 + Q_{\mathrm{E}}^2 - I_{\mathrm{L}}^2 - Q_{\mathrm{L}}^2\right). \tag{6.12}$$

The dot product uses all outputs,

$$D_{\mathrm{DLL3}} = I_{\mathrm{P}}(I_{\mathrm{E}} - I_{\mathrm{L}}) + Q_{\mathrm{P}}(Q_{\mathrm{E}} - Q_{\mathrm{L}}). \tag{6.13}$$

Figure 6.15 shows DLL discriminators. Figures 6.13 and 6.15 show idealized discriminator behavior for zero-rise-time base band signal. The early-minus-late discriminators are important for a software receiver, because they do not require output from the prompt branch. Each of the branches requires a number of operations proportional to the number of samples. The calculation of I and Q values presents a computational bottleneck for the software receiver. Thus removing one of the branches significantly decreases computational load.

If we have more than three replicas we can devise more sophisticated discriminators to account for changes in autocorrelation function shape caused for example by multipath.

The correlator spacing can be adjusted during receiver operation as a function of signal quality.

A receiver can use more than three replicas. Figure 6.16 shows an iPRx receiver correlator window with graphical output from 15 correlators. The number of correlators in a straightforward implementation is limited by the DIF sampling rate.

The loop order can be defined as the highest power in the denominator of the loop transfer function. As the phase-locked loop's order is increased, it tends to compensate for an instantaneous change in the next higher derivative of the input. Code loop order is usually smaller than carrier loop, because a carrier loop provides internal aiding to the

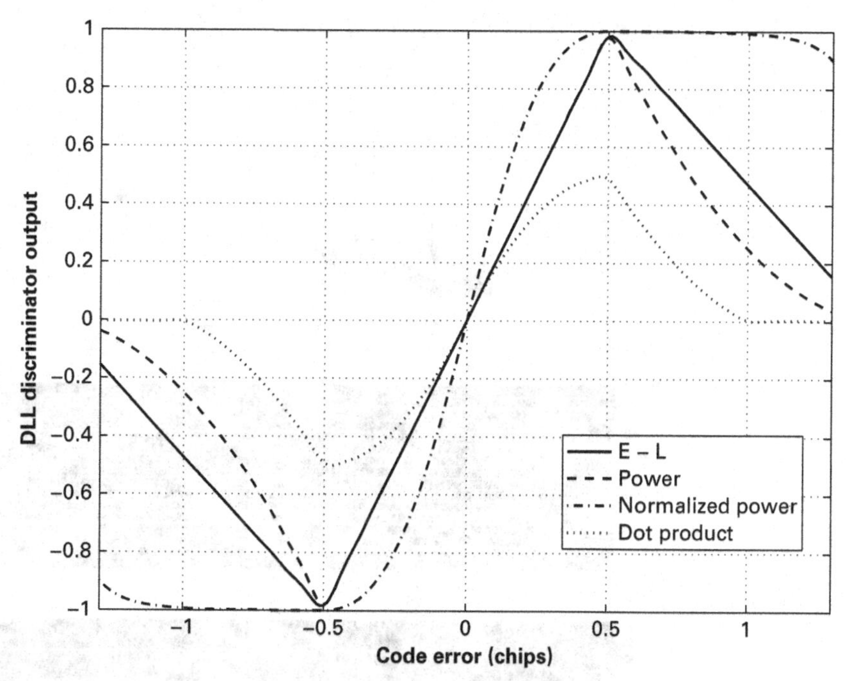

Figure 6.15 DLL discriminators.

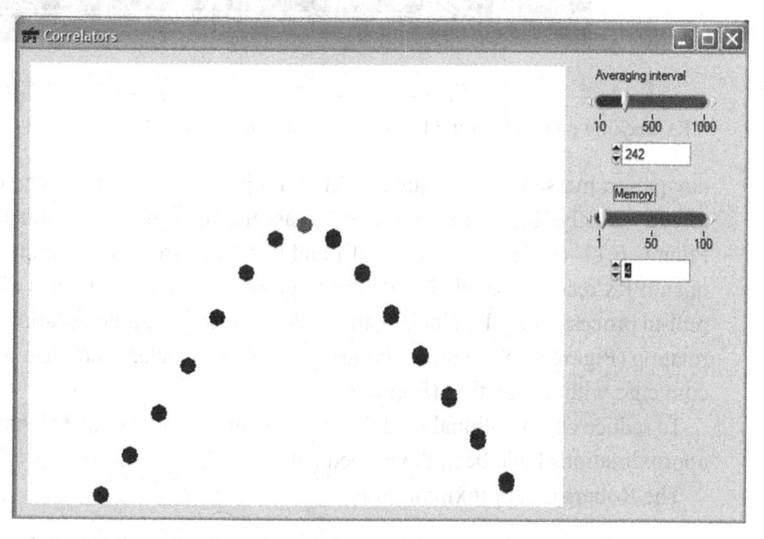

Figure 6.16 iPRx correlator panel : output from 15 correlators.

code loop and compensates for most of the dynamics in it. Thus code loop bandwidth is smaller than carrier loop bandwidth [3].

We can see an example of the behavior of I and Q outputs in the tracking panel of an iPRx receiver. The I-arm outputs the demodulated data symbol, which is in fact a

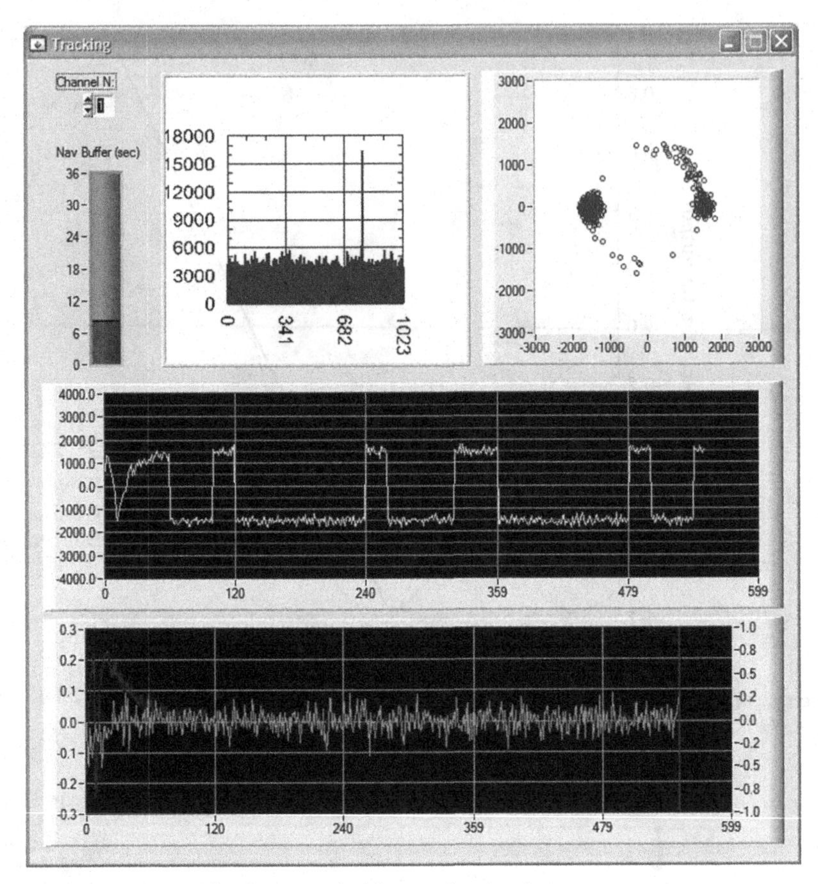

Figure 6.17 iPRx tracking panel : tracking loops are locking on the signal.

navigation message data sequence. In an IQ plot we can see that the code tracking loop affects mostly I-arm behavior, whereas the Q-arm shows carrier tracking error. Figures 6.17–6.19 shows output of I and Q signals from the prompt and early channels in an iPRx receiver panel. The iPRx I/Q graph shows the dynamics of the tracking loops pull-in process. If DLL is locked and PLL is not locked, then the spots in the IQ plot start rotating (Figure 6.18). If on the contrary, the PLL is locked and DLL is not, then the spots converge without rotation (Figure 6.19).

To reduce computational load for discriminators calculating a signal envelope, various approximations have been developed [3].

The Robertson approximation is

$$\sqrt{I^2 + Q^2} \simeq \mathrm{MAX}(|I| + 1/2|Q|, 1/2|I| + |Q|). \tag{6.14}$$

Receivers normally apply much more than three correlators. The price of correlators is so low that they are basically free for a hardware receiver. For a software receiver, extra correlators are extremely expensive in terms of computational load. However, their usefulness is limited to specific applications, such as multipath and interference mitigation [3].

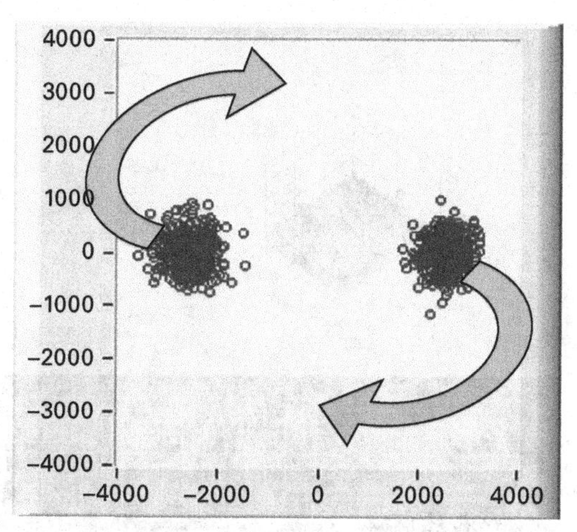

Figure 6.18 iPRx tracking panel : I/Q behavior for not locked PLL.

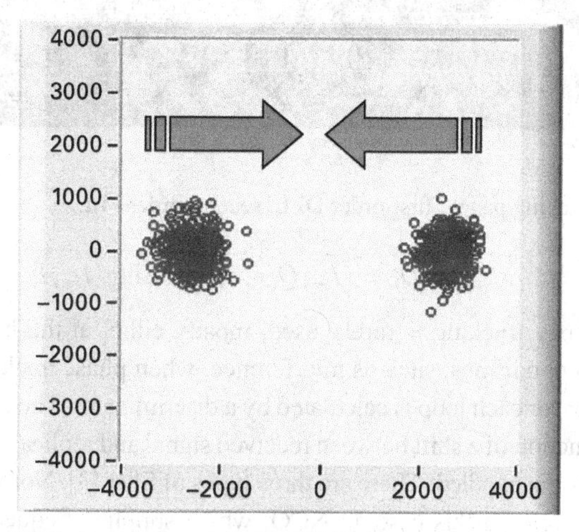

Figure 6.19 iPRx tracking panel : I/Q behavior for not locked DLL.

In hardware receivers correlators are also used in acquisition (software receivers usually implement acquisition using FFT). In this case multiple correlators are extremely useful for quick acquisition and high sensitivity and indoor receivers. This is especially if correlators are implemented on a mass scale. Receivers capable of indoor positioning may have tens of thousands of correlators. For indoor positioning the receivers are used in snapshot positioning and do not provide tracking.

A baseband processor can also implement frequency tracking, which can be considered as differential carrier phase tracking. Basically all frequency lock loop (FLL) discriminators suffer from a change of sign of the bits in a navigation message. The decision-directed cross-product discriminator [1] is created out of two discriminators to detect and compensate for the sign change:

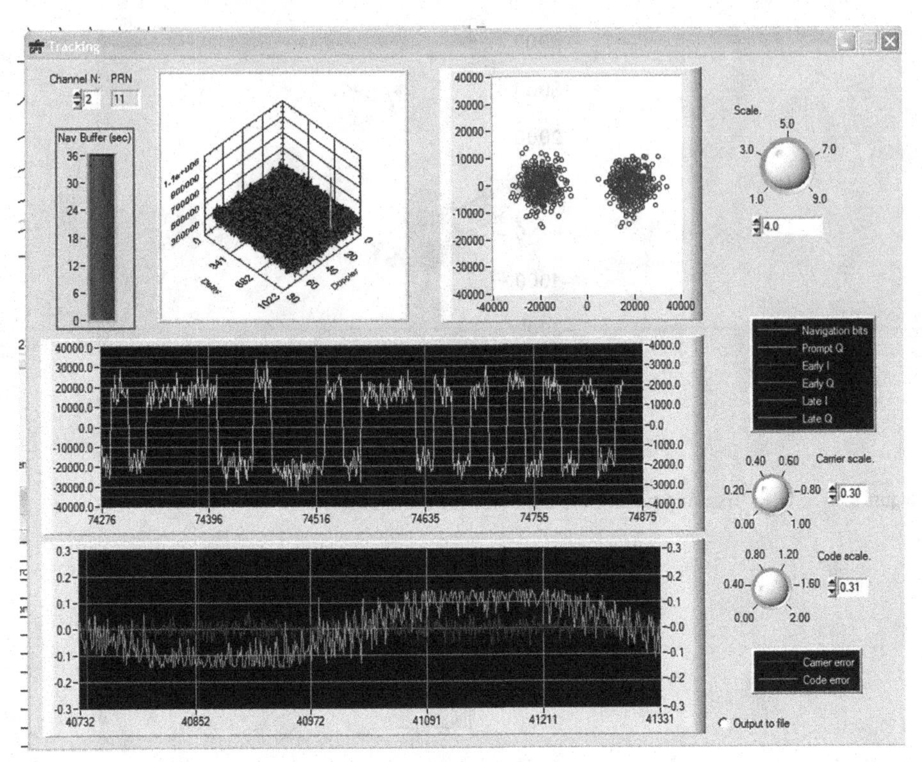

Figure 6.20 iPRx tracking panel : first-order DLL, second-order PLL.

$$D_f = (I_{i-1}Q_i - I_iQ_{i-1})\text{sign}(I_{i-1}I_i + Q_{i-1}Q_i). \tag{6.15}$$

Frequency tracking is rarely used, mostly either at the beginning of tracking or under special conditions, such as interference, when phase tracking is difficult.

Error for each loop is calculated by a discriminator. The discriminator is describing error as a function of a shift between received signal and replica. A loop filter is in charge of how this error is handled. There are three types of filter [3]. Normally receivers employ second-order loops. In this case an NCO, which supplies frequency for the replica's code and carrier, is controlled using information about error change. The first-order filter controls the NCO using information about error value. Beside second-order filters, an iPRx receiver also allows us to implement first-order DLL and third-order PLL. The first-order DLL can be used in INS aided mode. Figure 6.20 shows an iPRx tracking panel for first-order DLL. The bottom plot shows code and carrier errors. We can see that code error is compensated with a delay, because the loop does not use information about error rate.

The third-order filter controls the NCO using information about error rate of change. We should note that third-order loop bandwidth is limited to 18 Hz and less [3]. Due to the limited bandwidth it may also require external assistance such as from INS in the case of high dynamics, because of jerks, as this loop is not sensitive to accelerations. Figure 6.21 shows an iPRx tracking panel for third-order PLL. We can see that there is a constant uncompensated carrier phase error, also noticeable from the I/Q plot as a skew angle. An FLL shows similar behavior.

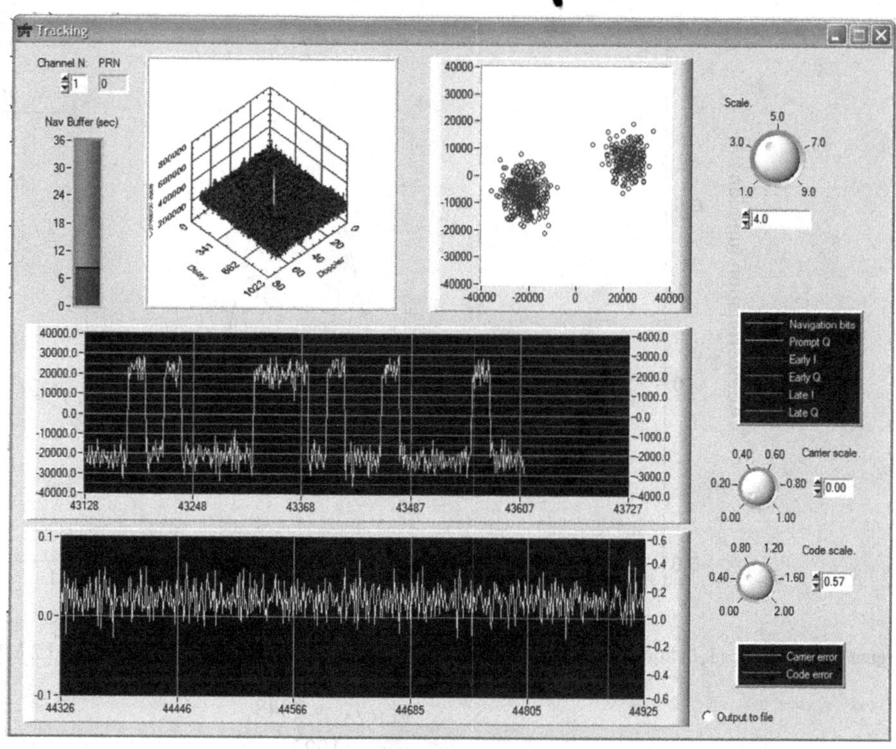

Figure 6.21 iPRx tracking panel : second-order DLL, third-order PLL.

Wide bandwidth allows tracking in high dynamics, but decreases tracking accuracy. It is possible to change the bandwidth of the loops on-the-fly in a software receiver. This can be done in accordance with signal-to-noise measurements, using simple rules of thumb.

6.5 Tracking and acquisition of other GNSS signals

6.5.1 Tracking and acquisition of GLONASS signals

In order to search for a GPS L1 C/A signal we need to go through a number of shifts defined by code length and length of shift in chips,

$$n = \frac{N}{d}, \tag{6.16}$$

where d is correlation spacing. For acquisition of BPSK we can conduct a search choosing d equal to a maximum value 0.5. The number of chips for GPS L1 C/A is equal to 1024, and for GLONASS to 511. Therefore, the initial search area is as follows:

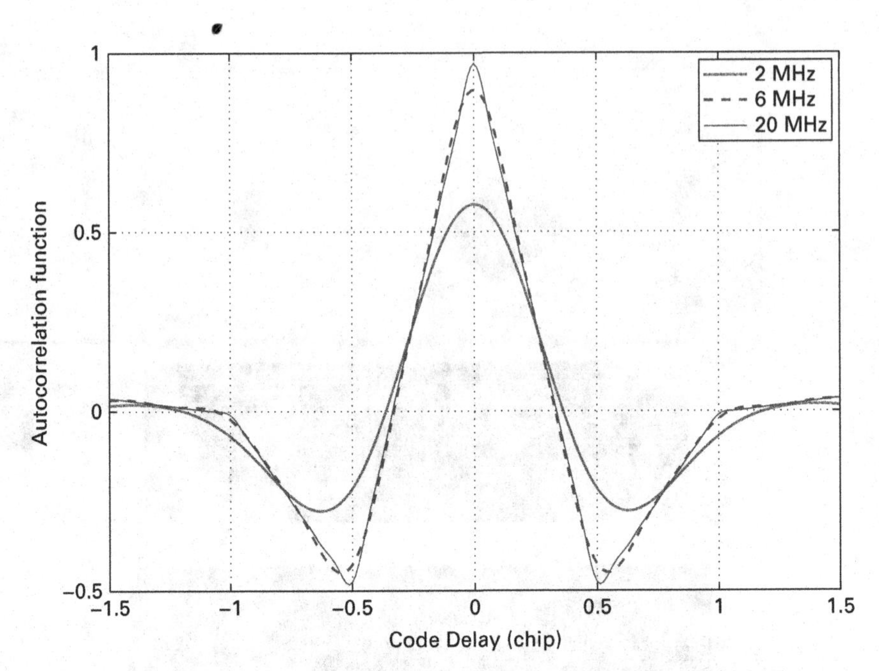

Figure 6.22 BOC (1,1) autocorrelation function for front-end bandwidth equal to 20, 6, and 2 MHz.

$$\begin{aligned} n_{\mathrm{L1C/A}} &= 2048, \\ n_{\mathrm{SP}} &= 1022, \end{aligned} \tag{6.17}$$

where $n_{\mathrm{L1C/A}}$ is searching space fr GPS, n_{SP} is searching space as for GLONASS, the same for L1 and L1 SP signals.

6.5.2 Tracking and acquisition of BOC signals

For acquisition of BOC signals the same method of circular correlation can be used. The difference is that a BOC signal may give multiple peaks and require more Doppler bins to search.

An ideal BOC(1,1) autocorrelation function is depicted in Figure 6.7. We can see that half of a chip distance, which was applied in the case of GPS L1 BPSK code, is not applicable here. Figure 6.22 shows the BOC(1,1) autocorrelation function for front-end bandwidth equal to 20, 6, and 2 MHz. The wider is the bandwidth, the sharper the correlation. We can see that 2 MHz bandwidth is not wide enough for BOC(1,1), since the main lobe has 4 MHz. The shape of the autocorrelation function provides better accuracy and multipath resistance (see more in Chapter 7).

6.6 Lock detectors

Lock detectors play a very significant role in receiver design. In order to be able to start using results from the tracking loop and start collecting measurements, we need to make sure that the tracking loop is actually locked on a satellite signal.

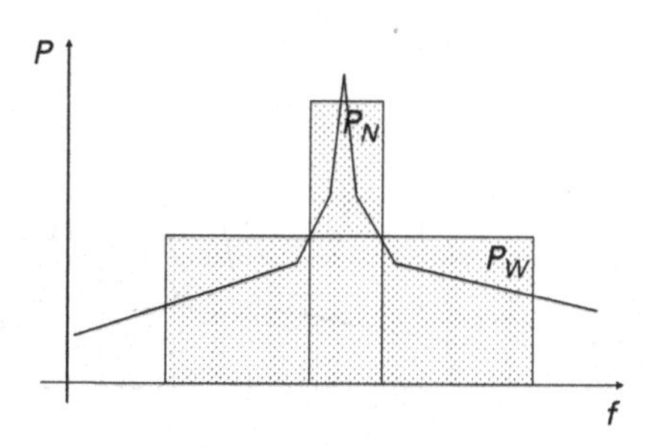

Figure 6.23 Tracking loop VCO spectrum in case of a locked loop.

It is relatively easy to see if a signal is locked, by watching the I and Q components in the receiver panel (see Figures 6.17–6.19). There are various ways to automate this process. We will follow [2] to devise such an algorithm. The lock detector is designed by analysis based on $C/CN_0 - N_0$ estimation. This is because a code lock is a necessary condition to achieve good $C/CN_0 - N_0$. In our design, code lock is impossible without phase lock. A lock detector may be defined as follows:

$$\hat{\mu}_P = \frac{1}{K}\sum_{i=1}^{K} P_i, \tag{6.18}$$

where P_i is normalized power for the ith chip length interval, and K is the total number of intervals. For C/A GPS signal lock detector measurement, $\hat{\mu}_P$ is taken over 50 intervals, which gives a total 1 second length of averaging interval.

If we compare signal power over two different bandwidths, it can give us an indication of the total signal-to-noise ratio. Figure 6.23 shows a schematic presentation of a power spectrum for a tracking loop NCO over time. If the tracking loop is locked, then the spectrum has a clear peak, and narrow bandwidth power is increasing. If the loop is not locked, then the peak is sliding and narrow bandwidth power is decreasing (Figure 6.24).

The normalized power P_i at each interval is calculated as the ratio of narrow-band power to wide-band power,

$$P = \frac{P_{\mathrm{N}}}{P_{\mathrm{W}}}, \tag{6.19}$$

where wide-band power can be defined as

$$P_{\mathrm{W}} = \sum_{j=1}^{M}\left(I_j^2 + Q_j^2\right), \tag{6.20}$$

and narrow-band power

$$P_{\mathrm{N}} = \left(\sum_{j=1}^{M} I_j\right)^2 + \left(\sum_{j=1}^{M} Q_j\right)^2, \tag{6.21}$$

both computed over M samples.

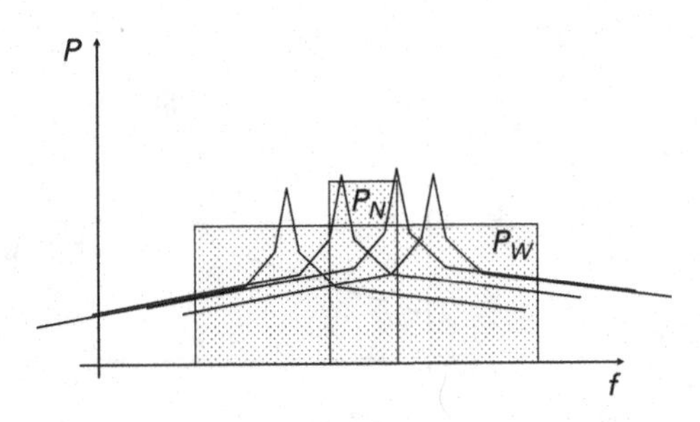

Figure 6.24 Tracking loop VCO spectrum in case of an unlocked loop.

The signal lock detector measurement $\hat{\mu}_P$ is compared against a threshold and a decision about lock is made.

The lock detectors continue to work all the time after the tracking loops are locked in order to indicate in time when the lock is lost.

The same $\hat{\mu}_P$ measurements can be used to estimate C/N_0 as follows:

$$\frac{\widehat{C}}{N_0} = 10\log_{10}\left(\frac{1}{T}\frac{\hat{\mu}_P - 1}{M - \hat{\mu}_P}\right). \tag{6.22}$$

6.7 Bit synchronization

After the tracking loop is locked, it starts to output data for measurements. The purpose of bit synchronization is to align with navigation message data bits, in order to be able to read a binary navigation message from the signal. We describe here a method given in [2].

We create a cell counter with the number of cells equal to the number of chips in one bit of a navigation message. When the tracking loops are locked, the cell counter counts each time an in-phase component changes its sign. We can actually see navigation message bits as outputs of in-phase component. Cell counters switch cells in intervals equal to code length, so change of sign of the in-phase component can happen only between cells. Every time the change of sign of an in-phase component happens, the counter increases number in the corresponding cell. The resulting bit synchronization histogram shows the number of such changes at each cell. If the tracking loops are properly locked, then all changes should happen and be counted in the same cell (see Figure 6.25). There are two thresholds set. The bit synchronization is successful if one cell exceeds the upper threshold. The bit synchronization has failed if two cells exceed the bottom threshold, or the tracking loop lock detectors indicate loss of lock. If bit synchronization is successful, then the bit synchronization counter should not be used further. After that it should just be checked continuously, that all changes of in-phase output sign occur in the bit indicated by synchronization counter timing.

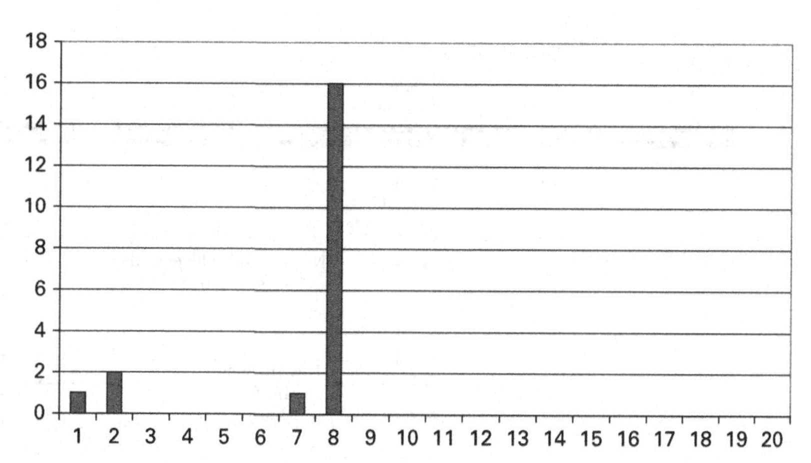

Figure 6.25 Bit synchronization histogram.

6.8 Measurements

A baseband processor should provide the following measurements:

- code phases (pseudoranges),
- Doppler,
- signal-to-noise measurements,
- carrier phase,
- decoded navigation message.

We have already described in Section 6.6 how to estimate signal-to-noise ratio by (6.22).

Carrier frequency estimates are provided by a carrier tracking loop. Doppler is calculated as a difference between those estimates for each satellite and a central IF signal frequency. These Doppler estimates include front-end clock drift. If we want to exclude this drift, then further adjustments should be done after a positioning algorithm estimates antenna coordinates along with the clock error.

Carrier phase measurements also come directly from carrier tracking loops. Code phase measurements, however, can be calculated normally only after at least one frame of navigation message is decoded.

We start to collect navigation bits after bit synchronization (Section 6.7) is achieved. After 6 seconds in the case of GPS L1 C/A we can decode a part of the message equal in length to one frame. From these data we can find an embedded time mark. This time mark is a Z-count, which is included at each 6 seconds of GPS navigation message. The Z-count shows time since the beginning of a GPS week in 1.5 second intervals. After we have decoded a Z-count, we need to pinpoint the current position of the tracking loop pointer to the time axis, in particular to an instant after the received Z-count and before the next one. This can be done using the GPS message structure. The position of the tracking loop pointer on the time axis gives a pseudorange measurement. This pseudorange measurement includes the front-end clock error, so it is useful only in a set with

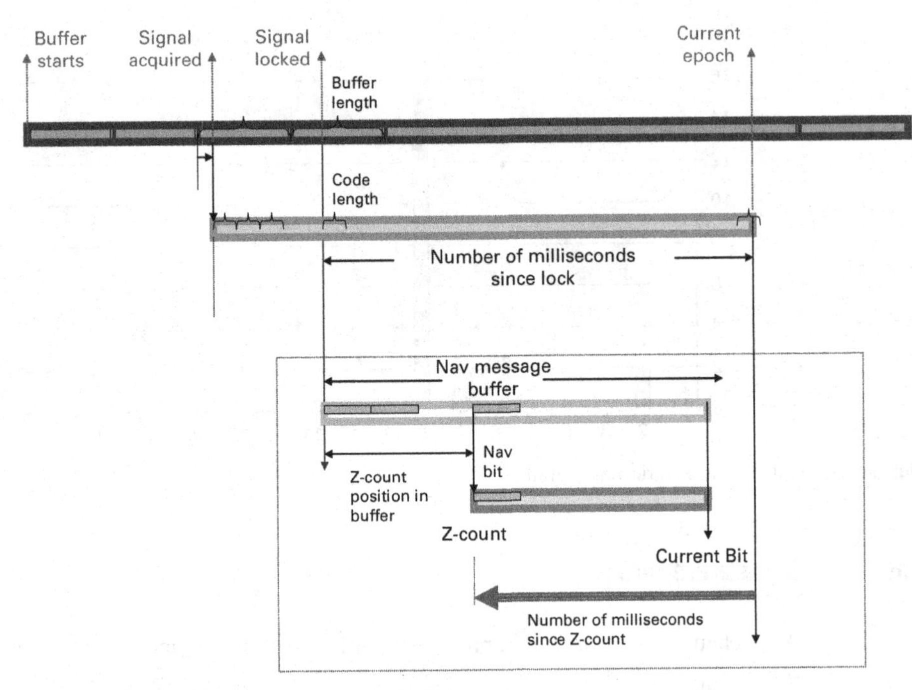

Figure 6.26 Concept of code phase measurements calculation.

pseudoranges from other satellites. Figure 6.26 shows how code phase measurements are calculated in a baseband processor.

The position of the tracking loop pointer on the time axis also allows us to pinpoint a satellite position at the time of transmission. This is required to be done at the positioning calculation step.

6.9 Real-time implementation

A real-time software (RTS) receiver structure is presented in Figure 6.27. The RTS receiver has to operate in real-time and for an unlimited period of time. Therefore it is bound to have acquisition, tracking, and navigation (if there is one) functions in separate threads. The front end and GUI also take separate threads. It makes an implementation of a RTS receiver more cumbersome, especially on a PC under Windows OS.

The front-end logs DIF signal into a memory buffer. Acquisition and tracking threads take data from this memory buffer (see Figure 6.28). Their operation must be organized in such a manner that there will be no conflicts between threads. This requires special care in a RTS receiver because at least three threads are trying to access the same shared data.

Baseband measurements should be transmitted on a regular basis from tracking to navigation threads. The acquisition thread should provide the tracking thread with initial code phase and frequency data for a particular channel. In the tracking thread a particular channel may start just by getting these data from an acquisition thread, while other

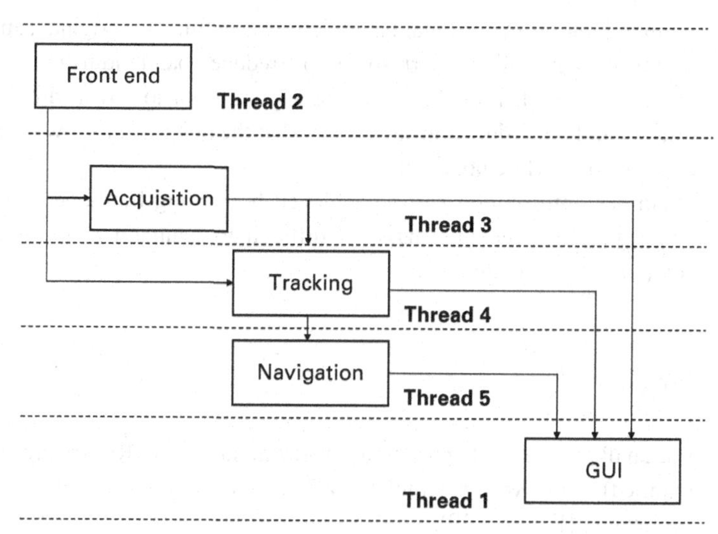

Figure 6.27 Real-time software receiver implementation.

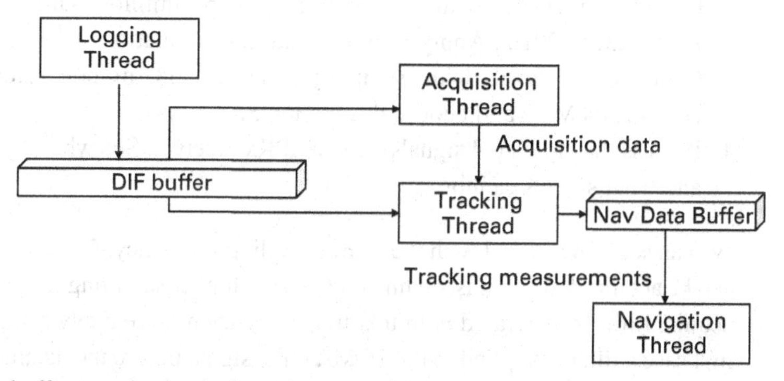

Figure 6.28 Memory buffer organization in the software receiver.

channels already provide tracking measurements to the navigation thread. All these functionalities require an introduction of thread-safe variables and vectors, and elements of thread-safe data sharing, such as semaphores.

Real-time operation also requires especial attention to a few places in the baseband processing, which consume most of the processing time. There are several ways to reduce calculation load on a processor. In this section we look at them briefly.

The most time-consuming parts in the software receiver implementation are tracking loop I,Q accumulators for prompt, early, and late replicas. The number of operations is proportional to the number of samples and to the number of accumulators. One way to reduce computational load is to choose discriminators, which require fewer accumulators. Another way is to decimate the digital signal. We just need to remember that in order for the signal to be useful, the signal bandwidth should be less than half of the new sampling rate. If the original bandwidth was defined by original sampling rate, then the decimated signal should correspond to a reduced bandwidth. Therefore filtering may be required. A bandpass filter should be applied prior to down-sampling. The decimation filter should be designed in such a way that the noise of the down-sampled signal can be considered as white noise.

Further optimization can be achieved by writing bottle-neck portions in assembly language. However, this makes it difficult to maintain code, because it depends on computer architecture.

6.10 Project

Use an iPRx receiver to process a signal simulated with ReGen (light version of ReGen is required). The ReGen simulator allows us to choose sampling frequency, IF, and decimation (Figure 6.29).

1. Generate a signal with sampling rate 16.368 million samples per second and IF = 4.1304 MHz.
2. Generate a signal with sampling rate 5.456 million samples per second and IF = –1.3256 MHz. Apply front-end filtering simulation.
3. Generate a signal with sampling rate 16.368 million samples per second, IF = 4.1304 MHz and choose decimation 3.
4. Process the simulated signals with an iPRx receiver. See what signal can be acquired and what signals cannot.

We can see that a signal with the same sampling frequency, 5.456 Msamples per second, can be acquired, when it is decimated from the higher sampling frequency, and cannot be acquired if it is generated with this frequency even if we apply a front-end filter. In the first case a filter is applied to the 16 Msample signal before the decimation. In the second case the filter is applied to the 5 Msample signal.

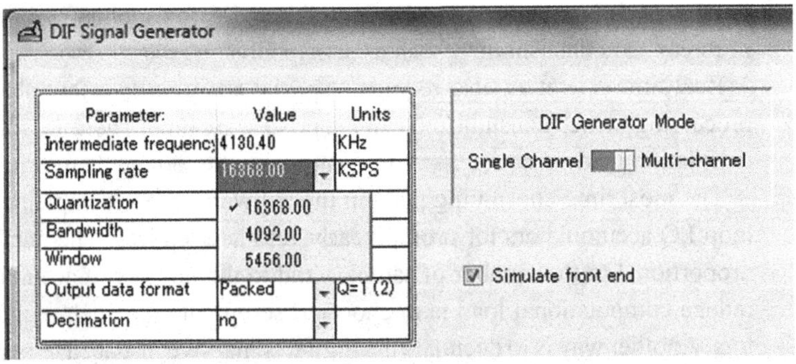

Figure 6.29 Signal generation options in ReGen DIF generation panel.

References

[1] J. J. Spilker, *Fundamentals of Signal Tracking Theory*, in *Global Positioning System: Theory, and Applications*, Vol. I, B. W. Parkinson and J. J. Spilker (editors), Washington, DC, American Institute of Aeronautics and Astronautics Inc., 1996.

[2] A. J. Van Dierendonck, *GPS Receivers*, in *Global Positioning System: Theory, and Applications*, Vol. I, B. W. Parkinson and J. J. Spilker (editors), Washington, DC, American Institute of Aeronautics and Astronautics Inc., 1996.

[3] P. W. Ward and J. W. Betz, *Satellite Signal Acquisition, Tracking, and Data Demodulation*, in *Understanding GPS, Principles and Applications*, E. Kaplan, and C. Hegarty (editors), 2nd edition, Norwood, MA, Artech House, 2006.

[4] J. Tsui, *Fundamentals of Global Positioning System Receivers: A Software Approach*, 2nd edition, Hoboken, NJ, John Wiley and Sons, Inc., 2005.

[5] T. Pany, *Navigation Signal Processing for GNSS Software Receivers*, Norwood, MA, Artech House, 2010.

[6] P. Rinder and N. Bertelsen, *Design of a Single Frequency GPS Software Receiver*, Aalborg, Aalborg University, 2004.

[7] A. Oppenheim and R. Schäfer, *Discrete-Time Signal Processing*, NJ, USA, Prentice-Hall, 1999.

[8] N. I. Ziedan, *GNSS Receivers for Weak Signals*, Norwood, MA, Artech House, 2006.

[9] S. J. Orfanidis, *Optimum Signal Processing*, 2nd edition, New York, NY, McGraw-Hill Publishing Company, 1988.

[10] E. V. Appleton, The automatic synchronization of triode oscillators, *Proc. Camb. Phil. Soc.*, **21**, 1922–1923, 231–248.

[11] R. E. Best, *Phase-Locked Loops: Design, Simulation, and Applications*, 5th edition, New York, NY, McGraw-Hill, 2003.

[12] F. M. Gardner, *Phaselock Techniques*, Second Edition, New York, NY, John Wiley and Sons, Inc., 1979.

[13] E. D Kaplan and C. J. Hegarty (editors), *Understanding GPS, Principles and Applications*, second edition, Boston, MA, Artech House, 2006.

[14] P. S. Jorgensen, *Ionospheric Measurements from NAVSTAR Satellites*, Report SAMSO-TR-79–29 (Space and Missile Systems Organization), Air Force Systems Command, December 1978.

Exercise

Exercise 6.1. Change PLL and DLL loop bandwidth in an iPRx receiver to see the I/Q behavior when PLL and DLL lose a lock on the signal (see more on this topic in Chapter 12).

7 Multipath

In this chapter we consider multipath error. The difference from other GNSS errors we have considered so far is that this error affects a baseband processor. The multipath error is created by signal coming to an antenna through different paths. These additional paths result from signal reflections from various surfaces. It is easy to see the difference between multipath error and other errors when we consider how this error is simulated in a signal simulator. All other errors, besides scintillation, are created when we make a scenario. A RINEX observation file, which can be generated at this stage, reflects all these errors, including tropospheric and ionospheric delays, clock errors, and orbit errors. All these errors are just added to the code and carrier phases, calculated for the satellite to antenna signal path. Multipath error is created when we generate a signal from this scenario in a DIF generator. In fact we generate an additional signal for each additional signal path. The simplest multipath error is created in a DIF generator by duplicating a simulated satellite signal with a phase shift and attenuation in amplitude.

7.1 Multipath error and its simulation

A GNSS signal may come to a receiver antenna by more than one way. It may come from a satellite directly or after it has been reflected by some surfaces. We can describe a GPS signal as follows (see Chapter 3):

$$A = A_0 \sin(\omega t + \varphi) \cdot D, \tag{7.1}$$

where D is a C/A code. We can omit a navigation message here, because the signal is affected by multipath only through code. The multipath signal can then be expressed in the following way:

$$A_{\mathrm{M}} = k_{\mathrm{M}} \cdot A_0 \cdot \sin(\omega t + \varphi + \varphi_{\mathrm{M}}) \cdot D(\varphi_{\mathrm{M}}), \tag{7.2}$$

where the signal is attenuated by k_{M} and delayed by φ_{M}. The total signal coming to a receiver can then be expressed as a sum of the signals,

$$\begin{aligned} A &= A_{\mathrm{D}} + \sum_{i=1}^{n} A_{\mathrm{M}_i} = A_0 \left(\sum_{i=0}^{n} \left[k_{\mathrm{M}_i} \cdot \sin(\omega t + \varphi + \varphi_{\mathrm{M}_i}) \cdot D(\varphi_{\mathrm{M}_i}) \right] \right), \\ k_{\mathrm{M}_0} &= 1, \end{aligned} \tag{7.3}$$

where A_{D} is a direct signal.

In general, multipath affects signal power, code delays, carrier phases, Doppler, and signal scintillation parameters.

Multipath effects can be described using mostly geometrical optics. Besides the reflection, we may need to consider effects of diffraction and signal scattering [1].

In the case of simulation these parameters can be defined either directly in a deterministic model for each satellite similar to (7.3) or based on environmental models. Environmental models can be implemented in two different ways in a signal simulator.

1 **Deterministic models**

A deterministic model can be based on geometrical optics ray-tracing. It may also include a reflected object description, including parameters of material and roughness. Material can be assigned to the object with different electromagnetic properties, i.e. permittivity and conductivity. A simple model for a time delay can be calculated in the case of reflection from a vertical surface as follows [2]:

$$\tau = \frac{2m}{c} \cos\alpha \cos E, \tag{7.4}$$

where m is distance between the antenna and the surface, c is the speed of light, α is the angle between the satellite azimuth and the azimuth of a vector normal to the surface, and E is elevation angle. The phase delay is calculated from the delay as follows:

$$\varphi_{\mathrm{M}_i} = 2\pi f\tau + \varphi_{\mathrm{R}_i}, \tag{7.5}$$

where f is the carrier frequency, φ_{R_i} is the phase shift caused by a reflector and the antenna phase pattern.

Equations (7.4) and (7.5) give a multipath model for corresponding code and carrier phases, which are implemented in the signal simulated by (7.3) for each ray–surface pair.

In the case of an aircraft the multipath delay can be calculated as follows [1]:

$$\tau = \frac{2h}{c} \sin E, \tag{7.6}$$

where h is aircraft altitude.

2 **Stochastic models**

A stochastic model can be based on a typical environment, characterized by typical statistical parameters and by the user motion model. A statistical fading process can be simulated by multiple echoes based on the typical environment scenario. In this case a model should provide realistic delay and phase correlation for multiple rays. A scattering model and statistical fading can be generated for vegetation as a function of motion.

In a receiver baseband processor, instead of one signal, correlators encounter two or more signals with the same spread code. The shape of a correlation plot for two or more signals is different, and the receiver tries to follow the envelope of the signals, rather than a direct signal. Figures 7.1 and 7.2 show multi-correlator output from an iPRx single-channel receiver in the case of a normal signal and one distorted by multipath.

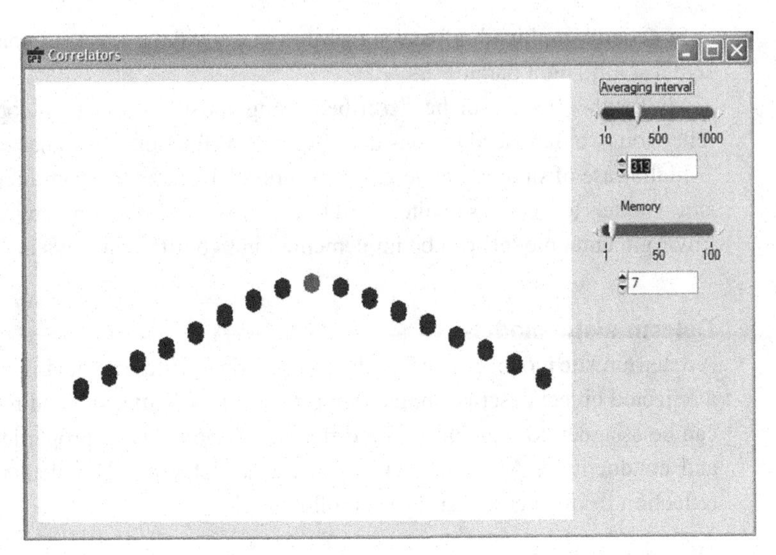

Figure 7.1 iPRx single-channel receiver correlator output panel. Correlation with an undistorted signal.

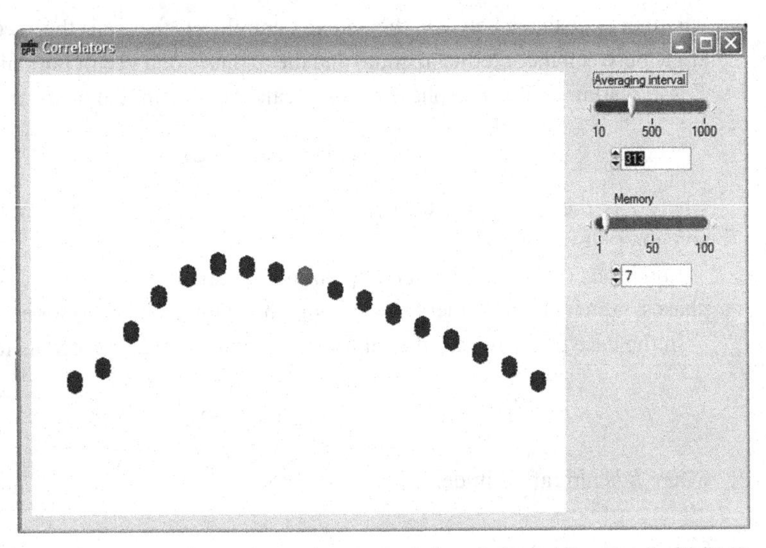

Figure 7.2 iPRx single-channel receiver correlator output panel. Correlation with a signal affected by multipath.

The reflected signal is not always a parasitic signal. In the case of indoor positioning, reflected signals are sometimes the only ones available. Therefore indoor positioning is often realized as a snapshot positioning using chunks of code, sometimes reflected from various surfaces.

It is also possible that a receiver acquires only reflected signal. There are interesting special cases for multipath generation. For example, when a reflected signal is stronger than the direct signal and $k_M > 1$. In real life such situations can happen, for example, in urban canyons.

7.2 Case study: multipath effect on BPSK and BOC(1,1) signals

In this section we consider how multipath affects BPSK and BOC signals. As we described in Chapter 3, one of the reasons behind introducing BOC signals was to improve GNSS signal resistance to multipath. As a case study we look at a test which allows us to compare BOC(1,1) and BPSK resistance to multipath. We use the Quasi-Zenith Satellite (QZS) -like signal for the test. QZS transmits BOC(1,1) in L1 band which is compatible with GALILEO and modernized GPS. We use the QZS-like signal transmitted from a helicopter-based pseudolite in order to have a controlled test environment.

First, we describe the theoretical range error due to multipath and then show the test results. In Chapter 6, we considered autocorrelation functions of BPSK and BOC(1,1). The middle peak of BOC(1,1) is steeper than the correlation function of BPSK and therefore the range error due to multipath should be smaller. The purpose of tracking is to find the peak of the correlation function. If we imagine a climber who seeks to reach the highest point, then he can most likely spot it if it is the peak of a mountain rather than a gentle hill.

The peaks of both BPSK and BOC correlation functions are ideally sharply pointed. However, the real correlation functions in the receiver despreading process are rounded due to the bandwidth of the front end and front-end bandpass filters (see Chapter 6). Figure 7.3

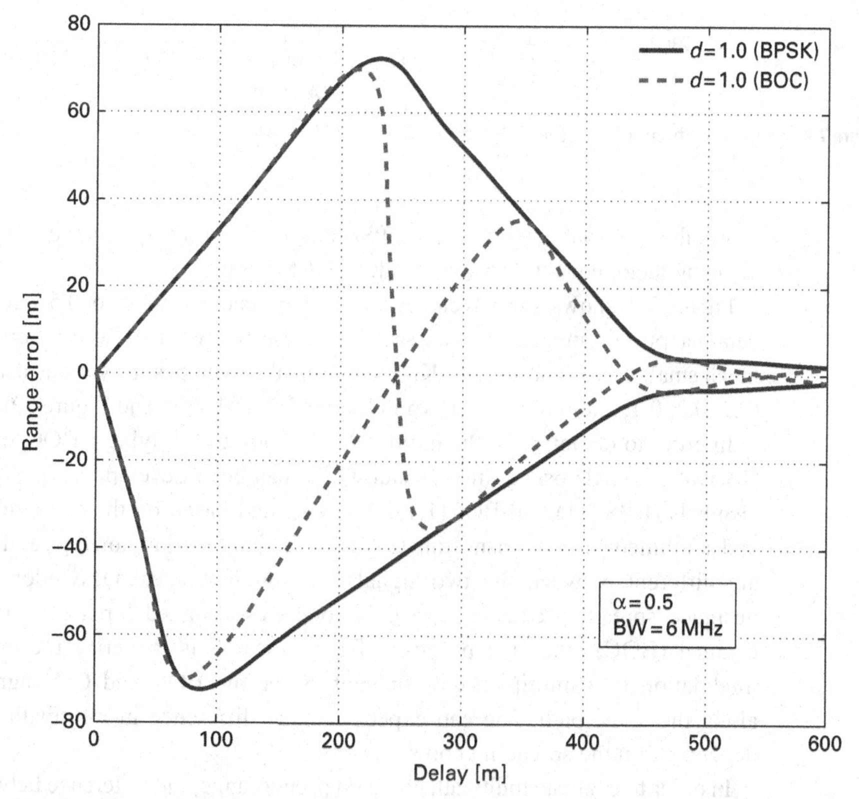

Figure 7.3 Multipath envelopes for BPSK and BOC (d=1.0chips).

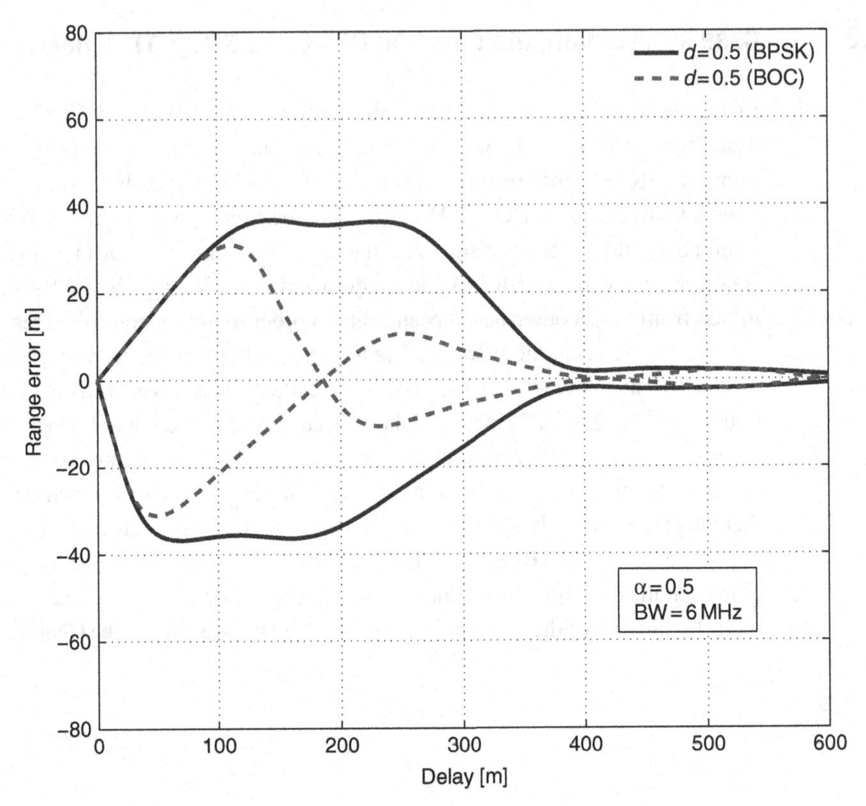

Figure 7.4 Multipath envelopes for BPSK and BOC ($d = 0.5$ chips).

shows the multipath envelopes for BPSK and BOC for a chip spacing equal to 1.0 chip, damping factor equal to 0.5, and bandwidth 6 MHz.

Figure 7.4 shows the envelopes for a chip spacing equal to 0.5 chips with other parameters unchanged. One can see from these figures that the envelope for BOC is much smaller than that for BPSK. The multipath envelopes for various damping factors (0.3, 0.2, 0.1, and 0.05) are shown in Figure 7.5 for BPSK and Figure 7.6 for BOC.

In order to demonstrate the multipath reduction by applying a BOC signal, a BOC/CA two channels pseudolite (Pseudo-QZS) has been developed [3]. Signals of both channels (BPSK(1) and BOC(1,1)) are generated based on the same rubidium clock and combined before transmitting (Figure 7.7). Only modulation type and PRN number are different between the two signals. Although the real QZS adopts a specially designed spreading code, a GPS C/A code was assigned for the pseudolite second channel (BOC) since the purpose of this research is to verify the effect of BOC modulation in a multipath environment. Since the BOC and C/A signal propagate along the same path, one can expect that the difference in multipath error will be dependent on the spreading code only.

In order to evaluate multipath effect on pseudorange, the difference between pseudorange and carrier phase or ***code-minus-carrier (CMC)*** is computed. Common errors such

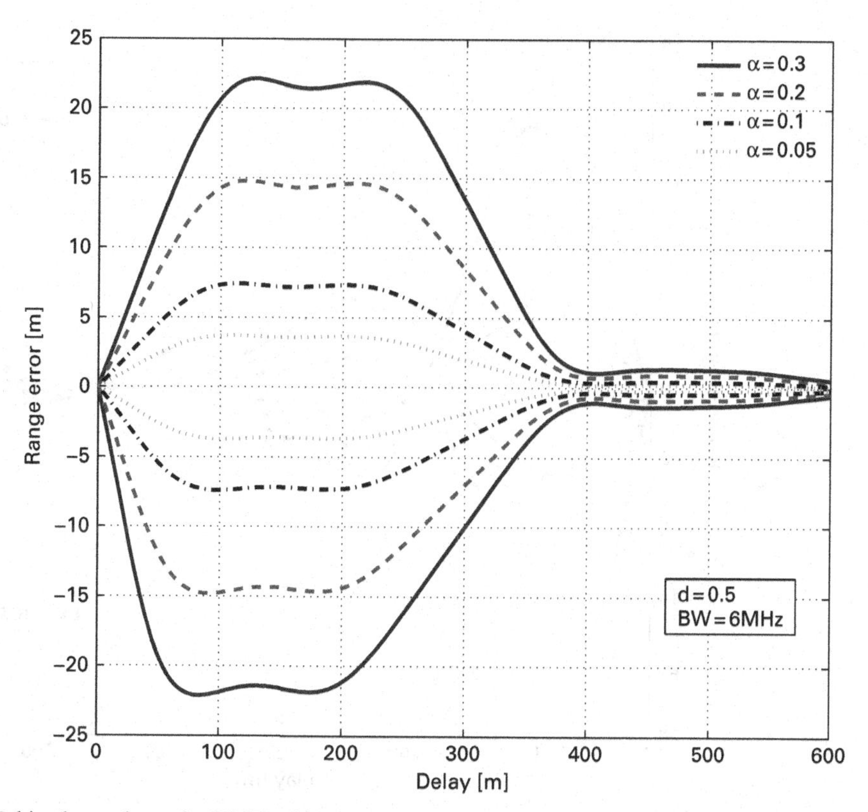

Figure 7.5 Multipath envelopes for BPSK with various damping factors.

as pseudolite/receiver clock error and tropospheric error cancel and multipath error and measurement noise remain in CMC. Although the CMC of a GPS signal suffers from ionospheric delay, the CMC of a pseudolite is not affected by ionospheric delay, since the pseudolite signal does not propagate in the ionosphere. Therefore, multipath error in a pseudolite can be evaluated correctly.

Before evaluating the multipath effect, the pseudorange noise levels of BPSK and BOC are verified. Figure 7.8 shows the variations of CMC from their averages for BPSK and BOC. Also, the standard deviations of CMC are given in the figures. Since the data were recorded in an area where there were not many buildings, the CMC represented mostly the pseudorange noise.

The theoretical standard deviation of pseudorange noise is given by (7.7), since coherent tracking is adopted in the software receiver:

$$\sigma_{\text{BPSK}} = cT_{\text{C}}\sqrt{\frac{d}{4(C/N_0)T}}. \tag{7.7}$$

On the other hand, the standard deviation of BOC pseudorange is given by (7.8), since the slope of the correlation function is three times steeper than for BPSK:

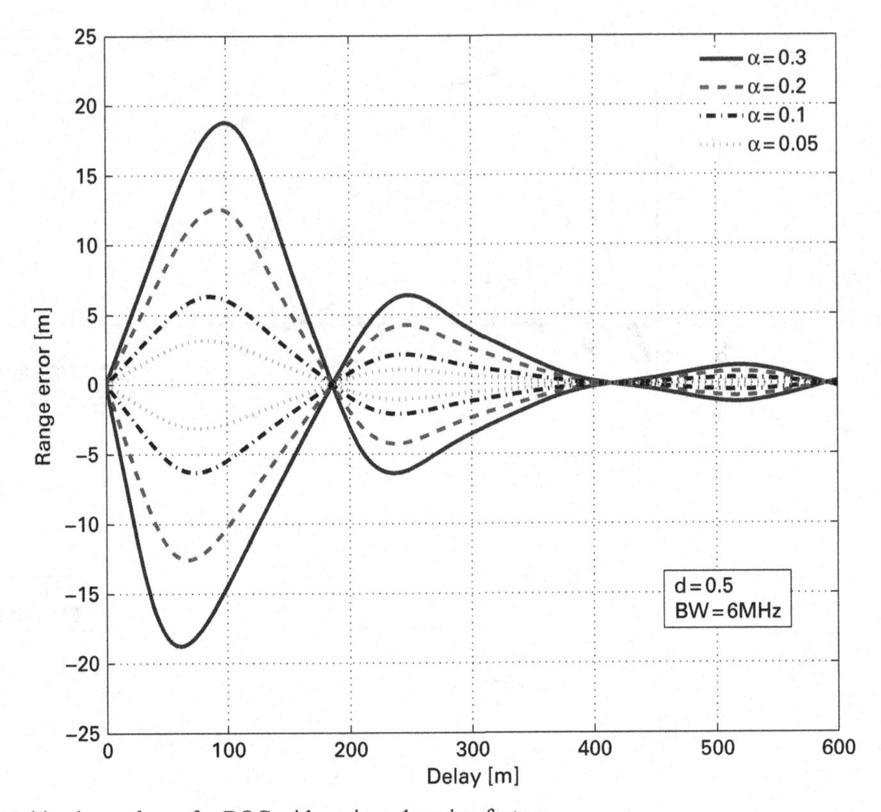

Figure 7.6 Multipath envelopes for BOC with various damping factors.

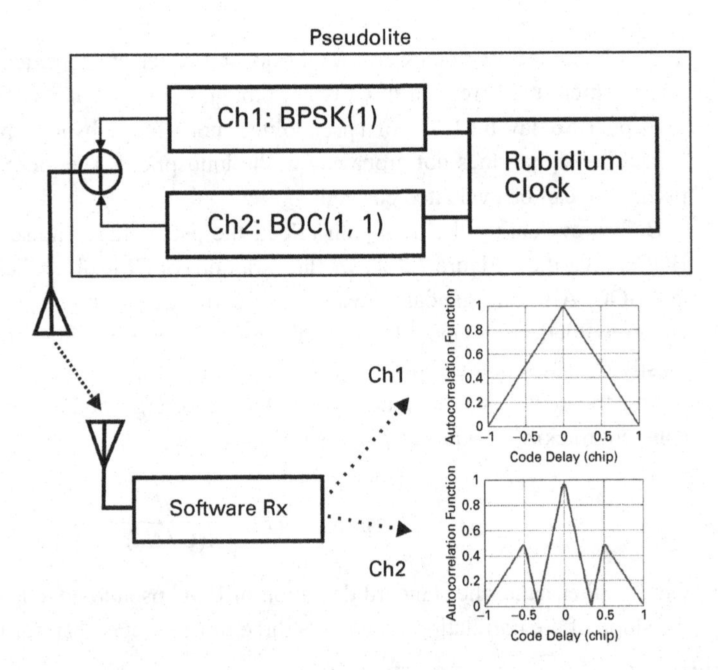

Figure 7.7 Test set-up.

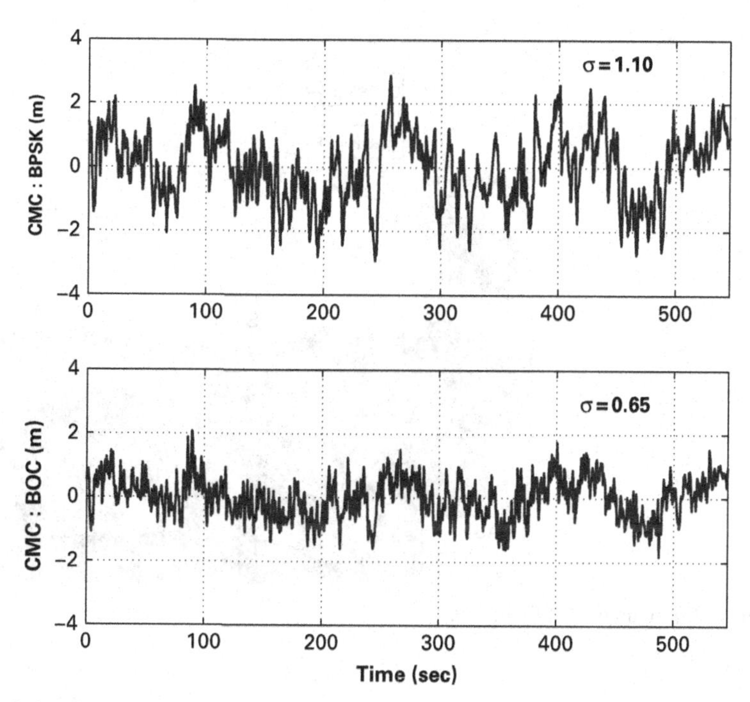

Figure 7.8 CMC variation for BPSK (top) and BOC (bottom) in a less-multipath environment.

$$\sigma_{\mathrm{BOC}} = cT_{\mathrm{C}}\sqrt{\frac{d}{4(C/N_0)T}}\cdot\frac{1}{\sqrt{3}}. \tag{7.8}$$

The chip width (T_C) is common for BOC and BPSK. Though integration time (T) and correlator spacing (d) are changeable, equivalent values are used for both BPSK and BOC signal tracking. Signal to noise ratio was slightly different between the two channels. However, the ratio of standard deviations shown in Figure 7.8 is 1.69, which is close to the theoretical ratio ($\sqrt{3}$).

Next, the flight test result is shown. The pseudolite was installed in the cabin of the JAXA's experimental helicopter MuPAL-e (Mitsubishi MH-2000A). The helicopter hovered at an altitude of 1500–2000 ft (457–610 m) above the control building as shown in Figure 7.9. A DGPS/INS, which is an item of standard equipment of MuPAL-e, was also installed and the position, velocity, and attitude data as well as a 1pps/time-tag were sent to the pseudolite. Then pseudolite ephemeris was computed and encoded as a navigation message, and sent to a ground user. The pseudolite antenna is mounted underneath the helicopter. Base station and rover are also shown in the figure. The large hangar, which was 30 meters high, was used as a reflector and multipath measurement tests were carried out. In this test, the rover moved along the nearby hangar very slowly.

Figure 7.10 shows the C/N_0 of the BOC signal. It is clear that C/N_0 changes drastically, and therefore the multipath error is likely to change accordingly. Since the rover moved near to the hangar, the amplitude of reflected signal seemed large.

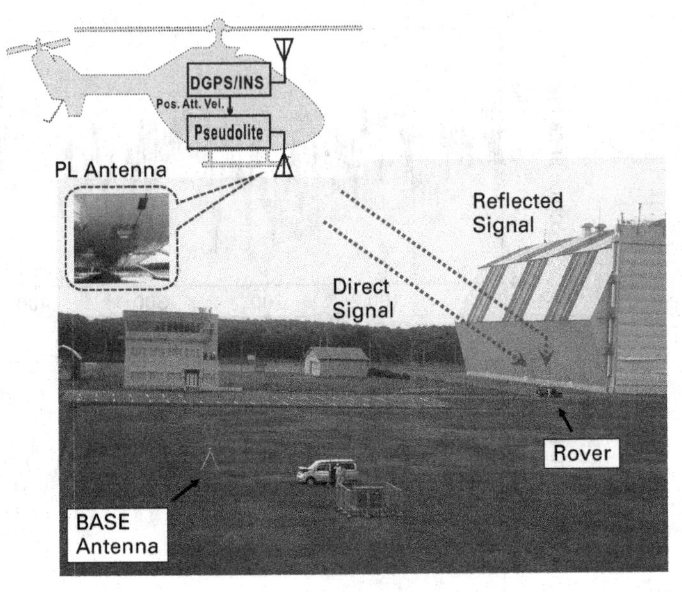

Figure 7.9 Configuration of flight test.

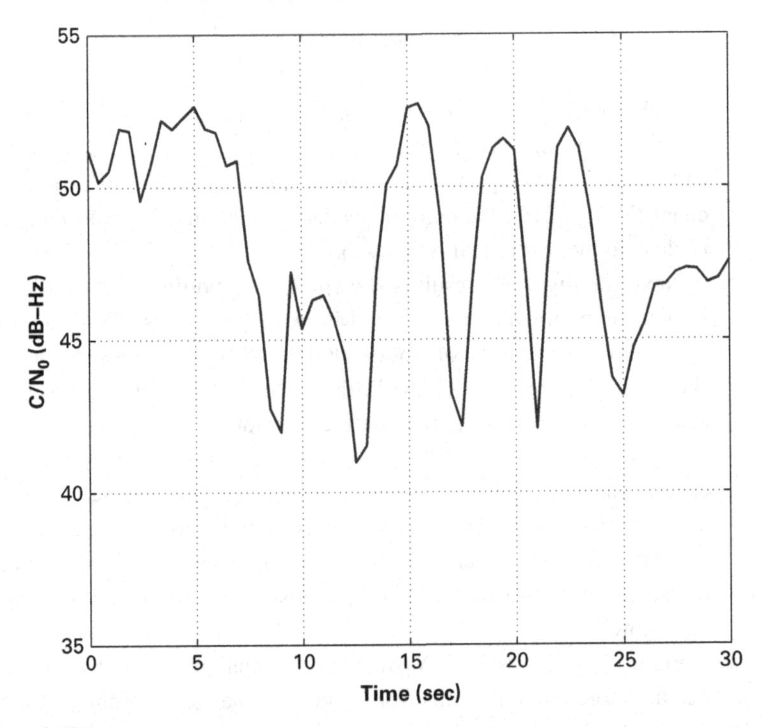

Figure 7.10 Variation of C/N_0.

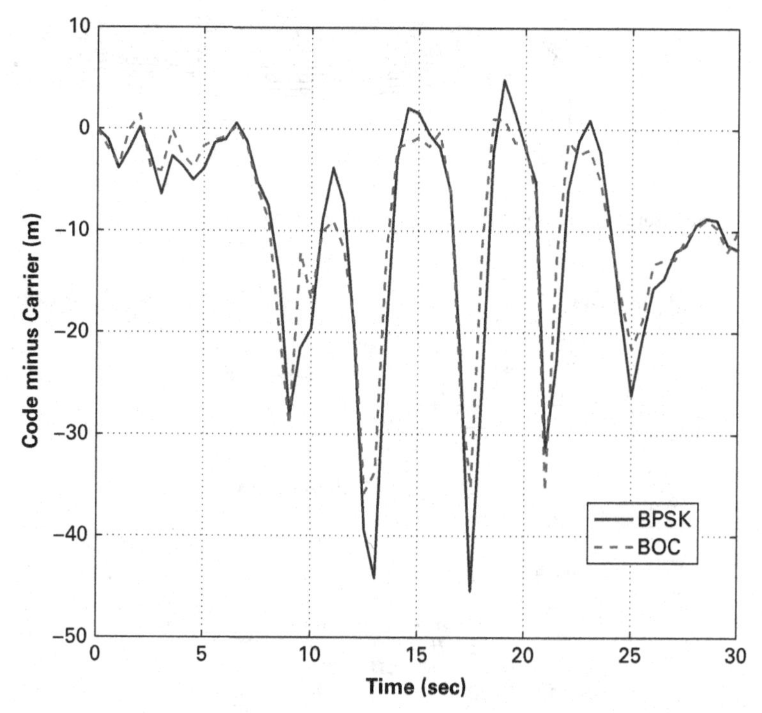

Figure 7.11 Difference between pseudorange and carrier phase for C/A (top) and BOC(bottom) channel.

Differences between pseudorange and carrier phase (CMC) for BPSK and BOC channels are shown in Figure 7.11, and the multipath reduction by BOC modulation is demonstrated. The large multipath fluctuations due to the slow movement of the receiver are also seen. Compared with Figure 7.10, it is clear that there is a strong correlation between C/N_0 and multipath error.

7.3 Multipath mitigation

7.3.1 Antennas

Antennas often implement special designs to mitigate multipath. Figure 7.12 shows schematically a structure applied for multipath mitigation. Metal rings located around the antenna core act as receiving and transmitting antennas. Secondary signal emissions from these rings are induced by a signal coming from a low angle. A superposition of the incoming signal and signals generated by the rings is designed to give almost zero power. A similar principle is embedded in interference mitigation by beam forming. Nullification of an interference signal coming from a not distant point can be done by using an array of antennas, forming a signal in such a way that the signal from a specific angle and azimuth can be nullified. An antenna array in general allows us to modify an antenna gain pattern on-the-fly.

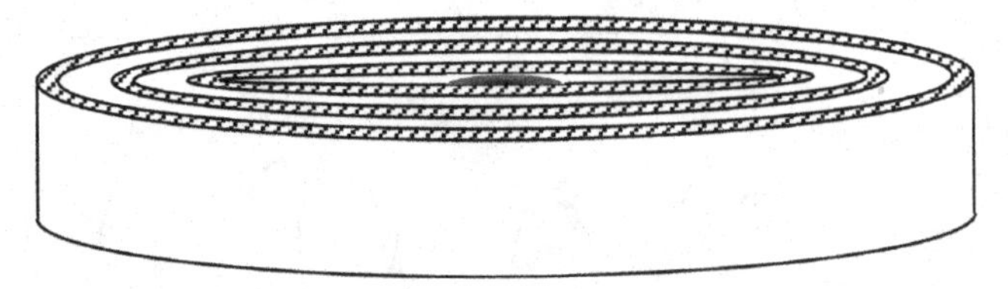

Figure 7.12 Structure for multipath mitigation in the antenna.

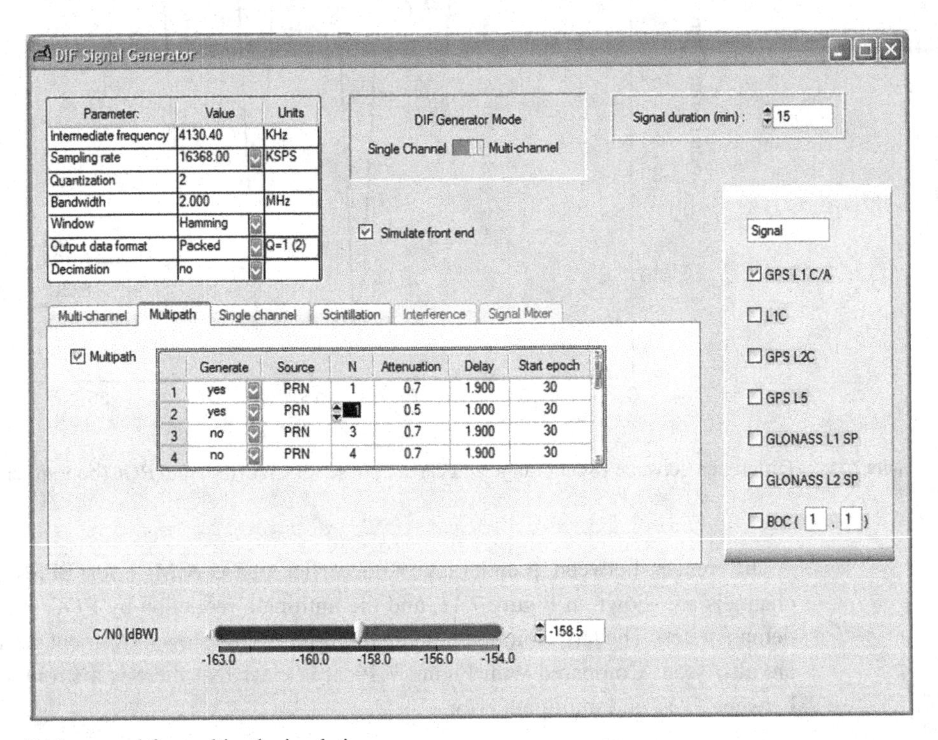

Figure 7.13 ReGen panel for multipath simulation.

7.3.2 Multi-correlator receiver

Inside a baseband processor it is possible to look at a finer representation of the signal autocorrelation function. This is possible by using a multi-correlator. An iPRx single-channel receiver allows use of up to 17 correlators (see Figure 7.2). There are several techniques which use multiple correlators to mitigate multipath [1].

7.4 Project: investigation of multipath effects on receiver performance using multipath simulation

1. Generate a satellite signal with multipath error as recommended below using a ReGen simulator (professional version is required). Set up an attenuation, phase, and start epoch for one channel, according to the ReGen User Manual (see Figure 7.13).

Set additional multipath signals with various start time, delay, and attenuation settings.

2. Investigate the results of signal tracking by a single-channel iPRx receiver with 3 and 17 correlators. Observe the tracking results and S/N.

References

[1] G. J. Bishop and J. A. Klobuchar, Multipath effects on the determination of absolute ionospheric time delay from GPS signals, *Radio Science*, **20**, (3), 1985, 388–396.

[2] R. Van Nee, *Multipath and Multi-Transmitter Interference in Spread-Spectrum Communication and Navigation Systems*, Delft, Delft University Press, 1995.

[3] T. Tsujii, H. Tomita, Y. Okuno, *et al.*, Development of a pseudo quasi zenith satellite and multipath analysis using an airborne platform, *Journal of Global Positioning Systems*, **6**, (2), 2007, 126–132.

8 Optimization of GNSS observables

In this chapter we consider in detail how one can compensate errors in GNSS observables. GNSS observables can be constructed from code and carrier measurements made on different frequencies in various ways. Some of these observables are less affected by particular errors than others. We also consider how the usage of measurements from other receivers located at known positions allows elimination of some errors (Figure 8.1).

8.1 Error budget of GNSS observables

As explained in previous chapters, we can describe GNSS errors in the form of an error budget. The budget consists of the following errors:

- DOP arising from geometrical properties of the satellite constellation (see Chapter 1);
- Satellite-related errors, including satellite clock error, satellite orbit errors, satellite transmitter errors, including biases (see Chapter 3);
- Propagation errors in a dispersive medium due to the ionosphere (see Chapter 4);
- Propagation errors in a non-dispersive medium due to the troposphere (see Chapter 4);
- Receiver-related errors including noise and hardware biases (see Chapter 5);
- Multipath (see Chapter 7).

We have described the errors related to signal propagation in the atmosphere in Chapter 4. Some R&D tasks may require more specific models to be implemented, especially where development of new algorithms is concerned. One example is spatially correlated ionospheric errors. Algorithm development related to a virtual reference station (VRS), network RTK, or ionospheric research may require an ability to generate a spatially correlated ionospheric model. In that case the signal is generated for more than one receiver and ionospheric errors are properly correlated. The specific fluctuations of TEC distribution can be added on top of nominal TEC distribution. These nominal TEC distribution values come from true Klobuchar, NeQuick, or IGS models. The other example is to specify an anomalous ionospheric gradient with parameters in accordance, for example, with WAAS Super Truth data analysis and simulate a moving slope for the ionospheric gradient. For details of the Threat Model of an anomalous ionosphere gradient see [1]. In this case some specific fluctuations or slope can be implemented on top of the regular total electron count (TEC) distribution model.

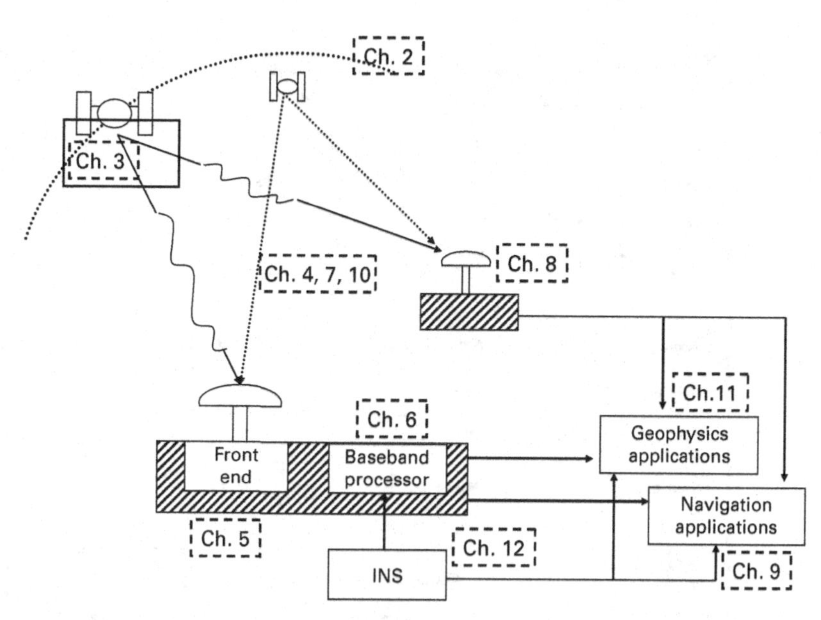

Figure 8.1 Subject of Chapter 8.

Such advanced spatially correlated ionospheric models may be required if a user wishes to test local area augmentation system (LAAS) algorithms. Simulation of such errors allows us to verify their correct implementation. We use LGF (LAAS ground facility) tests emulation software to test some LAAS algorithms. This software allows us to analyze spatial correlations between ionospheric errors from distantly located receivers. We have processed both measurements from real receivers and receiver measurements simulated by ReGen with spatially correlated ionospheric errors.

This chapter looks at the various methods to process various combinations of observables from one or more receivers. At this stage we can use observables either from real satellites, which would require a receiver, or those simulated with a ReGen simulator as described in Chapter 3, or observables from a software receiver using signal simulated by a ReGen simulator. A ReGen software simulator allows simulation of multiple observables for multiple receivers and multiple frequencies (see Figure 8.2). Usage of a simulator allows us to see how these observables are simulated and how they manifest themselves, so we can better understand how to utilize them.

8.2 Forming observables

8.2.1 Carrier smoothed code phase observables

Positioning can be done (as described in Chapter 1) using code phase observables. These observables have large uncompensated errors. These observables can be combined with carrier phase observables using various filters. We consider here a simple filter for combining code and carrier observables as proposed by R. Hatch (see [2]). Smoothed

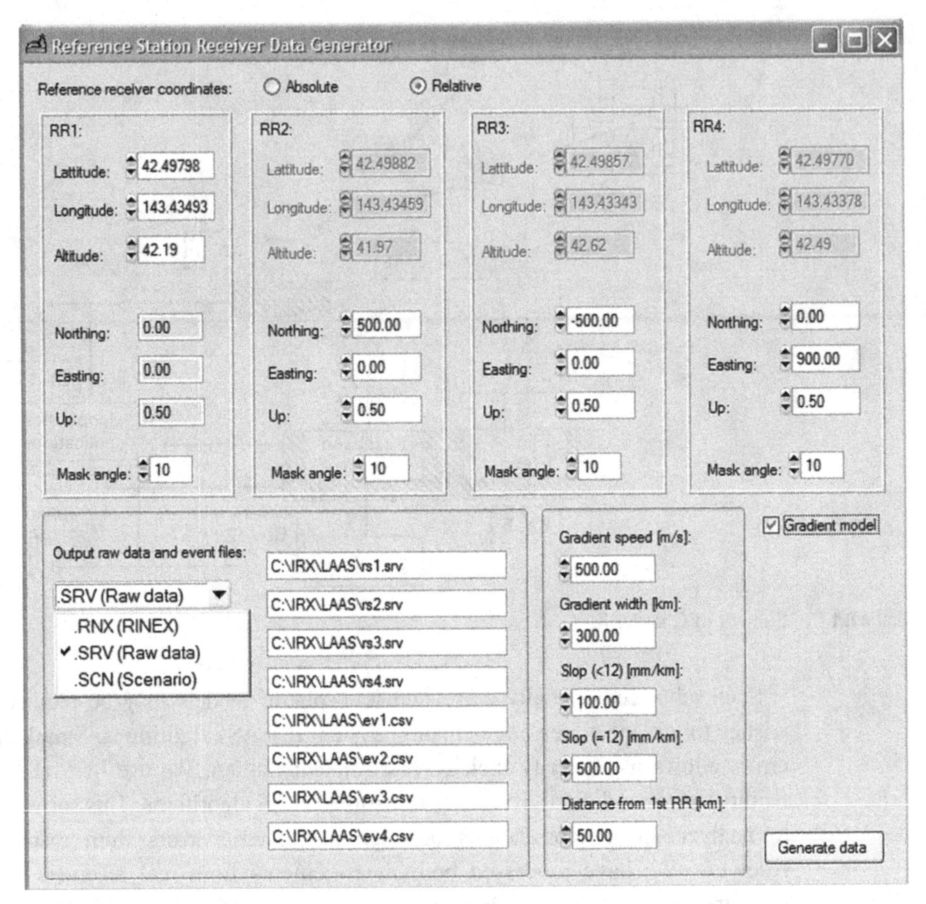

Figure 8.2 ReGen panel for spatially correlated signal generation for reference station receivers.

code phase observables for the ith satellite $\hat{\rho}_{s_i}$ can be calculated as a weighted sum of code (ρ_{s_i}) and carrier phase (ϕ) increment:

$$\hat{\rho}_{s_i}(t_k) = W_\rho \cdot \rho_{s_i}(t_k) + W_\phi \cdot \left[\hat{\rho}_{s_i}(t_{k-1}) + \phi(t_k) - \phi(t_{k-1})\right]. \tag{8.1}$$

The weight coefficients W_ρ and W_ϕ are defined as follows:

$$\begin{cases} W_\rho = 1 - \frac{k-1}{N_s}, \ W_\phi = \frac{k-1}{N_s} & (k = 1, 2, \cdots N_s), \\ W_\rho = \frac{1}{N_s}, \ W_\phi = 1 - \frac{1}{N_s} & (k > N_s), \ N_s = \tau_s / T_s, \end{cases} \tag{8.2}$$

where τ_s is smoothing time constant, T_s is sampling interval. For example, if the time constant is 100 seconds and the sampling rate is 1 Hz, then N_s is 100. After 100 seconds, the weight coefficient for code phases (W_ρ) is 0.01 and weight coefficient for carrier phases (W_f) is 0.99.

If for the particular application, the multipath error and the ionospheric error can be considered to be small, then the pseudorange noise is effectively reduced and more precise range information can be obtained. However, if there is a significant multipath

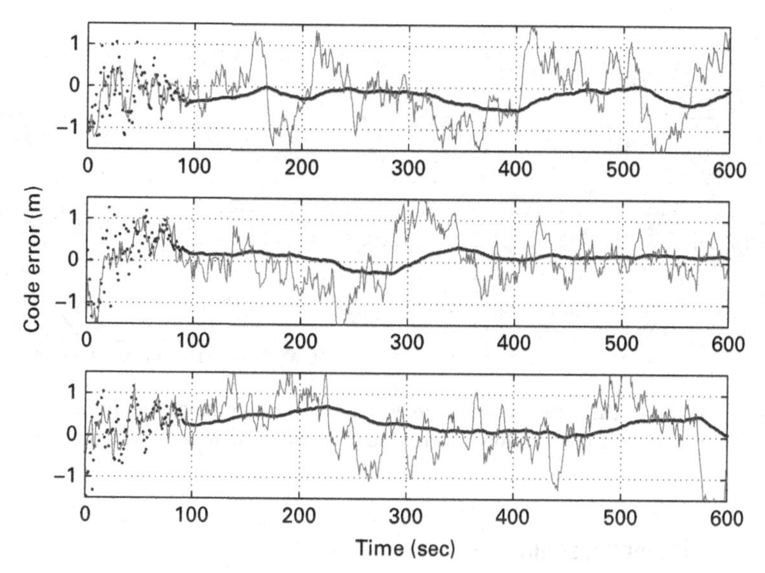

Figure 8.3 Smoothed code phase errors (dots) and multipath errors (thin lines) for three satellites.

error, the corresponding range error remains for a long time since the multipath error of the pseudorange is much larger than that of the carrier phase and they never cancel. Figure 8.3 shows an example of smoothed pseudorange error for three satellites in which the smoothing process starts at the origin of the time axis. It is clear that the pseudorange noise is significantly reduced by smoothing. A rather small multipath error simulated by the first-order Gauss-Markov method is included in each raw pseudorange, and therefore the range errors are not approaching zero.

An ionospheric delay also affects the smoothing process since the sign of the delay for pseudorange and carrier is opposite. In order to appreciate its importance, consider an example of a single-frequency receiver undergoing a carrier-smoothing test.

We use an iPRx software receiver for this test. A correct signal is generated by a ReGen simulator. We are also able to generate an intentionally incorrect signal, in which the sign of the ionospheric error is the same in both the code and carrier observables using a ReGen simulator. We model both code and carrier ionospheric errors as delays, because a possible mistake would be just to ignore that the carrier error has a negative value and to model it as all other errors as a delay. We can see that having code and phase incorrectly simulated would directly affect the results of an approach to designing a positioning algorithm. We use a simple smoothing algorithm without a reset. The code–carrier divergence becomes evident after about 30–60 minutes of filtering depending on receiver noise and multipath errors. The test results for two cases are depicted in Figure 8.4. We can see that the results of the smoothing algorithm would vary significantly depending on how the ionospheric error is accounted for.

The same effect can be used for detecting an ionospheric anomaly by using a combination of carrier smoothed code phase measurements with different time constants. Such a technique is implemented in ground-based augmentation systems (GBAS) [3].

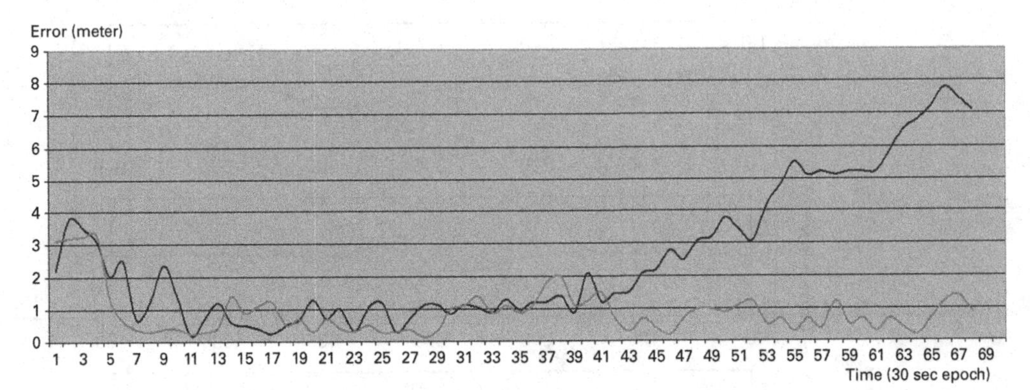

Figure 8.4 Effect of carrier and code divergence on carrier smoothing algorithm. Correctly simulated ionospheric error on top. Carrier and code ionospheric error with the same sign on bottom.

8.2.2 Differential solution

If we have another receiver installed in the vicinity of our rover receiver, then we can assume that some errors are similar to the errors in our ***rover receiver***. The coordinates of this ***reference receiver*** antenna should be surveyed in advance. Then we can use measurements from the reference receiver in order to compensate for some errors in our receiver. As distance to the reference receiver increases, the correlation between reference receiver and rover receiver error budget decreases. In order to compensate for that, we can use a few distant reference receivers instead of one. An extreme case is to use a global reference network, such as IGS, or its products. Below we consider how we can implement measurements from a reference receiver. We combine the observables from the receivers on a satellite by satellite basis.

8.2.2.1 Single difference

At first, we consider ***single difference*** observables. When two receivers and one satellite are considered, the single difference of code and carrier phase for the ith satellite is defined as the difference between measurements from receiver-1 and receiver-2 (Figure 8.5):

$$\Delta \rho_{1,2}^{i} = \rho_{1}^{i} - \rho_{2}^{i}, \tag{8.3}$$

$$\Delta \phi_{1,2}^{i} = \phi_{1}^{i} - \phi_{2}^{i}, \tag{8.4}$$

where subscripts denote the number of the receiver.

Carrier phase measurements can be expressed as follows:

$$\phi = \rho - \frac{f_2}{f_1} I + d_{\text{trop}} + b - b_{\text{SV}} + d_{\text{eph}} + d_{\text{m,phase}} + \lambda N + \varepsilon, \tag{8.5}$$

where d_{trop}, b, b_{SV}, d_{eph} $d_{\text{m,phase}}$ N, ε are tropospheric delay, receiver clock bias, satellite clock bias, ephemeris error, multipath error, ambiguity number, and noise, respectively.

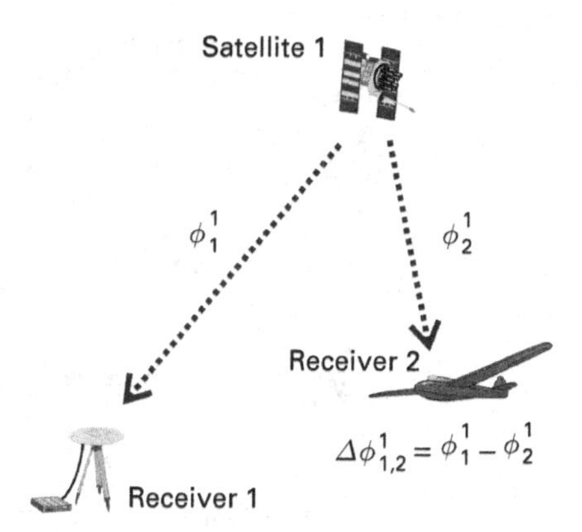

Figure 8.5 Single difference observables.

The second term of the right-hand side represents the ionospheric delay for L1 frequency, and the parameter I is given by (see Chapter 4, Equations (4.45), (4.46)):

$$I = 40.3 \frac{\int N_e(s)\, ds}{f_1 f_2}, \tag{8.6}$$

where N_e is the density of electrons, and f_1, f_2 are L1, L2 frequency. The electron density is integrated along the signal propagation path and higher orders are neglected.

Then single difference observables (8.4) for the ith satellite can be rewritten as follows:

$$\Delta\phi = \Delta\rho - \frac{f_2}{f_1}\Delta I + \Delta d_{\text{trop}} + \Delta b + \Delta d_{\text{eph}} + \Delta d_{\text{m,phase}} + \lambda \Delta N + \Delta\varepsilon, \tag{8.7}$$

where ΔN is an unknown single difference ambiguity number. This value should be estimated along with other parameters, such as receiver coordinates.

The subscripts and superscripts are omitted here to simplify the equation. The single difference of satellite clock bias can be expressed as follows:

$$b_{\text{SV}}(t_{r_2}) - b_{\text{SV}}(t_{r_1}) \cong \dot{b}_{\text{SV}}(t_{r_2}) \cdot (t_{r_2} - t_{r_1}), \tag{8.8}$$

where t_{rj} is the signal reception time by the jth receiver. Since the satellite clock drift $\dot{b}_{\text{SV}}$ is normally smaller than 10^{-3} (m/s) and the difference of reception time is usually less than a few milliseconds, the satellite clock biases cancel each other.

If the baseline length between two receivers is relatively short, then most of the orbit-related errors are canceled. If the signal propagation paths for both receivers are similar then propagation errors are also canceled. The rule of thumb is that ionospheric errors are well compensated if the reference receiver is within 10 km from a rover. The tropospheric delay errors for the receiver are much less well correlated, especially if there is a height difference between two receivers.

8.2.2.2 Double difference

Now we consider two receivers 1, 2 and two satellites i and j. We can form two single differences for satellite i and satellite j. The ***double difference*** is defined as the difference

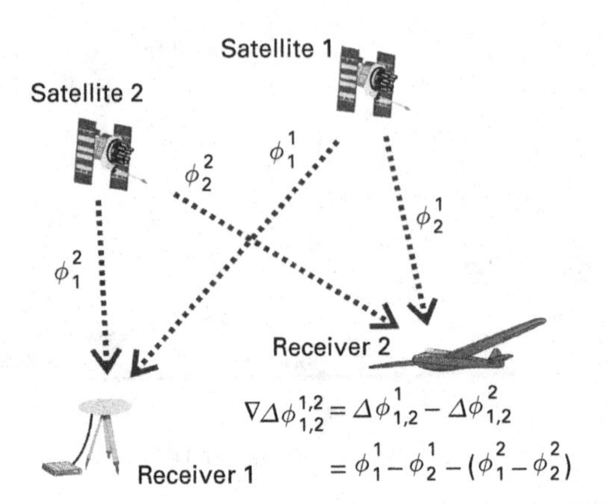

Figure 8.6 Double difference observables.

between these single differences (Figure 8.6). Denoting a double difference operator by $\nabla\,\Delta$, we can express it as follows for code and carrier phase measurements:

$$\nabla\Delta(\cdot) = (\cdot)_1^i - (\cdot)_1^j - (\cdot)_2^i + (\cdot)_2^j. \tag{8.9}$$

The double difference of the carrier phase is given as follows:

$$\nabla\Delta\phi = \nabla\Delta\rho - \frac{f_2}{f_1}\nabla\Delta I + \nabla\Delta d_{\text{trop}} + \nabla\Delta d_{\text{eph}} + \nabla\Delta d_{\text{m,phase}} + \lambda\nabla\Delta N + \nabla\Delta\varepsilon, \tag{8.10}$$

where $\nabla\Delta N$ is an unknown double difference ambiguity number.

The receiver clock biases are canceled because the reception time of satellite i and j signals is the same, so clock biases are also the same.

8.2.2.3 Triple difference

Taking the difference between the sequential double difference measurements, we can remove the ambiguity number $\nabla\Delta N$ from the state vector. Time differenced double differences are called ***triple differences***:

$$\delta\nabla\Delta\phi = \delta\nabla\Delta\rho - \frac{f_2}{f_1}\delta\nabla\Delta I + \delta\nabla\Delta d_{\text{trop}} + \delta\nabla\Delta d_{\text{eph}} + \delta\nabla\Delta d_{\text{m,phase}} + \delta\nabla\Delta\varepsilon. \tag{8.11}$$

When the observation rate is high and the ionosphere and troposphere are stable, the propagation delays may be negligible. If the initial position of a vehicle is known, positioning with triple difference is very easily carried out because it is not necessary to resolve ambiguities, i.e. the initial ambiguity resolution process can be omitted. However, the positioning accuracy is degraded gradually because the estimated position at each epoch depends on the position at the previous epoch, therefore the positioning error is accumulating. Furthermore, once a ***cycle slip*** (sudden jump of ambiguity in carrier phase measurements) occurs and the number of visible satellites becomes less than four, positioning cannot be performed any more.

8.3 Multi-frequency observables

In this section we consider how measurements made on more than one frequency can compensate for some errors. It makes sense to use the types of linear combination described below mostly when data from reference receivers are available. Therefore these linear combinations are normally formed using single, double, and triple difference observables.

8.3.1 Widelane linear combination

When a dual frequency GNSS receiver is used, we can form various linear combinations of GNSS measurements. These combinations can provide us with some error reductions and are advantageous for ambiguity resolution and cycle slip detection [4]. The following linear combination of L1 and L2 carrier phases, called ***widelane***, can be formed as follows:

$$\phi_{\mathrm{W}} = \left(\frac{\phi_1}{\lambda_1} - \frac{\phi_2}{\lambda_2}\right)\frac{c}{f_1 - f_2}, \tag{8.12}$$

where subscripts 1, 2, and W denote L1, L2, and widelane observables respectively.

Widelane wavelength accordingly yields:

$$\lambda_{\mathrm{W}} = \frac{c}{f_1 - f_2} \cong 86.2\,\mathrm{cm}. \tag{8.13}$$

When we form our carrier phase widelane observables, the ambiguity number is expressed through the ambiguity numbers in the L1 and L2 observables:

$$N_{\mathrm{W}} = N_1 - N_2. \tag{8.14}$$

If a reference receiver is available, then we can form the double differenced widelane observables:

$$\nabla\Delta\phi_{\mathrm{W}} = \nabla\Delta\rho + \nabla\Delta I + \nabla\Delta d_{\mathrm{trop}} + \nabla\Delta d_{\mathrm{eph}} + \nabla\Delta d_{\mathrm{m,W}} + \lambda_{\mathrm{W}}\nabla\Delta N_{\mathrm{W}} + \nabla\Delta\varepsilon_{\mathrm{W}}. \tag{8.15}$$

As we see in the next chapter, since the effective wavelength of widelane is about four times larger than the L1 wavelength, to resolve widelane ambiguity is easier than to resolve L1 ambiguity. However, the measurement noise becomes about three times as large as the L1 noise and ionospheric delay error is enlarged by a factor of about 1.3 due to a coefficient $f_2/f_1 = 60/77 \cong 0.78$. Therefore, the widelane observables are not suited to the positioning solution, and often used for initial ambiguity resolution.

8.3.2 Narrowlane linear combinations

We can form a ***narrowlane*** observable using L1 and L2 measurements as follows:

$$\phi_{\mathrm{N}} = \left(\frac{\phi_1}{\lambda_1} + \frac{\phi_2}{\lambda_2}\right)\frac{c}{f_1 + f_2}, \tag{8.16}$$

with wavelength

$$\lambda_N = \frac{c}{f_1 + f_2} \cong 10.7\,\text{cm}. \tag{8.17}$$

The narrowlane ambiguity number is expressed through the ambiguity numbers in the L1 and L2 observables:

$$N_N = N_1 + N_2. \tag{8.18}$$

The double differenced narrowlane is therefore

$$\nabla\Delta\phi_N = \nabla\Delta\rho - \nabla\Delta I + \nabla\Delta d_{\text{trop}} + \nabla\Delta d_{\text{eph}} + \nabla\Delta d_{\text{m,N}} + \lambda_N \nabla\Delta N_N + \nabla\Delta\varepsilon_N. \tag{8.19}$$

Since the effective wavelength is about half of the L1 wavelength, to resolve ambiguity is more difficult. However, the measurement noise is about half of the L1 noise, so the narrowlane solution may be a final solution for short baseline applications when the enlarged ionospheric delay error is sufficiently small. Note that the sign of the ionospheric delay is opposite to that of the widelane observable according to the definition of both observables (8.12) and (8.16).

8.3.3 Ionosphere-free observables

The purpose of creating ionosphere-free observables is to remove effects of the ionosphere by combining signals on different frequencies. Which frequency to use depends on which frequencies are available for a particular satellite system. We rewrite here Equation (4.45) derived in Chapter 4:

$$d_I = \frac{K_x}{2} \cdot \text{TEC} \cdot \left(\frac{1}{f}\right)^2, \tag{8.20}$$

where d_I is a code delay or phase advance due to the ionosphere, f is the signal frequency. Then

$$\frac{K_x}{2} \approx 40.3 \cdot 10^{16} \left[\frac{\text{m}}{\text{TECU} \cdot \text{sec}^2}\right], \tag{8.21}$$

where TEC is total electron content in TECU.

The ionosphere-free observables can be constructed using (8.20) in such a way that ionospheric delay is completely removed from a new linear combination as follows:

$$L_3 = \kappa_{1,3} L_1 + \kappa_{2,3} L_2, \tag{8.22}$$

where the coefficients are defined as

$$\kappa_{1,3} = \frac{f_1^2}{f_1^2 - f_2^2}, \qquad \kappa_{2,3} = \frac{f_2^2}{f_1^2 - f_2^2}. \tag{8.23}$$

For GPS L1 and L2 the coefficients are

$$\kappa_{1,3} \approx 2.546, \quad \kappa_{2,3} \approx -1.546. \tag{8.24}$$

Ionosphere-free observables cannot, however, remove higher terms components. The ray bending for two frequencies is different (see Chapter 4), which also introduces an error. This error depends on satellite elevation.

The main disadvantage of using ionosphere-free observables is that though they remove most of the ionospheric errors, the noise of these observables is approximately three times higher than the noise of an L1 or L2 signal [5]. Thus, $\sigma(L_1) \approx (L_2)$, and

$$\sigma(L_3) = \sqrt{\kappa_{1,3}^2 \sigma^2(L_1) + \kappa_{2,3}^2 \sigma^2(L_2)} = \sqrt{\kappa_{1,3}^2 + \kappa_{2,3}^2}\sigma(L_1) \approx 3\sigma(L_1). \tag{8.25}$$

The ionosphere-free observable (L_3) can also be formed from the widelane and narrowlane observable as follows:

$$\phi_3 = \frac{1}{2}(\phi_\mathrm{W} + \phi_\mathrm{N}). \tag{8.26}$$

The double differenced ion-free observable is then

$$\nabla\Delta\phi_3 = \nabla\Delta\rho + \nabla\Delta d_\mathrm{trop} + \nabla\Delta d_\mathrm{eph} + \nabla\Delta d_\mathrm{m,ion} + \frac{1}{2}(\lambda_\mathrm{W}\nabla\Delta N_\mathrm{W} + \lambda_\mathrm{N}\nabla\Delta N_\mathrm{N}) + \nabla\Delta\varepsilon_3. \tag{8.27}$$

Since the ionospheric delay cancels, this observable is suited to a long baseline application, though the measurement noise is enlarged by a factor of about 3. Finally, we introduce the ***ionospheric signal*** as follows:

$$\phi_I = \phi_\mathrm{N} - \phi_\mathrm{W} = -2I + \lambda_\mathrm{N} N_\mathrm{N} - \lambda_\mathrm{W} N_\mathrm{W} + (d_\mathrm{m,N} - d_\mathrm{m,W}) + (\varepsilon_\mathrm{N} - \varepsilon_\mathrm{W}). \tag{8.28}$$

When the ambiguities of widelane and narrowlane are resolved and multipath errors are sufficiently small, the amount of ionospheric delay can be evaluated using this ionospheric signal.

There are several other linear combinations. In geodetic applications, a geometry-free linear combination or a Melbourne–Wübbenna linear combination is often used. The type of observables which are used for data processing depends also on the network, in particular on the length of its baselines. In the case of long baselines, it is recommended to use the ionosphere-free linear combination. For short baselines it is more advantageous to process a geometry-free linear combination, because it allows removal of errors related to the geometrical distribution of the satellite constellation.

8.4 Improving observables with augmentation systems

In this chapter we consider a few examples of how the methods of error compensation described in the previous chapter are implemented. Augmentation systems basically supply a user with correction data, which encapsulate information collected from reference stations. These data are used to improve accuracy and reliability of user positioning with GNSS.

The local area differential GPS (LADGPS) described in the previous section uses a ground reference station which provides the ranging correction. The distance between user and reference station is limited to achieve high accuracy since the ranging error, such as ionospheric/tropospheric delay, depends on the location of the receiver. However, a user can be supplied with information coming either from a single reference station or from a network. Networks can also differ depending on their size and distribution. The data from a network can be processed and then supplied to a user through various communication links for real-time processing or post-processing. The networks providing real-time information are either commercial or governmental. The information from a global network such as IGS for example is available free of charge for post-processing through the Internet. The Internet can provide a data link for real-time systems as well, though the correction latency due to Internet delays can be a problem in the case of positioning with carrier phase observables. There are local reference station networks in many countries which provide users with correction data.

A LADGPS designed for aircraft precision approach is called a ***ground-based augmentation system (GBAS)***. GBAS is different from other reference system solutions in that they also provide an integrity service. Multiple reference receivers are installed in the vicinity of an airport, and the correction data as well as reliability information are broadcast though a ground transmitter.

The user positioning accuracy of LADGPS is degraded when the distance from the reference station increases. The reason is that the LADGPS provides the sum of various ranging errors (satellite clock, satellite ephemeris, ionospheric/tropospheric delay). Though the satellite clock error is identical for all users, other errors are dependent on the user position relative to the reference station. If such user dependent errors are properly modeled for a wider area and requisite parameters are available for users in the area, a wide area DGPS (WADGPS) can be established. The correction data of WADGPS is called ***vector correction***, while it is called ***scalar correction*** for LADGPS. The WADGPS utilizes the information of multiple reference stations in order to compute the correction parameters and send them to users by a communication link.

A WADGPS designed for civil aviation is called a ***space-based augmentation system (SBAS)***, which provides integrity information as well as correction parameters. Such data are supplied by geostationary satellites.

The GBAS and SBAS developed by the US are called ***local area augmentation systems (LAAS)*** and ***wide area augmentation systems (WAAS)***. The terms LAAS/WAAS are now sometimes used in a broader sense as general terms to describe similar augmentation systems.

8.4.1 Satellite-based augmentation system (SBAS)

8.4.1.1 SBAS objectives

A ***SBAS (satellite-based augmentation system)*** is a WADGPS which uses geostationary satellites as a communication link. SBAS is defined as a system to augment satellite

Table 8.1 GNSS signal-in-space performance requirements (according to [6]).

	Accuracy (95%) / Alert limit					
Operation	Horizontal (m)	Vertical (m)	Integrity	Time-to-alert	Continuity	Availability
En-route	3700/7400	–	$1-10^{-7}$/hr	5 min	$1-10^{-4}$/hr to $1-10^{-8}$/hr	0.99–0.99999
En-route, terminal	740/1850–3700	–	$1-10^{-7}$/hr	15 s	$1-10^{-4}$/hr to $1-10^{-8}$/hr	0.99–0.99999
NPA	220/556	–	$1-10^{-7}$/hr	10 s	$1-10^{-4}$/hr to $1-10^{-8}$/hr	0.99–0.99999
APV-I	220/556	20/50	$1-2\times10^{-7}$/ approach	10 s	$1-8\times10^{-6}$/15 s	0.99–0.99999
APV-II	16/40	8/20	$1-2\times10^{-7}$/ approach	6 s	$1-8\times10^{-6}$/15 s	0.99–0.99999
Cat-I	16/40	4–6/10–15	$1-2\times10^{-7}$/ approach	6 s	$1-8\times10^{-6}$/15 s	0.99–0.99999

navigation core systems (e.g. GPS and GLONASS) for the operation of civil aviation from en-route to approach. The requirements for SBAS are specified in the SARPs (Standards And Recommended Practices) of ICAO [6]. The most important function of SBAS is to provide integrity of satellite navigation. ***Integrity*** is an ability to alert users within a specified time, called ***time to alert (TTA)***, if a problem occurs and the navigation system signal should not be used for navigation solution. The performance requirements of a GNSS signal, including its integrity, are specified in SARPs for each aircraft operational phase (Table 8.1).

Continuity is the probability that the system will continue to be available through a phase of operation, assuming that the service was available at the beginning of the phase. If the continuity is lost, a go-around must be executed. However, it does not generate hazardous misleading information (HMI) since the failure has been detected. On the other hand, if the integrity is lost, the failure may not be detected and may cause generation of HMI. Therefore, the integrity requirement must be stringent, especially for approach phases. ***Availability*** is the percentage of time that the system is available to support the operation, and it depends on the time, user position, satellite constellation, etc.

8.4.1.2 SBAS messages

SBAS provides the information to ensure the integrity as well as the range correction parameters to improve positioning accuracy. Table 8.2 summarizes the typical broadcast message types [6]. The data block format of all message types is shown in Figure 8.7, which includes 8 bits preamble, 6 bits message type identifier, 212 bits data field, and 24 bits CRC parity. Although the data rate is bits per second, the coding rate is 500 symbols per seconds, since a forward error correction (FEC) with the rate of 1/2 is applied.

Table 8.2 SBAS broadcast message types.

Type	Contents	Update interval (s)
0	SBAS test mode	6
1	PRN mask	120
2–5	fast corrections (FC)	60
	user differential range error indication (UDREI)	6
6	integrity information, UDREI	6
7	fast correction degradation factor	120
9	GEO ranging function parameters	120
10	degradation parameters	120
12	SBAS network time/ UTC offset parameters	300
17	GEO satellite almanacs	300
18	ionospheric grid point (IGP) masks	300
24	fast corrections (FC)	60
	UDREI	6
	long-term satellite error corrections (LT) (satellite ephemeris and clock corrections)	120
25	long-term satellite error corrections (LT)	120
26	ionospheric delay corrections	300
27	SBAS service message	300
28	clock-ephemeris covariance matrix message	120

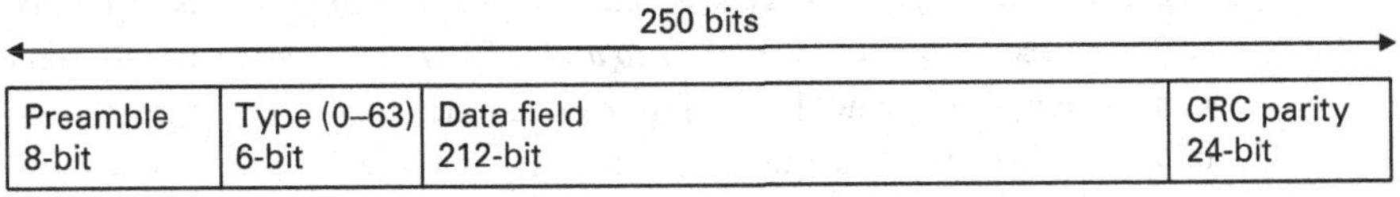

250 bits

Preamble 8-bit	Type (0–63) 6-bit	Data field 212-bit	CRC parity 24-bit

Figure 8.7 SBAS data block format.

The fast corrections (FC) are mostly clock errors and applied to correct pseudorange errors. The message types 24 and 25 provide the satellite ephemeris corrections as well as clock biases. Ionospheric delay correction is carried out by using message type 26, which includes the ionospheric grid point (IGP) delay estimate and grid ionospheric vertical error indicator (GIVEI). The IGP delay estimate is the vertical ionospheric delay estimated at the grid point, which is broadcast by message type 18. The ionospheric delay correction for a specific user is computed at the ionospheric pierce point (IPP), which is the intersection of the signal path and the ionosphere with the maximum electron density. The ionospheric delay at IPP is computed by the interpolation method using the delay estimates of four surrounding IGPs, as shown in Figure 8.8.

The vertical ionospheric delay at IPP is computed as follows:

$$I_{\mathrm{IPP}}^{\mathrm{V}} = (1-x)(1-y)I_{\mathrm{IGP1}}^{\mathrm{V}} + x(1-y)I_{\mathrm{IGP2}}^{\mathrm{V}} + xyI_{\mathrm{IGP3}}^{V} + (1-x)yI_{\mathrm{IGP4}}^{\mathrm{V}}, \quad (8.29)$$

where

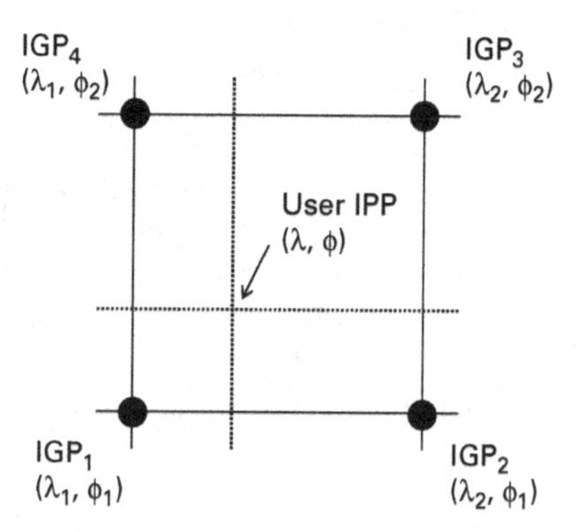

Figure 8.8 Relationship between IPP and surrounding IGPs.

$$x = \frac{\lambda - \lambda_1}{\lambda_2 - \lambda_1}, \quad y = \frac{\phi - \phi_1}{\phi_2 - \phi_1}. \tag{8.30}$$

The variance of user ionospheric vertical error (σ^2_{UIVE}) is computed similarly by using the variance of grid ionospheric vertical error(σ^2_{GIVE}).

The slant ionospheric delay error and variance at the user are then computed by multiplying the obliquity factor (Fpp) as follows (see Figure 8.9):

$$I_{\mathrm{IPP}} = F_{\mathrm{PP}} \cdot I^{\mathrm{V}}_{\mathrm{IPP}}, \tag{8.31}$$

$$\sigma^2_{\mathrm{UIRE}} = F^2_{\mathrm{PP}} \cdot \sigma^2_{\mathrm{UIVE}}. \tag{8.32}$$

8.4.1.3 SBAS integrity service

The integrity is monitored on board by comparing the protection level computed at the user and the alert limit which is defined for the executing operation (Table 8.1). The horizontal and vertical protection levels (HPL, VPL) are computed by the following equations:

$$\mathrm{HPL} = k_{\mathrm{H}} \cdot d_{\mathrm{major}}, \tag{8.33}$$

$$\mathrm{VPL} = k_{\mathrm{V}} \cdot d_{\mathrm{V}}, \tag{8.34}$$

where the constant $k_{\mathrm{H}} = 6.18$ for non-precision approach (NPA) and $k_{\mathrm{H}} = 6.0$ $k_{\mathrm{V}} = 5.33$ for APV (approach with vertical guidance)-I/II and Category-I approach.

The d_{major} and d_{V} are derived from the covariance matrix of position error when a least squares positioning is carried out:

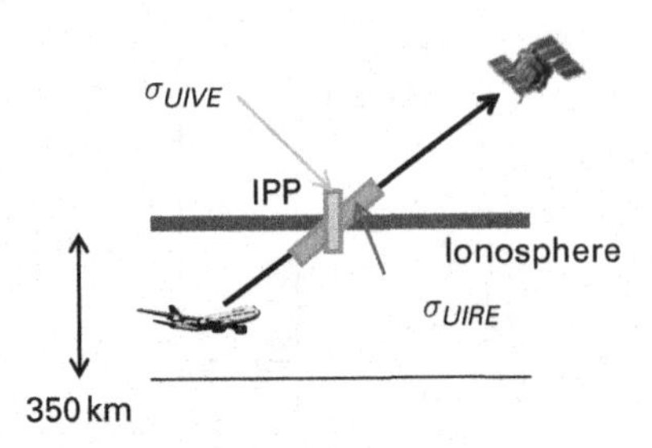

Figure 8.9 Relationship between IPP, σ_{UIVE}, and σ_{UIRE}.

$$\begin{bmatrix} d_{\text{E}}^2 & d_{\text{EN}} & \cdot & \cdot \\ d_{\text{EN}} & d_{\text{N}}^2 & \cdot & \cdot \\ \cdot & \cdot & d_{\text{V}}^2 & \cdot \\ \cdot & \cdot & \cdot & d_{\text{T}}^2 \end{bmatrix} = [S][W]^{-1}[S]^T, \tag{8.35}$$

where matrix $[S]$ can be expressed as follows:

$$[S] = \left([G]^T[W][G]\right)^{-1}[G]^T[W], \quad [W] = [\text{diag}(\sigma_i^{-2})]. \tag{8.36}$$

The ith row of geometry matrix $[G_i]$ is given as follows:

$$[G_i] = \left[-\cos el^i \sin az^i \quad -\cos el^i \cos az^i \quad -\sin el^i \quad 1\right], \tag{8.37}$$

where el^i and az^i are elevation and azimuth of the ith satellite, respectively.

The d_{V} represents the error uncertainty in the vertical direction while d_{major} is the uncertainty along the semi-major axis of the error ellipse:

$$d_{\text{major}} = \sqrt{\frac{d_{\text{E}}^2 + d_{\text{N}}^2}{2} + \sqrt{\left(\frac{d_{\text{E}}^2 - d_{\text{N}}^2}{2}\right)^2 + d_{\text{EN}}^2}}. \tag{8.38}$$

The range error variance for the ith satellite is represented as follows:

$$\sigma_i^2 = \sigma_{i,\text{flt}}^2 + \sigma_{i,\text{UIRE}}^2 + \sigma_{i,\text{air}}^2 + \sigma_{i,\text{trop}}^2, \tag{8.39}$$

where $\sigma^2_{i,\text{flt}}$, $\sigma^2_{i,\text{UIRE}}$, $\sigma^2_{i,\text{air}}$, $\sigma^2_{i,\text{trop}}$ are variance of fast/long-term correction residuals, ionospheric delay (8.32), airborne receiver error, and tropospheric error, respectively. The methods to calculate these terms are described in Appendix J of DO-229D [7] for cases both with and without a SBAS message. The ionospheric delay error variance usually bears the most important information in this message.

There are three SBAS which are currently operational: US WAAS (Wide Area Augmentation System), European EGNOS (European Geostationary Navigation Overlay Service), and Japanese MSAS (MTSAT Satellite-based Augmentation System) where MTSAT stands for multi-functional transport satellite.

SBAS consists of ground reference stations, central processing stations, satellite uplink stations, and geostationary satellites. WAAS comprises multiple wide area reference stations (WRSs), wide-area master stations (WMSs), ground uplink stations

(GUSs), and geostationary satellites. EGNOS system architecture is similar; however, the names of the ground facilities are different, e.g., ranging and integrity monitoring stations (RIMS), mission control centers (MCCs), and navigation land earth stations (NLES). The MSAS ground facility consists of four ground monitor stations (GMS): two master control stations (MCS) located in Japan, and two monitor ranging stations (MRS) in Hawaii and Australia (Figure 8.10). Two MTSAT satellites are in orbit as space components of MSAS.

In contrast to the WAAS and EGNOS coverage, the shape of Japan's mainland is elongated. Therefore, the interpolation method to estimate ionospheric correction might not provide sufficient performance with a limited number of ground reference stations. Figure 8.11 shows an example of σ_{UIVE} around Japan in February 2008. The value of σ_{UIVE} increases drastically when the location is separate from the mainland, and the σ_{UIRE} may become very large for a low elevation satellite since the obliquity factor is multiplied.

8.4.1.4 Case study: integrity monitoring using MSAS

Here we consider an example of positioning and integrity monitoring using MSAS. A flight experiment was conducted at Hachijojima Island, Tokyo, Japan. The trajectory of the aircraft is shown in Figure 8.12. The Beech 65 took off from Hachijojima airport and flew south east, then conducted counter clockwise circling twice and clockwise circling twice. Next, it ascended to 1200 meters, descended to 600 meters, then landed at Hachijojima airport. The height of the airport above the ellipsoid (WGS84) is about 130 meters. The attitude of the aircraft is shown in Figure 8.13. Since roll angles in the four circling stages were over 25°, the satellite at a lower elevation might be blocked by the aircraft itself.

Figure 8.14 shows the number of satellites observed by the onboard receiver and the horizontal protection level computed by the SBAS method with and without MSAS

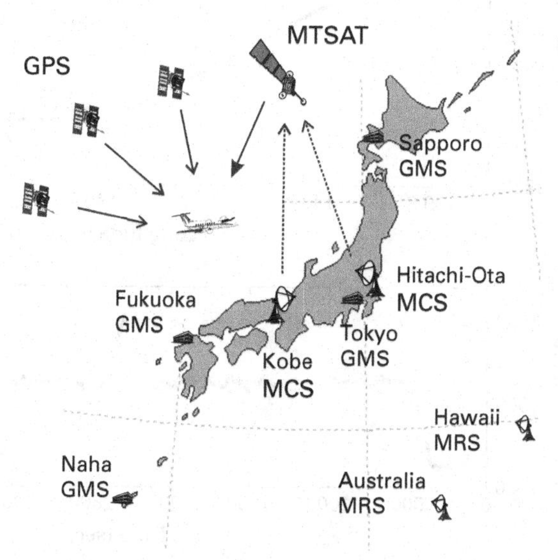

Figure 8.10 MSAS configuration.

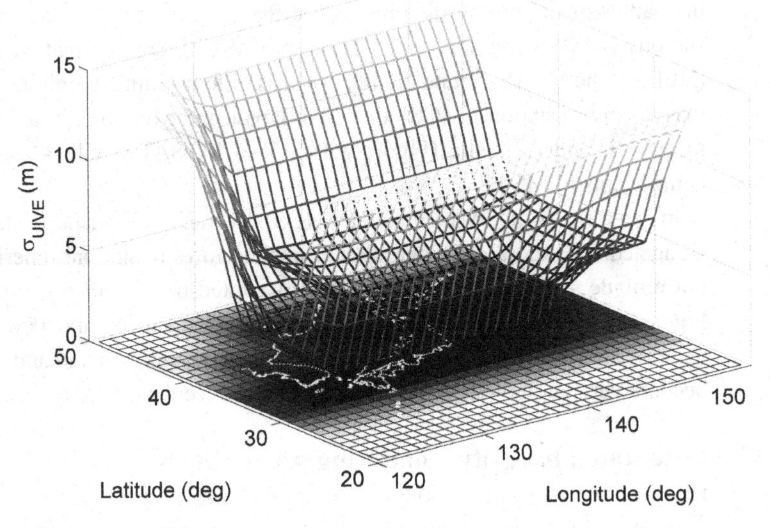

Figure 8.11 Example of σ_{UIVE} around Japan.

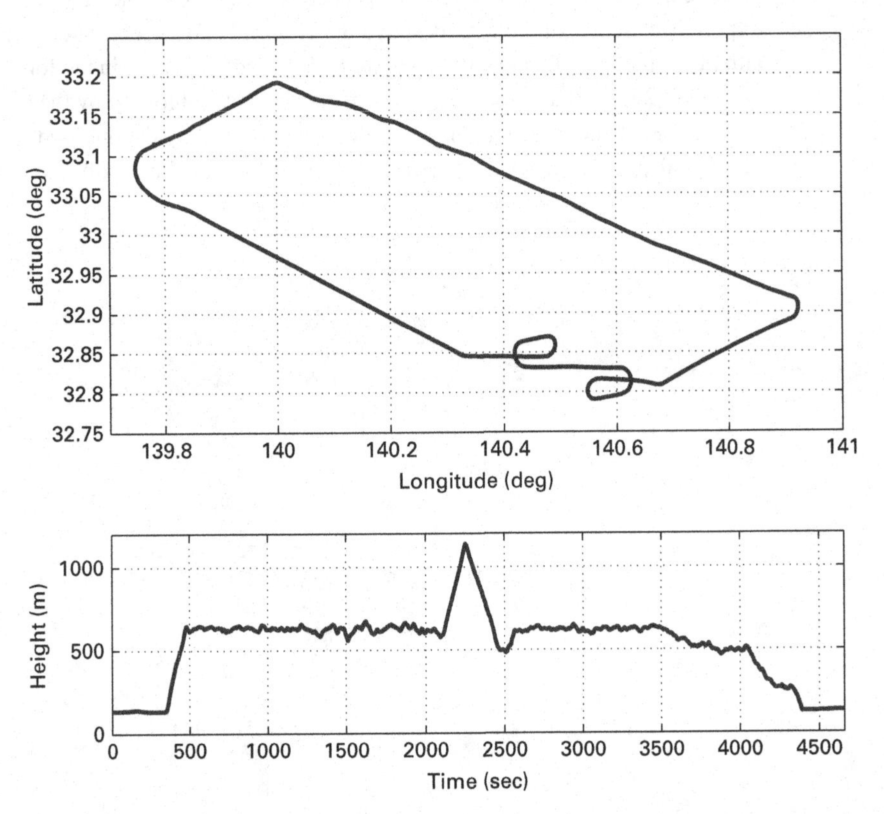

Figure 8.12 Flight trajectory.

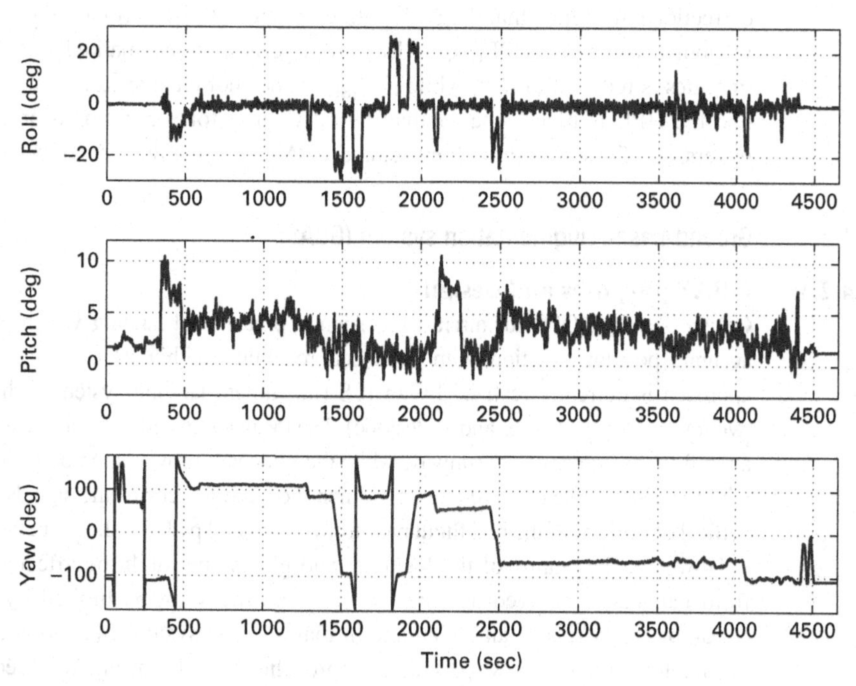

Figure 8.13 Aircraft attitude.

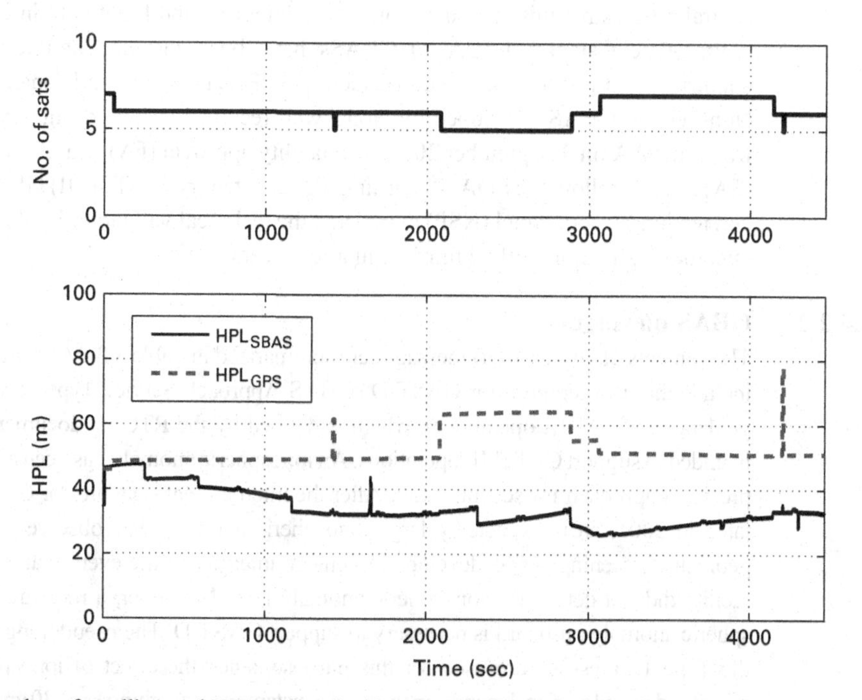

Figure 8.14 Number of satellites and HPL with and without MSAS corrections.

corrections. It is clear that the HPL with corrections (HPL_{SBAS}) is smaller than the HPL without corrections (HPL_{GPS}). Also, HPL_{GPS} is affected markedly if the number of satellites is reduced to five, while HPL_{SBAS} does not change so much. It is noteworthy that the HPL is much smaller than the alert limit for a non-precision approach (556 meters, see Table 8.1) in both cases, and NPA can be conducted safely.

8.4.2 Ground-based augmentation system (GBAS)

8.4.2.1 GBAS purposes and design

GBAS is a LADGPS for aircraft precision approach and landing which provides a very accurate position solution with enhanced reliability within its coverage. The precision approach is currently realized by an ILS (instrument landing system), which consists of two transmitters (localizer and glideslope). The localizer provides the signal which guides the aircraft to the center of the runway, while the glidepath provides the descending flight path with the usual angle of three degrees. When the aircraft receives the signals, the lateral and vertical deviations of the real flight path from the guided path are shown in the flight director.

However, the beams of the localizer and glideslope might be affected by multipath from hangars, a snowed-up runway, etc., and the siting is limited by the terrestrial conditions. GBAS has an advantage in that a single system can provide the approach service for multiple runways and for approaches from both runway directions. Also, it has the potential to realize a curved approach by which the aircraft operator can avoid noise-conscious areas. GBAS comprises multiple (usually four) reference receivers, a central processing facility, and a VHF data broadcast (VDB) facility (Figure 8.15). The central processing unit computes correction, integrity, and flight path information.

Prototype Category-I (CAT-I) GBASs have been installed and tested in several countries such as the US, Australia, Germany, Brazil, Spain, and Japan. And a CAT-I Non-Federal GBAS developed by Honeywell received system design approval (SDA) from the FAA on 3 September 2009. The facility approval (FA) and the service approval (SA) have to follow the SDA. Regarding higher category (CAT-II/III) GBASs, the ICAO Navigation System Panel (NSP) completed the technical validation in May 2010 and the operation validation will be finished in a few years.

8.4.2.2 GBAS messages

The contents of GBAS VDB messages are summarized in Table 8.3. Message types 2 and 11 include the data required for GAST D (GBAS Approach Service Type D, which provides guidance for CAT-III operations), which is defined by the RTCA's documents [8,9] and is intended to support CAT II/III operations. An ionospheric anomaly has been a major threat for precision approach for several years. After the international standard for CAT-I was formulated in 2001, an unexpectedly large ionospheric anomaly was observed. A technique – geometry screening – was developed to ensure integrity in the event that a GBAS ground facility did not detect any ionospheric anomaly [10]. Moreover, a measure to detect ionospheric anomaly on-board is necessary to support GAST D. The pseudorange corrections of the Type 11 message are used for this purpose, since the effect of ionospheric delay on smoothed pseudorange depends on the time constant for smoothing (e.g. 30 and 100 seconds).

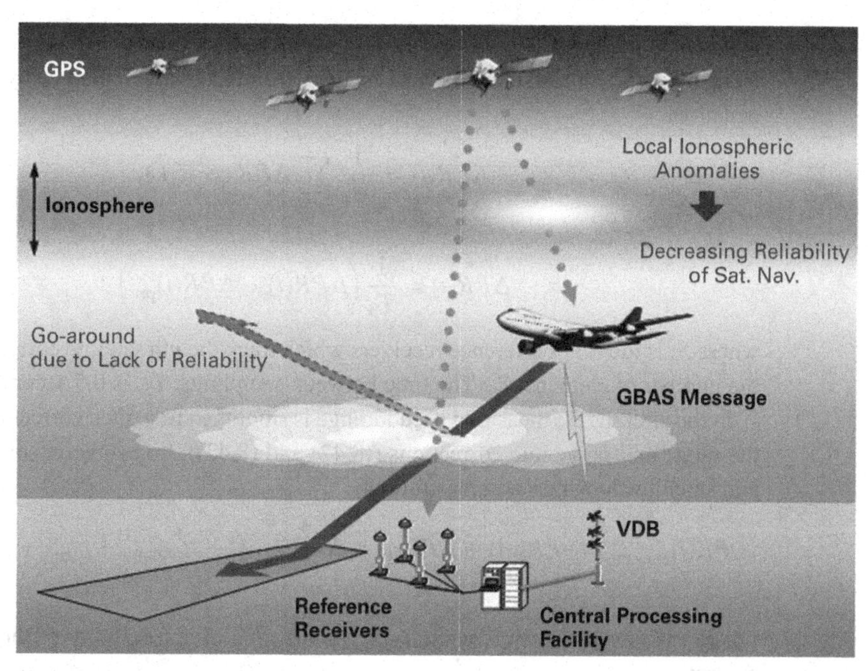

Figure 8.15 GBAS ground facility and potential ionospheric threat.

Table 8.3 GBAS VDB messages.

Message type	Message
1	pseudorange corrections
2	GBAS-related data
3	null
4	final approach segment (FAS) data
5	predicted ranging source availability (optional)
6–8	reserved
11*	pseudorange corrections (incl. 30 second smoothed PR)
101	GRAS pseudorange corrections
9,10,12–100, 101–255	spare

* *Not applicable for APV-I/II, and Category-I.*

The smoothed pseudorange for the *i*th satellite and *j*th receiver, $PR_S(i, j)$, is computed by using the Hatch filter (see section 8.2.1) with time constant 100 seconds. Then the correction for smoothed pseudorange is calculated as follows:

$$\Delta PR_S(i,j) = \rho(i,j) - PR_S(i,j) - c * \Delta t_{SV}(i), \tag{8.40}$$

where $\rho(i,j)$ and $\Delta t_{SV}(i)$ are the computed range and the clock correction for the satellite.

Next, the receiver clock adjusted pseudorange correction is obtained as follows:

$$\Delta PR_{S,\text{adj}}(i,j) = \Delta PR_S(i,j) - \frac{1}{N_C}\sum_{k\in S_c} \Delta PR_S(k,j), \tag{8.41}$$

where S_c is the set of satellites tracked by all reference receivers, and N_c is the number of elements of S_c. Finally, the broadcast correction and its rate are obtained by the following equations:

$$\Delta PR(i) = \frac{1}{M_n} \sum_{k \in S_n} \Delta PR_{S,\mathrm{adj}}(i,k), \tag{8.42}$$

$$\Delta \dot{PR}(i) = \frac{1}{T_S} \left(\Delta PR(i) - \Delta PR(i)_{\mathrm{pre}} \right). \tag{8.43}$$

where S_n is the set of reference receivers which track the nth satellite correctly, and M_n is the number of elements S_n. The time between samplings, T_S, is 0.5 seconds.

In the aircraft, the measured pseudorange is smoothed first, then corrected by applying the range and range rate corrections ((8.42) and (8.43)), tropospheric correction (ρ_{trop}), and satellite clock correction as follows:

$$PR(i)_{\mathrm{corrected}} = PR_S(i) + \Delta PR(i) + \Delta \dot{PR}(i) * (t - t_{Z\mathrm{count}}) + \rho_{\mathrm{trop}} + c * \Delta t_{SV}(i), \tag{8.44}$$

where t is the current time while $t_{Z\mathrm{count}}$ is the time of applicability of the pseudorange correction.

The integrity for precision approach is ensured by comparing the protection level and the corresponding alert limit. If the lateral protection level (LPL) or the vertical protection level (VPL) exceeds the lateral or vertical alert limit, the approach service has to be invalidated within 0.4 seconds.

The LPL and VPL are computed as maximum values among the protection levels for the H_0 and H_1 hypothesis (LPL_{H0}, HPL_{H1}, VPL_{H0}, VPL_{H1}) [8]:

$$\mathrm{LPL} = \max[\mathrm{LPL_{H0}},\ \mathrm{LPL_{H1}}], \tag{8.45}$$

$$\mathrm{VPL} = \max[\mathrm{VPL_{H0}},\ \mathrm{VPL_{H1}}]. \tag{8.46}$$

The protection levels for an H_0 hypothesis are given by

$$\mathrm{LPL_{H0}} = K_{\mathrm{ffmd}} \sqrt{\sum_{i=1}^{N} s_{\mathrm{lat},i}^2 \sigma_i^2} + D_L, \tag{8.47}$$

$$\mathrm{VPL_{H0}} = K_{\mathrm{ffmd}} \sqrt{\sum_{i=1}^{N} s_{\mathrm{vert},i}^2 \sigma_i^2} + D_V, \tag{8.48}$$

where K_{ffmd} is the coefficient to determine the probability of fault-free missed detection. The parameters D_L and D_V are zeros except for the GAST D. The $s_{\mathrm{lat},i}$ and $s_{\mathrm{vert},i}$ for the ith satellite are computed from the elements of the least squares projection matrix, S, which is used for SBAS positioning (see (8.35)) although the projection directions are different. For GBAS, the coordinate system is defined such that the x-axis is along-track, y-axis is cross-track, and the z-axis is vertically up:

$$s_{\mathrm{lat},\,i} = s_{y,i} \tag{8.49}$$

$$s_{\mathrm{vert},i} = s_{x,i} + s_{z,i} \cdot \tan\theta_{\mathrm{GPA}}, \tag{8.50}$$

where θ_{GPA} is the glide path angle. The range error variance for ith satellite is represented as:

$$\sigma_i^2 = \sigma^2_{\mathrm{pr\_gnd},\,i} + \sigma^2_{\mathrm{iono},\,i} + \sigma^2_{\mathrm{pr\_air},\,i} + \sigma^2_{\mathrm{trop},\,i}, \tag{8.51}$$

where $\sigma_{\mathrm{pr\_gnd},i}$ is computed based on the ground data for several days (up to a few weeks) and included in the Type 1 message for GAST C (CAT-I) and in the Type 11 for GAST D. The $\sigma_{\mathrm{iono},i}$ is calculated by using the data in the Type 2 message, and differently for GAST C and D, while the $\sigma_{\mathrm{trop},i}$ is calculated by using the Type 2 message and common for both approach service types. The $\sigma_{\mathrm{pr\_air},i}$ is the standard deviation of the fault-free airborne error, and is dependent on the performance of airborne equipment.

On the other hand, the protection levels for the H_1 hypothesis are computed based on the maximum values of the protection levels for each reference receiver denoted as j:

$$\mathrm{LPL}_{\mathrm{H1}} = \max[\mathrm{LPL}_{\mathrm{H1}}(j)] + D_{\mathrm{L}}, \tag{8.52}$$

$$\mathrm{VPL}_{\mathrm{H1}} = \max[\mathrm{VPL}_{\mathrm{H1}}(j)] + D_{\mathrm{V}}, \tag{8.53}$$

$$\mathrm{LPL}_{\mathrm{H1}}(j) = \left|B_{\mathrm{lat},\,j}\right| + K_{\mathrm{MD}}{}^{*}\sigma_{\mathrm{lat\_H1}}, \tag{8.54}$$

$$\mathrm{VPL}_{\mathrm{H1}}(j) = \left|B_{\mathrm{vert},\,j}\right| + K_{\mathrm{MD}}{}^{*}\sigma_{\mathrm{vert\_H1}}, \tag{8.55}$$

where K_{MD} is the coefficient to determine the probability of missed detection assuming that the reference receiver is faulted.

The $B_{\mathrm{lat},j}$, $B_{\mathrm{vert},j}$ are calculated by using the so called "B-values," which are included in the Type 1 message, as follows:

$$B_{\mathrm{lat},\,j} = \sum_{i=1}^{N} s_{\mathrm{lat},\,i} B(i,j), \tag{8.56}$$

$$B_{\mathrm{vert},\,j} = \sum_{i=1}^{N} s_{\mathrm{vert},\,i} B(i,j). \tag{8.57}$$

The B-value is computed by using (8.41) and (8.42):

$$B(i,j) = \Delta PR(i) - \frac{1}{M_n - 1} \sum_{\substack{k \in S_n \\ k \neq j}} \Delta PR_{\mathrm{S,adj}}(i,k). \tag{8.58}$$

The $\sigma_{\mathrm{lat\_H1}}$ and the $\sigma_{\mathrm{vert\_H1}}$ in (8.54) and (8.55) are obtained as follows:

$$\sigma_{\mathrm{lat\_H1}} = \sqrt{\sum_{i=1}^{N} s^2_{\mathrm{lat},\,i} \sigma^2_{i\_\mathrm{H1}}}, \tag{8.59}$$

* Similar to Equations (8.33) and (8.39).

$$\sigma_{\mathrm{vert\_H1}} = \sqrt{\sum_{i=1}^{N} s_{\mathrm{vert},\,i}^2 \sigma_{i\_\mathrm{H1}}^2}, \tag{8.60}$$

where

$$\sigma_{i\_\mathrm{H1}}^2 = \frac{M(i)}{U(i)} \sigma_{\mathrm{pr\_gnd},\,i}^2 + \sigma_{\mathrm{iono},\,i}^2 + \sigma_{\mathrm{pr\_air},\,i}^2 + \sigma_{\mathrm{trop},\,i}^2. \tag{8.61}$$

The $M(i)$ is the number of reference receivers whose pseudoranges are used to calculate the correction for the ith satellite, while the $U(i)$ is the same with respect to $M(i)$ but without counting the jth reference receiver.

8.4.2.3 Case study: navigation with GBAS

Here we describe an example of GBAS navigation performance. The flight experiments were conducted in October 2010 at Noto airport (N37.29, E136.96 degrees), Japan. Four GPS receivers with choke-ring antennas were installed in the airport, which made a square with 200 meter sides, and the recorded data were used to generate the correction data The data from the onboard receiver and the generated GBAS message data were used to compute lateral and vertical protection levels in the off-line mode. Several tens of approaches including curved approaches were conducted and the selected three trajectories are show in Figure 8.16. The curved approach had the benefits of avoiding obstacles in the terrain or reducing noise in the area of the nearby airport, and could be realized by using the TAP (terminal area path) data which could be included in the GBAS message

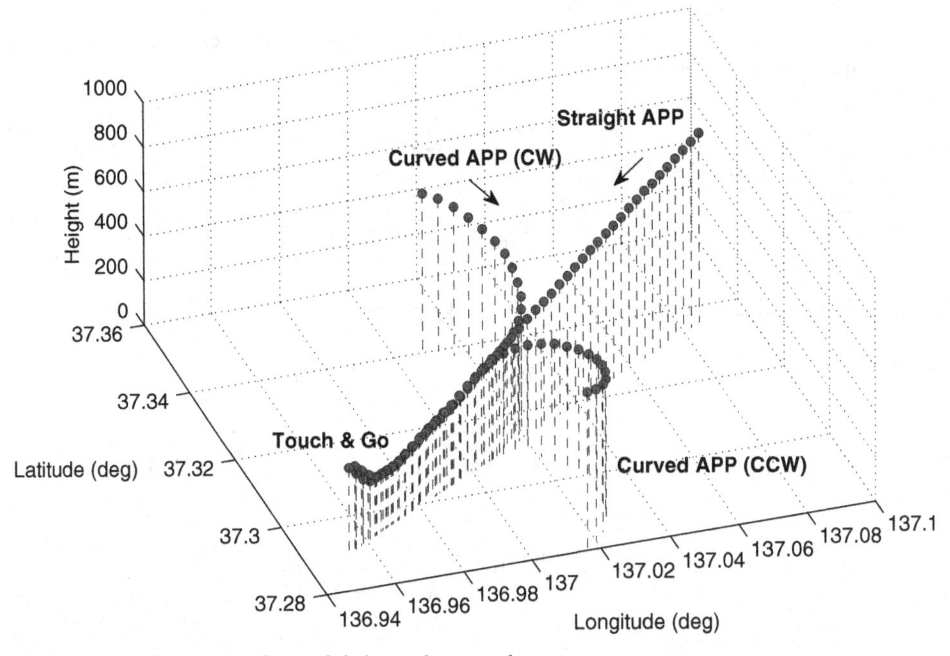

Figure 8.16 Aircraft trajectory for straight/curved approaches.

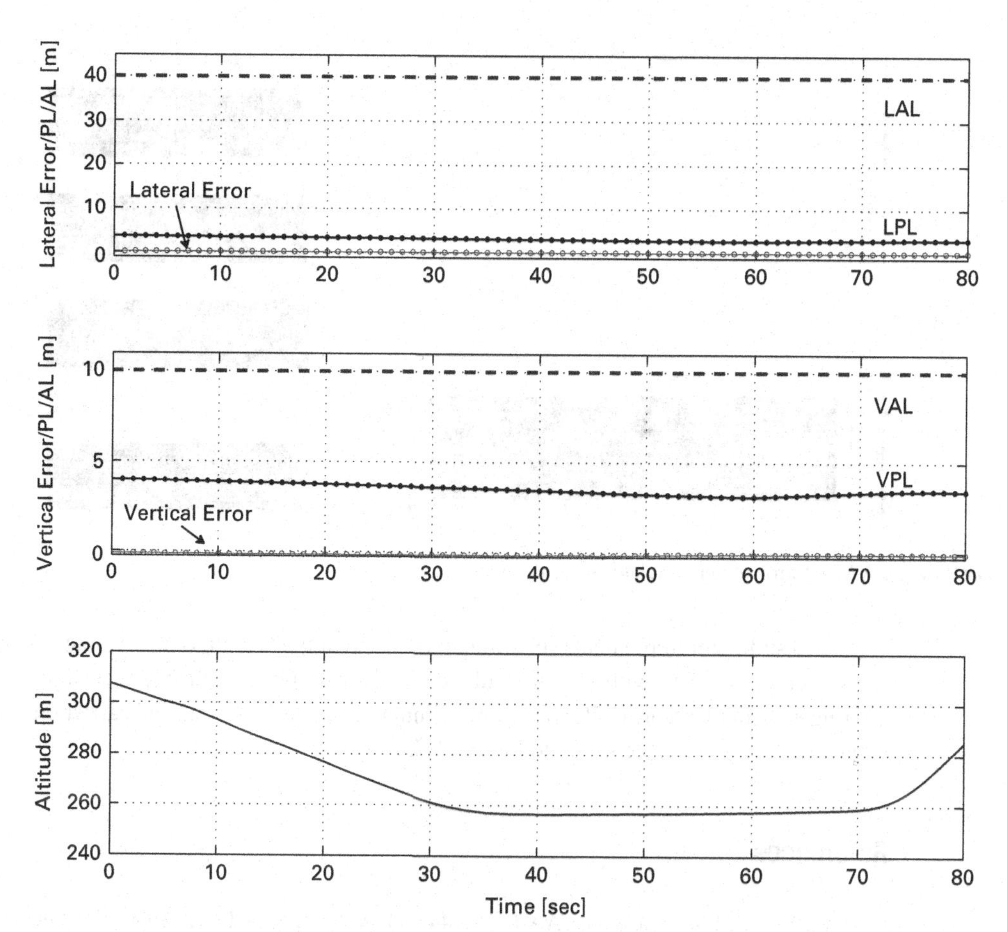

Figure 8.17 Lateral/vertical navigation errors and protection levels (top/middle), as well as the height profile (bottom) during a clockwise curved approach.

[9]. Since a commercial guidance system utilizing the TAP procedure is not available, an integrated display, which indicates a tunnel-like flight path to the pilot, is used for the operation of a curved approach. Figure 8.17 shows the computed CAT-I lateral/vertical protection levels as well as the navigation errors during the clockwise curved approach where the "true" trajectory is calculated by a carrier-phase-based DGPS. The lateral/vertical protection levels are well below the alert limits, therefore it is obvious that the CAT-I precision approach is available.

8.5 Project: simulation data for LAAS

1. Use ReGen to simulate output from four reference receivers. Generate RINEX observation files with spatially correlated ionospheric errors.
2. iP-Solutions LAAS simulator software (Figure 8.18) can be used to process receiver output files simulated with ReGen or from receivers working with live satellites. The

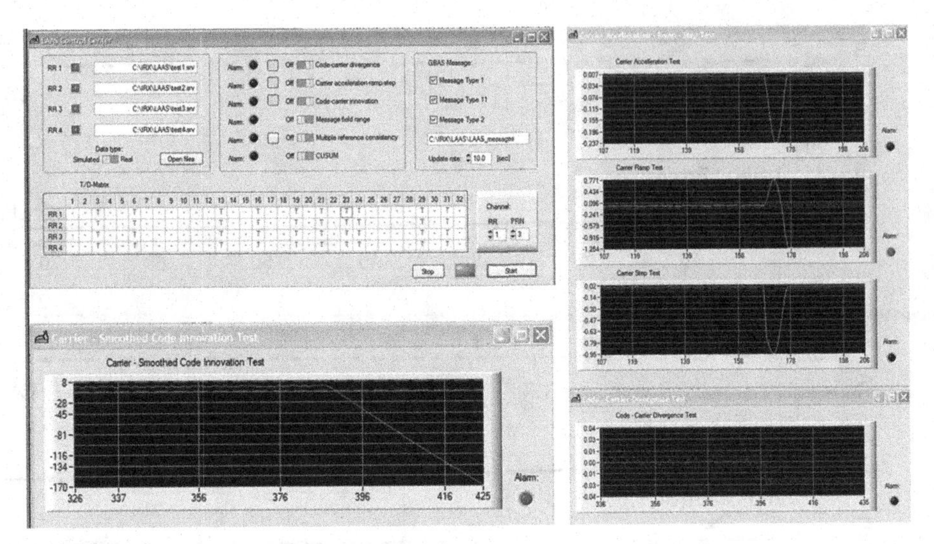

Figure 8.18 LAAS control center simulator screenshot.

LAAS simulator implements six different tests and allows output of GBAS messages. By generating data with various fault scenarios in ReGen, one can see what errors trigger particular alarms. ReGen allows simulation of various code and carrier errors, and fault scenarios according to [1] and [9].

References

[1] FAA Non-Fed Specification, FAA-E-AJW44-2937A, Category I Local Area Augmentation System Ground Facility, 2005.

[2] R. Hatch, Instantaneous ambiguity resolution, Proceedings of *IAG International Symposium No.107 on Kinematic Systems in Geodesy, Surveying and Remote Sensing*, New York, Springer Verlag, 1991, pp 299–308.

[3] D. V. Simili and B. Pervan, Code-carrier divergence monitoring for the GPS local area augmentation system, Proceedings of IEEE/ION PLANS 2006, San Diego, CA, 25–27 April 2006, pp. 483–493.

[4] G. Blewitt, An automatic editing algorithm for GPS data, *Geophysical Research Letters*, **17**, (3), 1990, 199–202.

[5] S. Schaer, *Mapping and Predicting the Earth's Ionosphere Using the Global Positioning System*, Volume 59 of Geodätisch-geophysikalische Arbeiten in der Schweiz, Schweizerische Geodätische Kommission, Institut für Geodäsie und Photogrammetrie, Eidg. Technische Hochschule Zürich, Zürich, Switzerland, 1999.

[6] ICAO, *International Standards and Recommended Practices, Aeronautical Telecommunications, Annex 10 to the Convention on International Civil Aviation*, Vol.I, Montreal, Canada, 2008. (http://www.icao.int)

[7] RTCA, *Minimum Operational Performance Standards for Global Positioning System/Wide Area Augmentation System Airborne Equipment, DO-229D*, Washington, DC, 2006. (http://www.rtca.org)

[8] RTCA, *Minimum Operational Performance Standards for GPS Local Area Augmentation System Airborne Equipment, DO-253C*, Washington, DC, 2008. (http://www.rtca.org)

[9] RTCA, *GNSS Based Precision Approach Local Area Augmentation System (LAAS) · Signal-in-Space Interface Control Document (ICD), DO-246D*, Washington, DC, 2008. (http://www.rtca.org)

[10] S. Ramakrishnan, J. Lee, S. Pullen, and P. Enge, Targeted ephemeris decorrelation parameter inflation for improved LAAS availability during severe ionosphere anomalies, Proceedings of ION 2008 National Technical Meeting, San Diego, CA, 28–30 Jan., 2008, pp. 354–366.

Exercise

Exercise 8.1. Consider two extreme cases for the carrier smoothing algorithm given in Section 8.2.1, one with weight coefficient for code phases $W_\rho = 0$ and another with weight coefficient for carrier phases $W_\varphi = 0$. What are the drawbacks of implementation in each case ?

9 Using observables in navigation-related tasks

In this chapter we consider a few examples of GNSS applications in navigation and look at how GNSS observables are used for various positioning and navigation-related tasks.

9.1 Precise positioning with carrier phase observables

We described positioning with code phase observables (pseudoranges) in Chapter 1. In the previous chapter we saw that carrier phase observables have a much lower error budget. However, it is necessary to introduce extra unknowns in the GNSS equations when using carrier phase measurements. A receiver can only measure the difference between phases of an incoming signal and a replica. These measurements are bound within a wavelength, which for example is about 19 cm for a GPS L1 signal. We need to add to these measurements an unknown ***ambiguity number*** of whole carrier waves between a satellite and receiver antenna phase center, so we can have measurements of total distance of about 20 000 km. These measurements will be different from the real distance to the satellite. The difference is because of satellite and receiver clock errors, as described in Chapter 1 for code phase observables.

Here we introduce the measurement equations for carrier phase observables. The carrier phase measured by a GNSS receiver is the difference between the phase of the incoming satellite signal as it was at transmission time ($\theta_{SV}(t_{SV})$) and the phase generated by the receiver at reception time ($\theta(t)$). These phases can be defined by the following equations [1],[2]:

$$\theta_{SV}(t_{SV}) = f \cdot t_{SV}, \tag{9.1}$$

$$\theta(t) = f \cdot t, \tag{9.2}$$

where f is the frequency of the carrier. In the above equations, the phases are defined to be zeros when the time epochs in satellite and receiver clocks are equal to zeros. Thus, the carrier phase Φ is described by

$$\Phi(t) = f(t - t_{SV}) + \varepsilon, \tag{9.3}$$

where ε is the measurement noise. However, the observed carrier phase at the start time of measurement t_0 is only a fraction of the full wave. So, the observed carrier phase Φ_m is written as follows at the initial time:

$$\Phi_m(t_0) = fr(\Phi(t_0)), \tag{9.4}$$

where $fr(\cdot)$ means that the fractional part of a wave is taken. Hence, the real carrier phase, which contains an unmeasured number of integer cycles (ambiguity number) N, is written as follows:

$$\Phi(t_0) = \Phi_m(t_0) - N. \tag{9.5}$$

Since the carrier phase is integrated continuously unless a cycle slip occurs, the measured carrier phase at time t is given by

$$\Phi_m(t) = \Phi(t) - \Phi(t_0) + \Phi_m(t_0). \tag{9.6}$$

Substituting (9.5) for (9.6), the measured carrier phase is

$$\Phi_m(t) = \Phi(t) + N. \tag{9.7}$$

Since $\Phi(t)$ can be treated like a pseudorange, the measurement equation of the carrier phase is given by multiplying the wavelength λ by (9.7) and rewriting $\lambda\Phi_m(t)$ to ϕ as follows:

$$\phi = \rho - \frac{f_2}{f_1} I + d_{\mathrm{trop}} + b - b_{\mathrm{SV}} + d_{\mathrm{eph}} + d_{\mathrm{m,phase}} + \lambda N + \varepsilon, \tag{9.8}$$

where N is the integer unknown ambiguity number, $d_{\mathrm{m,phase}}$ is the multipath error of the carrier phase, which is smaller than a few centimeters. Note that the sign of ionospheric delay is negative, while it is positive for the pseudorange (see Chapter 3).

9.1.1 Ambiguity resolution

In order to use carrier phase observables we need to resolve ambiguity, i.e. find the unknown integer ambiguity number for each carrier observable. There is a number of algorithms for ambiguity resolution suitable for particular tasks [3–5]. In this section we describe a simple ambiguity resolution method based on the least squares search approach in order to introduce the concept following [6] and [7]. In the next section we introduce a more efficient LAMBDA method [5].

In the search algorithm, the initial vector between the user and reference receiver is determined first based on unambiguous code measurements. After that, the carrier phase ambiguities are determined by choosing the best fit to the measurements from a number of candidates. Figure 9.1 shows a search cube in the ambiguity space. The search cube shows the candidates for possible coordinates, which would result from chosing one or another ambiguity number. If a receiver antenna is static, then the ambiguity is resolved by making use of the change of satellite constellation with time. Figure 9.2 shows the changes of positions corresponding to each ambiguity candidate with change of satellite constellation. Since the position solution corresponding to the correct ambiguity does not

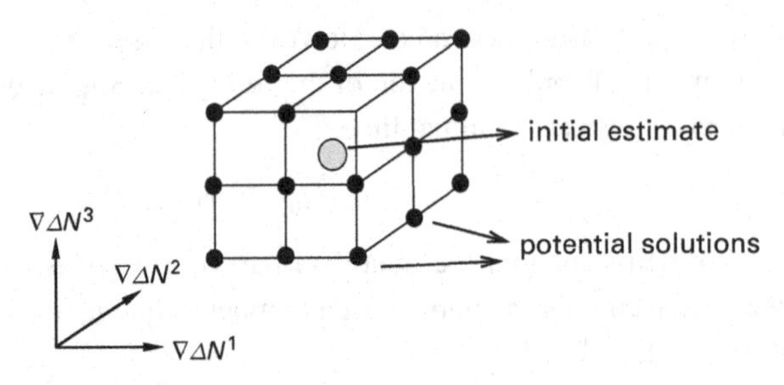

Figure 9.1 Search cube in ambiguity space.

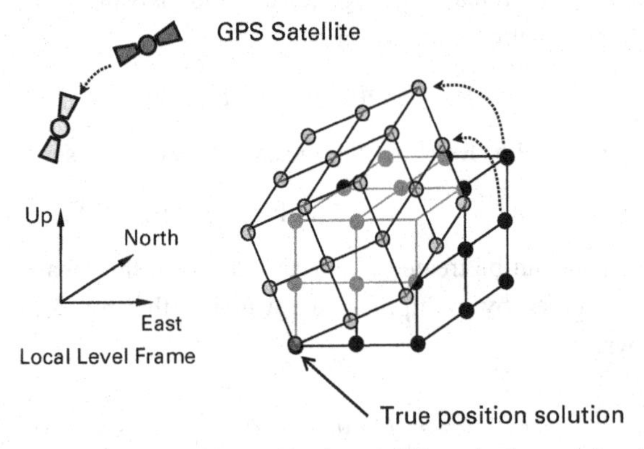

Figure 9.2 Changes of position calculated using the ambiguity candidates in the search cube.

change with time, the ambiguity will be resolved after a considerable change of satellite constellation. Therefore, it is necessary to receive the GNSS signal until the constellation changes sufficiently.

The time required for resolution depends on the baseline length or other environmental conditions such as the ionospheric errors and the multipath. Typically, 15 to 30 minutes are necessary for 10 km or shorter baselines while several hours are necessary for a few hundred kilometer baselines [8].

In addition to the geodetic survey, precise (centimeter-level accuracy) carrier phase positioning has a wide area of application in aerospace technology such as precision approach, taxi guidance, and rendezvous docking. The carrier phase positioning of moving platforms with one or more reference GPS receivers is called kinematic GPS (KGPS). At an early stage of the history of ambiguity resolution technology, the ambiguity was resolved by a batch process in static mode before moving, which is called "static initialization", and the KGPS was carried out as long as the receiver tracked four or more satellites continuously [9]. However, if some losses of lock (cycle slip) occur, or some satellites go out of sight, we cannot continue to perform the KGPS. So, ambiguity resolution ***on-the-fly (OTF)***, i.e. without static initialization, is necessary for KGPS.

A typical goal of ambiguity resolution is to determine the L1 ambiguity. However, if L2 measurements are available, then the widelane observable can be used to reduce the search space of the L1 ambiguity. Here, we give the double difference (DD) observation equations for the ambiguity resolution algorithm (see Chapter 8):

$$\nabla\Delta PR_1 = \nabla\Delta\rho + \nabla\Delta\frac{f_2}{f_1}I + \nabla\Delta d_{\text{trop}} + \nabla\Delta d_{\text{eph}} + \nabla\Delta d_{\text{m1}} + \nabla\Delta\varepsilon_{\text{PR}_1}, \tag{9.9}$$

$$\nabla\Delta\phi_{\text{W}} = \nabla\Delta\rho + \nabla\Delta I + \nabla\Delta d_{\text{trop}} + \nabla\Delta d_{\text{eph}} + \nabla\Delta d_{\text{m,W}} + \lambda_{\text{W}}\nabla\Delta N_{\text{W}} + \nabla\Delta\varepsilon_{\text{W}}, \tag{9.10}$$

$$\nabla\Delta\phi_1 = \nabla\Delta\rho - \nabla\Delta\frac{f_2}{f_1}I + \nabla\Delta d_{\text{trop}} + \nabla\Delta d_{\text{eph}} + \nabla\Delta d_{\text{m1,phase}} + \lambda_1\nabla\Delta N_1 + \nabla\Delta\varepsilon_1. \tag{9.11}$$

Using these observables sequentially, we can finally resolve the L1 ambiguity, and obtain the positioning solution. The first equation for code phase measurements does not contain an ambiguity number, because pseudoranges are not ambiguous at this stage. Code phases are also ambiguous, because they repeat with 300 km intervals, but their ambiguities are resolved with a navigation message time stamp (see Chapter 6). Measurement errors for two carrier phase observables are different (see Table 9.1).

These values are calculated using the accumulated data in a flight test where two Trimble 4000SSE receivers were used. Due to the test environment and aircraft maneuvers the data also contain multipath error. For example, during some maneuvers with large bank angles the aircraft wings may reflect the satellite signal and cause a multipath error.

The ambiguity resolution algorithm can be described as follows (see the flowchart in Figure 9.3):

(1) An initial position is estimated by using the carrier smoothed pseudorange.
(2) An initial estimate of widelane ambiguities is determined by the following equation, where the double differenced ranges from the receiver to satellites, $\nabla\Delta\rho$, are calculated using the initial position estimate from the first step:

$$\nabla\Delta N_{\text{W0}} = \text{idnint}\left(\frac{\nabla\Delta\phi_{\text{W}} - \nabla\Delta\rho - \nabla\Delta d_{\text{trop}}}{\lambda_{\text{W}}},\right) \tag{9.12}$$

where the notation "idnint" means the nearest integer.

We choose four satellites with the minimum RDOP (relative dilution of precision) as primary satellites among all of the observed satellites. The RDOP is a factor that is

Table 9.1 Example of measurement errors and corresponding position errors for constellation geometry with relative dilution of precision RDOP=3. Values are for a Trimble 4000SSE receiver and include multipath error.

Observable	Measurement error (cm)	Position error (cm)
DD of carrier smoothed pseudorange	65	195
DD of widelane carrier phase	4	12
DD of L1 carrier phase	1	3

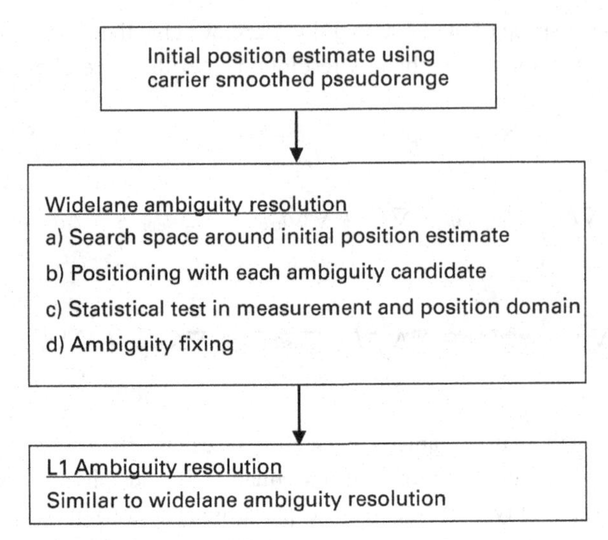

Figure 9.3 Flowchart of ambiguity resolution algorithm.

defined by the distribution of observed satellites in the sky, and given by the following equations:

$$\mathrm{RDOP} = \sqrt{\mathrm{trace}([H_r]^T[H_r])^{-1}}, \tag{9.13}$$

$$[H_r] = \begin{bmatrix} \frac{\partial DD_1}{\partial \vec{r}} & \cdots & \frac{\partial DD_{\mathrm{nsv}-1}}{\partial \vec{r}} \end{bmatrix}^T$$

$$= \begin{bmatrix} \frac{\vec{r}^T - \vec{r}^T_{\mathrm{SV1}}}{\rho^1_u} - \frac{\vec{r}^T - \vec{r}^T_{\mathrm{SV2}}}{\rho^2_{\mathrm{u}}} \\ \vdots \\ \frac{\vec{r}^T - \vec{r}^T_{\mathrm{SV1}}}{\rho^1_{\mathrm{u}}} - \frac{\vec{r}^T - \vec{r}^T_{\mathrm{SVnsv}}}{\rho^{\mathrm{nsv}}_{\mathrm{u}}} \end{bmatrix} \tag{9.14}$$

where $[H_r]$ is the measurement matrix, ρ^i_{u} is the distance from the user receiver to the ith satellite, DD_i is the double differenced observable between the first and i+1th satellite, in which the smoothed pseudorange, widelane, or L1 carrier phase should be inserted. Using the RDOP, the standard deviation of double differenced measurement error σ_{m} and positioning error σ_{p} satisfy the following relation:

$$\sigma_{\mathrm{p}} = \mathrm{RDOP} \cdot \sigma_{\mathrm{m}}. \tag{9.15}$$

The ambiguity search cube for these four satellites, which constitutes a set of three double differences, is constructed around the center defined by (9.12). When the position calculation is performed by using a set of three ambiguity candidates, the ambiguities for other satellites' double differences are calculated.

(3) Receiver position is computed with each ambiguity candidate, and the statistical tests are performed in measurement domain and positioning domain.

(3a) Statistical test in the measurement domain.

A chi-squared (χ^2) test is performed using the sum of measurement residuals. The candidates satisfying the following condition are rejected:

$$\frac{\vec{\nu}^T [C_W]^{-1} \vec{\nu}}{df} > \frac{\chi^2_{df, 1-\alpha}}{df}, \tag{9.16}$$

where $\vec{\nu}$ is the residual vector, C_W is the DD measurement error covariance matrix for a widelane linear combination, α is the significance level of the χ^2 test, and df is the degree of freedom, respectively:

$$df = N_{SV} - 1, \quad (N_{SV} : \text{number of satellites}). \tag{9.17}$$

(3b) Statistical test in the positioning domain.

Taking the differences between horizontal position computed using smoothed pseudoranges and horizontal position computed using each ambiguity candidate, the candidates which satisfy the following condition are rejected:

$$\left| \vec{r}^{PR} - \vec{r}^{W} \right|_H > k^W \sigma_H^{PR-W}, \tag{9.18}$$

where $\vec{r}^{PR}$ and $\vec{r}^{W}$ are the position vectors of antenna calculated using smoothed pseudorange and widelane, respectively, $| \cdot |_H$ is a horizontal norm, σ_H^{PR-W} is the standard deviation of the difference between pseudorange-based position and widelane-based position in the horizontal direction, k^W is an empirical parameter of tolerance, which is set to 2 or 3 depending on the measurement conditions. Theoretically, $k^W = 1, 2, 3$ corresponds to the significance level of 68, 95, and 99%. Although the 3D or vertical position can be chosen for the test in position domain, an evaluation of horizontal position difference usually gives better performance since horizontal positioning is less affected by the residual measurement errors.

(4) If only one ambiguity candidate set is retained, then it is considered as the solution. If two or more candidates are retained, similar statistical tests should be performed for the next epoch.

(5) Steps (3) and (4) are repeated until only one candidate is retained. If the number of total epochs exceeds a pre-determined threshold number, the algorithm is back to step (2).

(6) The initial values of L1 ambiguity are calculated from the widelane position.

(7) Steps similar to (3) and (4) are repeated until only one candidate is retained. If the number of total epochs exceeds a pre-determined threshold number, the process goes back to step (6).

We look at an example of an L1 ambiguity search for the case when five satellites are observed. The ambiguity search is performed for ±1 cycle search area from the initial estimate of the ambiguity, so the total number of candidates is 27. Figure 9.4 shows the sum of squared residuals (SOS) values of measurement residuals divided by the degree of freedom (the left side of Equation (9.16)) for all candidates. Since the degree of freedom is equal to one, the value of $\chi^2_{df, 1-\alpha}/df$ in the equation becomes 6.63 with a significance

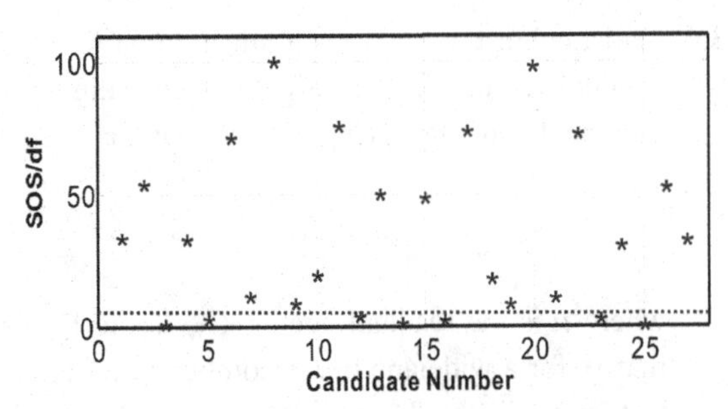

Figure 9.4 Sum of squared residuals/degree of freedom for each L1 phase ambiguity candidate. Solid line represents the threshold of 99%.

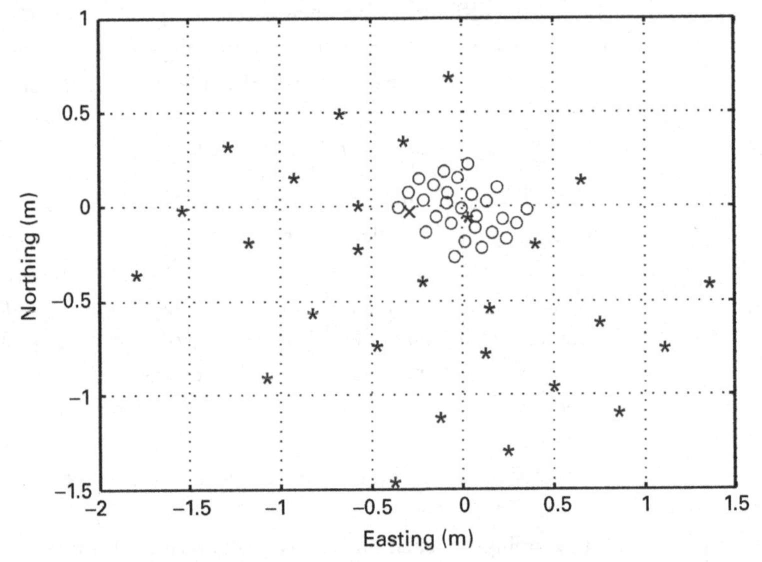

Figure 9.5 Horizontal position of the receiver, calculated for smoothed pseudorange(×), widelane ambiguity candidates(∗), and L1 phase ambiguity candidates(○).

level of 99%. The broken line in Figure 9.4 indicates the threshold for this test. Accordingly, seven candidates are retained. Although the correct ambiguity set is candidate No. 14, the SOS value is a minimum for candidate No. 25.

Now we look at the result of the test in the positioning domain for L1 ambiguity resolution. Figure 9.5 depicts the distribution of the positions calculated using the smoothed pseudorange, the widelane, and L1 ambiguity candidates. The position of the origin is calculated using a correct L1 ambiguity. In this case, the position corresponding to the correct widelane ambiguity is closest to the origin among all the widelane positions. If the difference between the position for correct widelane ambiguity and the position for an L1 candidate (○) were larger than the threshold value, the candidate would be rejected according to the test in the positioning domain. Figure 9.6 gives the horizontal, vertical, and three-dimensional differences between

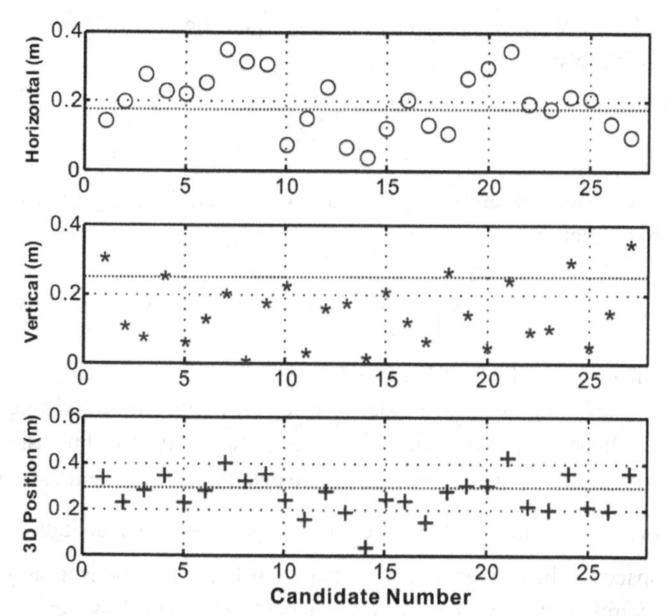

Figure 9.6 Difference between position calculated using the correct widelane ambiguity and positions calculated using each candidate of the L1 phase ambiguity.

the positions calculated using the correct widelane ambiguity and the positions calculated using each L1 candidate. Here the broken line indicates the threshold of the test in the positioning domain with a significance level of 99%. As a result, 10, 22, and 18 candidates pass the tests in horizontal, vertical, and three-dimensional positioning domain. If we compare the vertical or three-dimensional position (middle and bottom of Figure 9.6), all seven candidates which pass the test in the measurement domain also pass the test in the position domain. On the other hand, only one candidate is retained if the horizontal position is evaluated. This example indicates that the ambiguity can be resolved fast by evaluating the horizontal position in the positioning domain test.

9.1.2 LAMBDA method

The least-squares AMBiguity decorrelation adjustment (LAMBDA) is one of the most widely used ambiguity resolution methods which reduces the search space effectively by decorrelating double differenced ambiguity biases [10]. A generalized measurement equation for positioning is assumed as follows:

$$\vec{y} = [H]\vec{x} + [A]\vec{N} + \vec{\varepsilon}, \tag{9.19}$$

where $\vec{y}$ is the residual vector for double differenced (DD) carrier phase, $\vec{x}$ is the position adjustment from the initial estimate, $\vec{N}$ is the DD ambiguity vector, and $[H]$ and $[A]$ are corresponding design matrices. First, the position solution ($\hat{\vec{x}}$) and the float ambiguity solution $\hat{\vec{N}}$ are computed by the least squares method. The covariance matrix is denoted as follows:

$$[Q] = \begin{bmatrix} Q_{\hat{\vec{x}}} & Q_{\hat{\vec{x}}\hat{\vec{N}}} \\ Q_{\hat{\vec{N}}\hat{\vec{x}}} & Q_{\hat{\vec{N}}} \end{bmatrix}. \tag{9.20}$$

The purpose of ambiguity resolution is to find the integer vector $\vec{n}$ which minimizes the cost function

$$\chi^2 = \left\|\hat{\vec{N}} - \vec{N}\right\|^2_{\left[Q_{\hat{\vec{N}}}\right]} = (\hat{\vec{N}} - \vec{N})^T \left[Q_{\hat{\vec{N}}}\right]^{-1} (\hat{\vec{N}} - \vec{N}), \quad \vec{N} \in Z^{N_{SV}-1}, \tag{9.21}$$

where N_{SV} is the number of satellites.

The ambiguity search space is inside the hyper-ellipsoid defined by the above equation. The hyper-ellipsoid is reduced to an ellipsoid for the two-dimensional case. The size, shape, and directions of semi-major axes are determined by χ^2 and the eigenvalues (eigenvectors) of $\left[Q_{\hat{\vec{N}}}\right]^{-1}$. The search space can be elongated as shown in Figure 9.7 (left). Since the search space for the conventional ambiguity resolution method is usually a rectangle, the number of grid points to be searched is significantly reduced. However, the method to select the grid points is complicated. If the weight matrix ($\left[Q_{\hat{\vec{N}}}\right]^{-1}$) is diagonal, the cost function becomes the weighted sum of squares of the vector elements ($\chi^2 = \sum_{i=1}^{N_{SV}-1} \left[Q_{\hat{\vec{N}}}\right]^{-1}_{ii} (\hat{N}_i - N_i)^2$). Therefore, the solution is the integer nearest to the float solution. Hence, the next step is transformation (decorrelation) of the ambiguity vector. A transformation matrix, $[Z]$, and the transformed ambiguity vector, $\vec{z}$, are assumed as follows:

$$\vec{z} = [Z]\vec{N}. \tag{9.22}$$

Since the elements of the $\vec{z}$ and $\vec{N}$ are all integers, the elements of the transformation matrix are integers as well. Since $\vec{N}$ has been computed by back-transformation, the

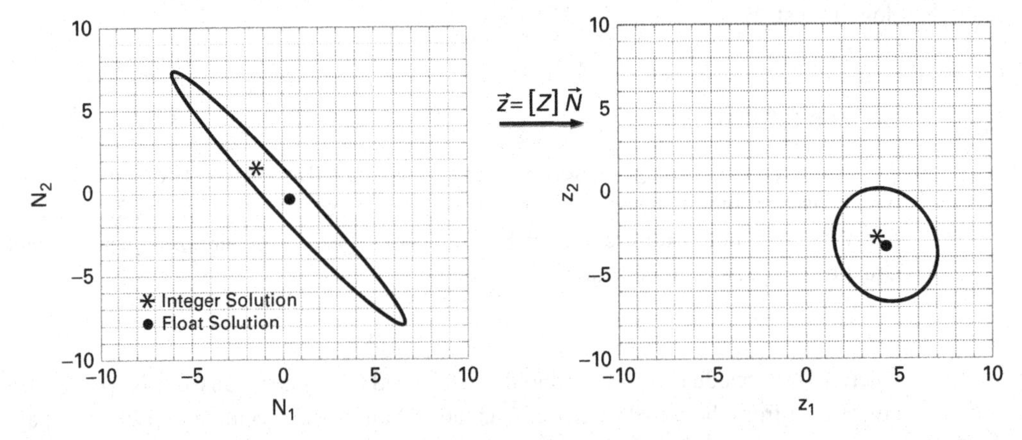

Figure 9.7 Ambiguity search ellipsoid in original space (left) and transformed space (right) (after [10]).

inverse matrix $[Z]^{-1}$ should exist and all elements of the matrix should be integers. Also, from the conditions above $|\det([Z])| = 1$.

The cost function (9.21) can now be expressed by using the transformed ambiguity vector:

$$\begin{aligned}\chi^2 &= ([Z]^{-1}\hat{\vec{z}} - [Z]^{-1}\vec{z})^T\left[Q_{\hat{\vec{N}}}\right]^{-1}([Z]^{-1}\hat{\vec{z}} - [Z]^{-1}\vec{z}) \\ &= (\hat{\vec{z}} - \vec{z})^T[Z]^{-T}\left[Q_{\hat{\vec{N}}}\right]^{-1}[Z]^{-1}(\hat{\vec{z}} - \vec{z}).\end{aligned} \tag{9.23}$$

If the matrix $[Z]^{-T}\left[Q_{\hat{\vec{N}}}\right]^{-1}[Z]^{-1}$ is diagonal, then the elements of the solution $(\vec{z})$ are the integers nearest to the elements of $\hat{\vec{z}}$. Generally, the matrixes cannot be strictly diagonal, but rather close to diagonal. Figure 9.7 shows the transformed ellipsoid on the right. The area of the ellipsoid is the same as that of the original since $|\det([Z])| = 1$. However, the shape is nearly a circle and the searching process can be easier than searching the elongated area. A method to obtain the transformation matrix based on LDL^{T} and UDU^{T} factorizations, as well as a method to reduce the search space, have been proposed in [11].

As the final step, the ambiguity vector in original space is obtained by back transformation (9.24), and used for the positioning:

$$\vec{N} = [Z]^{-1}\vec{z}. \tag{9.24}$$

The LAMBDA method is especially useful when many satellites are available. If we use GPS together with GLONASS, then the number of satellites can be more than 12 for most of the time. In this case it is very difficult to conduct a full grid search for real-time applications with conventional methods, because the search number increases exponentially. The LAMBDA method, however, allows us to reduce the search number dramatically, and correct ambiguity can be found instantaneously [12].

9.1.3 Cycle slip detection

Cycle slip occurs if the receiver loses phase lock of the satellite signal. The most frequent reason is signal obstruction due to trees, buildings, or vehicles themselves. Another reason is a low SNR due to bad ionospheric conditions, multipath, high receiver dynamics, or low satellite elevation. When a cycle slip occurs, the carrier phase jumps by an integer cycle while the fractional part of the phase remains unchanged. The cycle slip may be as small as a few cycles, or exceed millions of cycles. Cycle slips have to be detected because the corresponding measurements are not available for positioning until the new ambiguities are resolved. However, cycle slips can be easily detected for dual frequency by monitoring the ionospheric signal defined by (8.28). Taking the time difference of the ionospheric signal and denoting it as *SLIP*, the result without cycle slip is

$$SLIP \equiv \phi_I(t_n) - \phi_I(t_{n-1}) = -2\{I(t_n) - I(t_{n-1})\}, \tag{9.25}$$

where the multipath and measurement noise are omitted. If a cycle slip occurs in the L1 and L2 carriers as δN_1 and δN_2, the index will jump as

$$
\begin{aligned}
SLIP &= -2\{I(t_n) - I(t_{n-1})\} + \lambda_N(\delta N_1 + \delta N_2) - \lambda_W(\delta N_1 - \delta N_2) \\
&= -2\{I(t_n) - I(t_{n-1})\} - 75.5 \cdot \delta N_1 + 96.9 \cdot \delta N_2.
\end{aligned}
\tag{9.26}
$$

Assuming the measurement noise of the L1 and L2 carrier phase to be 0.1 radians, corresponding to 3 and 3.9 mm, respectively, the measurement noise of the ionospheric signal is 20 mm. Then the measurement noise in *SLIP* becomes 28 mm because the time difference is taken. Therefore, even a cycle slip of one cycle in the L1 or L2 carrier would be easily detected by using a threshold value of 8.4 cm (3σ). However, there are some special cases in which detection is very difficult, for example, 5-cycles slip in the L1 and 4-cycles in the L2 carrier, that cause only a 10 cm jump in *SLIP*. Nevertheless, since L1 and L2 cycle slips are independent and normally large numbers, we adopt this index to detect cycle slips. When a cycle slip is detected and the phase locks of more than three satellites are maintained, the corresponding ambiguity is computed using the receiver position obtained from the remaining satellites' carrier phases. And if the number of satellites maintaining the phase locks becomes less than four, the OTF is performed as the initialization. Instead of the initialization, some methods of cycle slip fixing have been proposed using a simple linear regression [2], or using Kalman filtering [13].

If an L1 single-frequency receiver is used, the cycle slip may be detected using the carrier smoothed pseudorange $\overline{PR_1}$ as follows [6]:

$$
\begin{aligned}
SLIP_{\mathrm{S}} &\equiv \{\phi_1(t_n) - \overline{PR_1}(t_n)\} - \{\phi_1(t_{n-1}) - \overline{PR_1}(t_{n-1})\} \\
&= -2\frac{f_2}{f_1}\{I(t_n) - I(t_{n-1})\},
\end{aligned}
\tag{9.27}
$$

or using the carrier Doppler $\dot{\phi}_1$, as

$$
SLIP_{\mathrm{D}} \equiv \{\phi_1(t_n) - \phi_1(t_{n-1})\} - \dot{\phi}_1(t_n) \cdot (t_n - t_{n-1}). \tag{9.28}
$$

However, detection of a few cycles slip is difficult because the accuracy of a smoothed pseudorange is normally worse than 50 cm and the Doppler changes significantly during the observation interval. If the smoothed pseudorange or carrier Doppler were sufficiently accurate, these indexes could be used for cycle slip fixing.

9.2 Case study: aircraft attitude determination

9.2.1 Algorithm

In addition to the precise position of a vehicle, the GPS interferometric technique can give attitude information if the vehicle has more than one antenna onboard. The attitude of a vehicle is defined by a rotation transformation between a coordinate system fixed in the body and a coordinate system fixed in the NED (North East Down) local frame. Here, an attitude determination algorithm for an aircraft with four GPS antenna is derived as an example. Suppose $\vec{r}_j^{\mathrm{B}} = (x_j^{\mathrm{B}}, y_j^{\mathrm{B}}, z_j^{\mathrm{B}})^T$, $j = 1 \sim 3$, are the antenna vectors of the jth antenna in the body frame, and $\vec{r}_j^{\mathrm{L}} = (N_j, E_j, D_j)^T$ are those in the local level frame.

Then the 3 × 3 matrix $\left[r^{\mathrm{B}}\right] = (\vec{r}_1^{\mathrm{B}}, \vec{r}_2^{\mathrm{B}}, \vec{r}_3^{\mathrm{B}})$ and $\left[r^{\mathrm{L}}\right] = (\vec{r}_1^{\mathrm{L}}, \vec{r}_2^{\mathrm{L}}, \vec{r}_3^{\mathrm{L}})$ satisfy the following relation:

$$\left[r^{\mathrm{B}}\right] = \left[R_{\mathrm{L}}^{\mathrm{B}}(\varphi, \theta, \psi)\right]\left[r^{\mathrm{L}}\right], \tag{9.29}$$

where $\left[R_{\mathrm{L}}^{\mathrm{B}}(\varphi, \theta, \psi)\right]$, φ, θ, ψ are a rotation matrix, roll, pitch, and yaw angle respectively, so that

$$\left[R_{\mathrm{L}}^{\mathrm{B}}(\varphi, \theta, \psi)\right] = \begin{bmatrix} \cos\theta\cos\psi & \cos\theta\sin\psi & -\sin\theta \\ \sin\varphi\sin\theta\cos\psi - \cos\varphi\sin\psi & \sin\varphi\sin\theta\sin\psi + \cos\varphi\cos\psi & \sin\varphi\cos\theta \\ \cos\varphi\sin\theta\cos\psi + \sin\varphi\sin\psi & \cos\varphi\sin\theta\sin\psi - \sin\varphi\cos\psi & \cos\varphi\cos\theta \end{bmatrix}. \tag{9.30}$$

If there is no structural flexure, the rotation matrix is calculated using the least squares method to minimize the following cost function:

$$J = \sum_{i,j} \left\{\left[r^{\mathrm{B0}}\right] - \left[R_{\mathrm{L}}^{\mathrm{B}}(\varphi, \theta, \psi)\right]\left[r^{\mathrm{L}}\right]\right\}_{i,j}^{2}. \tag{9.31}$$

For attitude determination of the aircraft, the importance of wing flexure modeling to obtain precise aircraft attitude has been reported by several authors [14],[15].

Additionally, the lateral flexure of the aircraft fuselage can be modeled, since the fuselage is likely to bend when yaw angle changes rapidly [16]. To estimate the vertical flexure of the fuselage is difficult because the direction of the vertical fuselage is the same as that of wing flexure. Supposing $\left[r^{\mathrm{B0}}\right]$ and $\left[r^{\mathrm{B}}\right]$ are antenna vector matrices when the aircraft is stationary and in motion, respectively. The structural flexure matrix $[B]$ is introduced as follows.

$$\left[r^{\mathrm{B}}\right] = \left[r^{\mathrm{B0}}\right] - [B], \tag{9.32}$$

$$[B] = \left[\vec{b}_1, \vec{b}_2, \vec{b}_3\right] = \begin{bmatrix} 0 & 0 & 0 \\ -fl & -fl & -fl \\ 0 & fw & fw \end{bmatrix}, \tag{9.33}$$

where fw is the wing flexure and fl is the lateral flexure of the fuselage as shown in Figure 9.8. The fw and fl are positive when the wing and the fuselage flex upward and rightward respectively. In this case, the cost function to be minimized is given by

$$J = \sum_{i,j} \left\{\left[r^{\mathrm{B0}}\right] - [B] - \left[R_{\mathrm{L}}^{\mathrm{B}}(\varphi, \theta, \psi)\right]\left[r^{L}\right]\right\}_{i,j}^{2}. \tag{9.34}$$

Next, the mathematical equations for attitude determination by the least squares method are given. The state vector to be estimated is defined as follows:

$$\vec{x}_a = [\varphi, \theta, \psi, fw, fl]^T. \tag{9.35}$$

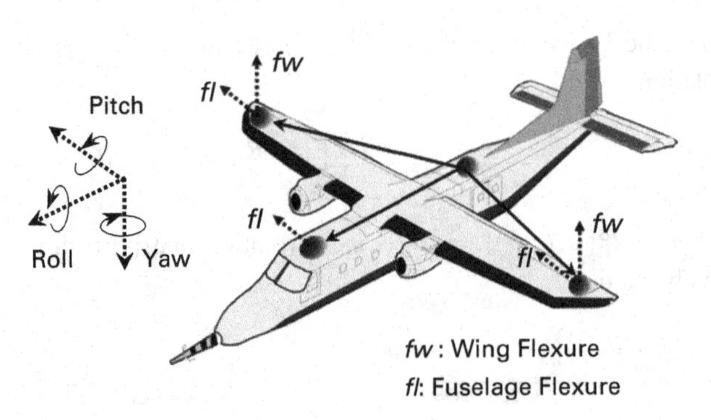

Figure 9.8 Structural flexure modeling.

Now, we introduce a fictitious measurement vector $\vec{y}_a$ (9 × 1) as follows:

$$\vec{y}_a \equiv \vec{h}(\vec{x}_a) = \begin{bmatrix} \vec{b}_1 \\ \vec{b}_2 \\ \vec{b}_3 \end{bmatrix} - \begin{bmatrix} [R_{\mathrm{L}}^{\mathrm{B}}]\vec{r}_1^{\,\mathrm{L}} \\ [R_{\mathrm{L}}^{\mathrm{B}}]\vec{r}_2^{\,\mathrm{L}} \\ [R_{\mathrm{L}}^{\mathrm{B}}]\vec{r}_3^{\,\mathrm{L}} \end{bmatrix}. \tag{9.36}$$

Denoting a priori the state vector by $\bar{\vec{x}}_a$, and computed measurement vector by $\bar{\vec{y}}_a$, the residual vector is given as

$$\delta\vec{y}_a \equiv \begin{pmatrix} \vec{r}_1^{\,\mathrm{B0}} \\ \vec{r}_2^{\,\mathrm{B0}} \\ \vec{r}_3^{\,\mathrm{B0}} \end{pmatrix} - \bar{\vec{y}}_a = \left.\frac{\partial \vec{h}_a(\vec{x}_a)}{\partial \vec{x}_a}\right|_{\vec{x}_a = \bar{\vec{x}}_a} (\vec{x}_a - \bar{\vec{x}}_a) = \left[H_a(\bar{\vec{x}}_a)\right]\delta\vec{x}_a. \tag{9.37}$$

Therefore, the least squares estimate which minimizes the cost function in (9.34) is given as

$$\delta\hat{\vec{x}}_a = \left([H_a]^T[H_a]\right)^{-1}[H_a]^T\delta\vec{y}_a. \tag{9.38}$$

The measurement matrix H_a (9 × 5) is written as

$$[H_a] = \begin{bmatrix}
\frac{\partial R_{1i}}{\partial\varphi} r_{i1} & \frac{\partial R_{1i}}{\partial\theta} r_{i1} & \frac{\partial R_{1i}}{\partial\psi} r_{i1} & 0 & 0 \\
\frac{\partial R_{2i}}{\partial\varphi} r_{i1} & \frac{\partial R_{2i}}{\partial\theta} r_{i1} & \frac{\partial R_{3i}}{\partial\psi} r_{i1} & 0 & -1 \\
\frac{\partial R_{3i}}{\partial\varphi} r_{i1} & \frac{\partial R_{3i}}{\partial\theta} r_{i1} & \frac{\partial R_{3i}}{\partial\psi} r_{i1} & 0 & 0 \\
\frac{\partial R_{1i}}{\partial\varphi} r_{i2} & \frac{\partial R_{1i}}{\partial\theta} r_{i2} & \frac{\partial R_{1i}}{\partial\psi} r_{i2} & 0 & 0 \\
\frac{\partial R_{2i}}{\partial\varphi} r_{i2} & \frac{\partial R_{2i}}{\partial\theta} r_{i2} & \frac{\partial R_{3i}}{\partial\psi} r_{i2} & 0 & -1 \\
\frac{\partial R_{3i}}{\partial\varphi} r_{i2} & \frac{\partial R_{3i}}{\partial\theta} r_{i2} & \frac{\partial R_{3i}}{\partial\psi} r_{i2} & 1 & 0 \\
\frac{\partial R_{1i}}{\partial\varphi} r_{i3} & \frac{\partial R_{1i}}{\partial\theta} r_{i3} & \frac{\partial R_{1i}}{\partial\psi} r_{i3} & 0 & 0 \\
\frac{\partial R_{2i}}{\partial\varphi} r_{i3} & \frac{\partial R_{2i}}{\partial\theta} r_{i3} & \frac{\partial R_{2i}}{\partial\psi} r_{i3} & 0 & -1 \\
\frac{\partial R_{3i}}{\partial\varphi} r_{i3} & \frac{\partial R_{3i}}{\partial\theta} r_{i3} & \frac{\partial R_{3i}}{\partial\psi} r_{i3} & 1 & 0
\end{bmatrix}. \tag{9.39}$$

where the R_{ij}, r_{ij} denote (i,j) elements of matrix $\left[R_{\mathrm{L}}^{\mathrm{B}}\right]$ and $\left[r^{\mathrm{L}}\right]$. The partial derivatives of R_{ij} with respect to attitude angles are given in the following equations:

$$
\begin{aligned}
\frac{\partial R_{11}}{\partial \varphi} &= 0, \\
\frac{\partial R_{11}}{\partial \theta} &= -\sin\theta\cos\psi, \\
\frac{\partial R_{11}}{\partial \psi} &= -\cos\theta\sin\psi,
\end{aligned}
$$

$$
\begin{aligned}
\frac{\partial R_{12}}{\partial \varphi} &= 0, \\
\frac{\partial R_{12}}{\partial \theta} &= -\sin\theta\sin\psi, \\
\frac{\partial R_{12}}{\partial \psi} &= \cos\theta\cos\psi,
\end{aligned}
$$

$$
\begin{aligned}
\frac{\partial R_{13}}{\partial \varphi} &= 0, \\
\frac{\partial R_{13}}{\partial \theta} &= -\cos\theta, \\
\frac{\partial R_{13}}{\partial \psi} &= 0,
\end{aligned}
$$

$$
\begin{aligned}
\frac{\partial R_{21}}{\partial \varphi} &= \cos\varphi\sin\theta\cos\psi + \sin\varphi\sin\psi, \\
\frac{\partial R_{21}}{\partial \theta} &= \sin\varphi\cos\theta\cos\psi, \\
\frac{\partial R_{21}}{\partial \psi} &= -\sin\varphi\sin\theta\sin\psi - \cos\varphi\cos\psi,
\end{aligned}
$$

$$
\begin{aligned}
\frac{\partial R_{22}}{\partial \varphi} &= \cos\varphi\sin\theta\sin\psi - \sin\varphi\cos\psi, \\
\frac{\partial R_{22}}{\partial \theta} &= \sin\varphi\cos\theta\sin\psi, \\
\frac{\partial R_{22}}{\partial \psi} &= \sin\varphi\sin\theta\cos\psi - \cos\varphi\sin\psi,
\end{aligned}
$$

$$
\begin{aligned}
\frac{\partial R_{23}}{\partial \varphi} &= \cos\varphi\cos\theta, \\
\frac{\partial R_{23}}{\partial \theta} &= -\sin\varphi\sin\theta, \\
\frac{\partial R_{23}}{\partial \psi} &= 0,
\end{aligned}
$$

$$\frac{\partial R_{31}}{\partial \varphi} = -\sin\varphi\sin\theta\cos\psi + \cos\varphi\sin\psi,$$
$$\frac{\partial R_{31}}{\partial \theta} = \cos\varphi\cos\theta\cos\psi,$$
$$\frac{\partial R_{31}}{\partial \psi} = -\cos\varphi\sin\theta\sin\psi + \sin\varphi\cos\psi,$$

$$\frac{\partial R_{32}}{\partial \varphi} = -\sin\varphi\sin\theta\sin\psi - \cos\varphi\cos\psi,$$
$$\frac{\partial R_{32}}{\partial \theta} = \cos\varphi\cos\theta i\sin\psi,$$
$$\frac{\partial R_{32}}{\partial \psi} = \cos\varphi\sin\theta\cos\psi + \sin\varphi\sin\psi,$$

$$\frac{\partial R_{33}}{\partial \varphi} = -\sin\varphi\cos\theta,$$
$$\frac{\partial R_{33}}{\partial \theta} = -\cos\varphi\sin\theta, \tag{9.40}$$
$$\frac{\partial R_{33}}{\partial \psi} = 0.$$

9.2.2 Flight test

In this section we consider an example of attitude determination using real flight data. Four L1 GPS antennas were mounted on an aircraft (Dornier Do-228). The antenna separations are 6.4, 8.4, and 8.2 m for the front, right, and left antennas from the aft antenna. The two on the aircraft fuselage were connected to Trimble 4000SSE receivers, while the two on the wing tips were connected to NovAtel GPSCards. GPS L1 carrier phase measurements at 2 Hz were recorded and analyzed after the flight. In this example, three antenna vectors relative to one reference antenna are determined independently by kinematic GPS positioning. Although the attitude can be computed by using a single difference between antennas, both receivers have to be precisely synchronized. A commercial attitude determination system may have one receiver in which its channels are assigned for multiple antennas separately. The time synchronization is assured for such a system; however, the antenna line biases should be precisely calibrated.

The least squares search method was adopted to resolve the ambiguity of the L1 carrier phase. In a kinematic positioning of a vehicle relative to a ground reference receiver, it is not easy to resolve carrier phase ambiguities if L1 single-frequency receivers are used. It requires from a few minutes to some tens of minutes to resolve ambiguities depending on the baseline length; since the GPS satellites' positions relative to the baseline have to change significantly. On the other hand, directions of baseline vectors in the attitude determination system change dramatically relative to the satellites due to changes of attitude. In other words, the observability of the GPS carrier phase is very high for the attitude. Besides that, the known antenna separations are available as a constraint to reduce the number of ambiguity candidates. Thus, L1 ambiguities can be resolved quite easily in a few seconds.

During the flight, more than six satellites were normally observed, and the satellites with an elevation higher than 10 degrees were used for processing. The GPS-based attitude is continuously determined, as shown in Figure 9.9, with the roll angle ranging from −40 to +35 degrees and the pitch ranging from -5 to 16 degrees. The estimated wing flexure as well as the aircraft height during landing are shown in Figure 9.10 where the height of runway is about 110 m (WGS84). It is clear that the wing flexure is nominally 12 cm in flight, and becomes near zero after landing.

GPS attitude was compared with the reference attitude that was provided by a strapdown ring laser gyro INS (Litton LTN-92), calibrated for misalignment between the INS and the body frame. The accuracy of the INS attitude is 0.05 degrees in pitch and roll, and 0.4 degrees in the yaw axis. The RMS values of roll, pitch, and yaw differences were 0.052, 0.060, and 0.075 degrees, respectively. The accuracy of attitude angle is approximately expressed by the value of the positioning accuracy divided by the baseline length. For example, when we adopt 0.5 cm as a nominal positioning accuracy and 6.41 m as a baseline length, the accuracy of angle will be 0.04 degrees. This value is slightly smaller than the computed RMS shown beforehand. However, considering that the calibrated INS attitude is described by the rotation angles from the local level to the nominal body frame and ignores the instantaneous structural flexure, this discrepancy can be thought to be acceptable.

In order to demonstrate the effect of modeling the lateral/wing flexure of the fuselage, aircraft maneuvers such as yaw reversals and pitch reversals were conducted. Figure 9.11

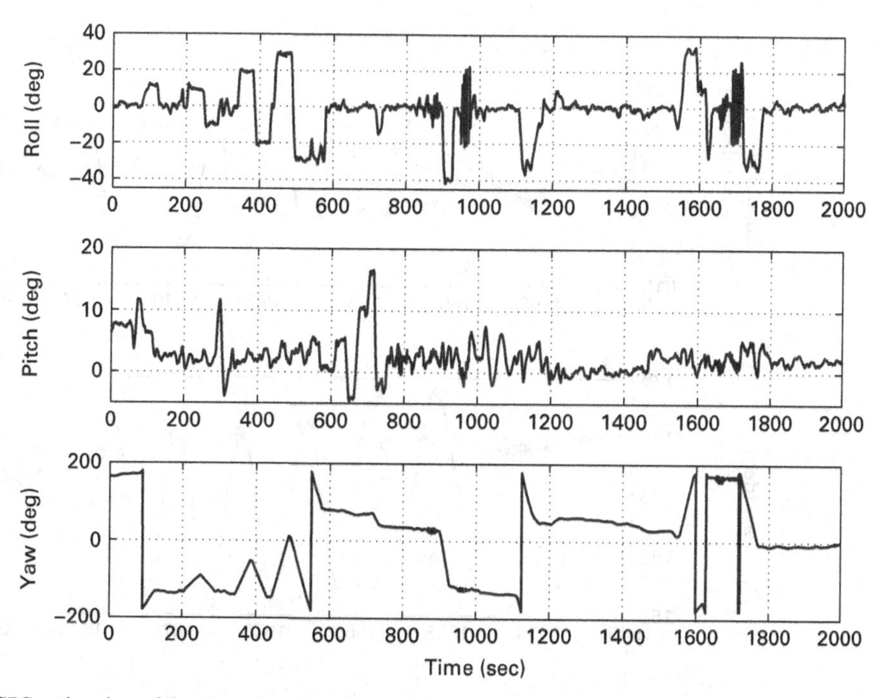

Figure 9.9 GPS estimation of the aircraft attitude.

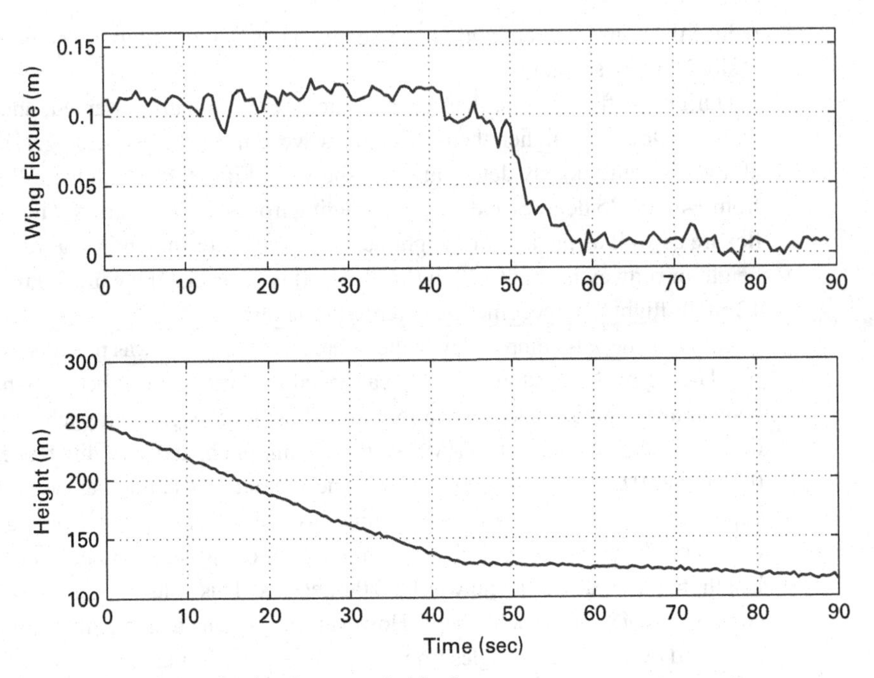

Figure 9.10 GPS estimation of aircraft wing flexure and altitude.

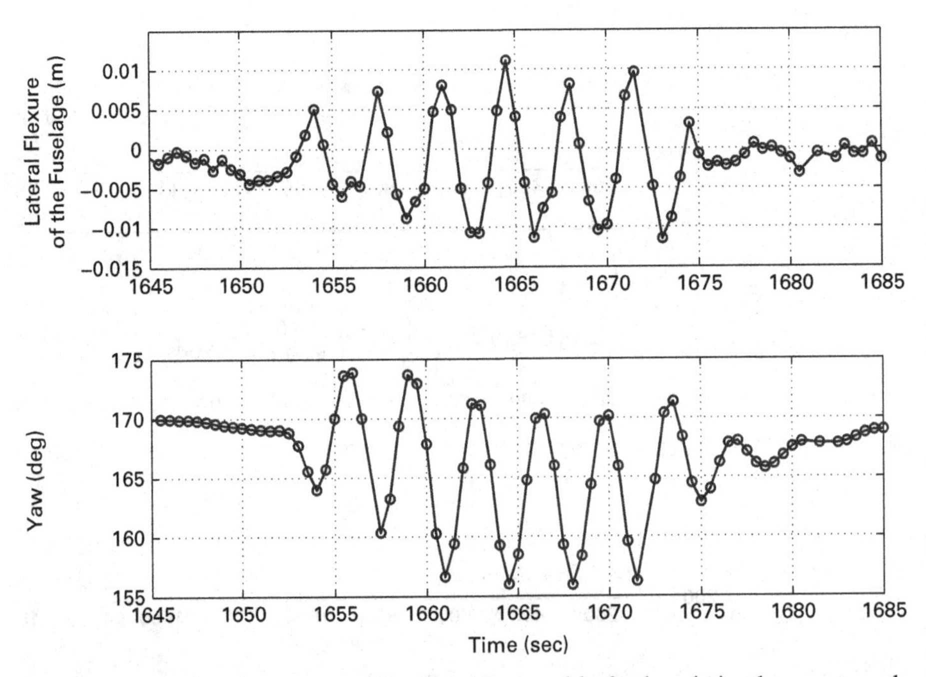

Figure 9.11 Change of yaw angle and the corresponding lateral flexure of the fuselage during the yaw reversals.

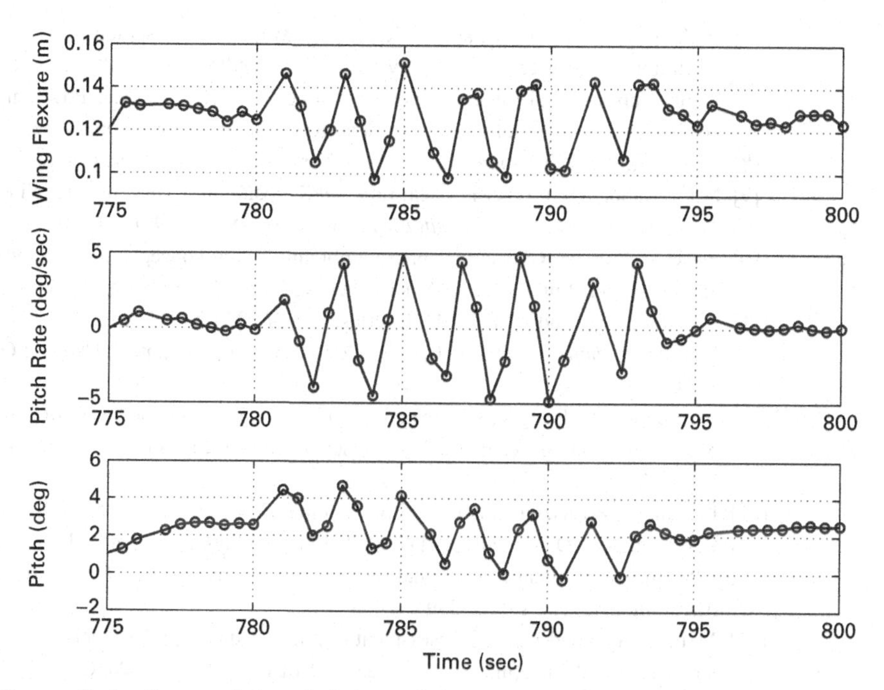

Figure 9.12 Change of wing flexure, pitch, and pitch rate during pitch reversals.

shows a change of yaw angle and the corresponding lateral flexure of the fuselage during the yaw reversals. The estimated lateral flexure of the fuselage reached up to ±1.1 cm due to the rapid maneuver of the aircraft, and it is clear that the flexure strongly correlates with the change of yaw angle in the period of yaw maneuver. A similar correlation between wing flexure and pitch is shown in Figure 9.12 during the pitch reversals.

References

[1] B. W. Remondi, Global positioning system: Description and use, *Bulletin Geodesique*, **59**, 1985, 361–377.

[2] G. L. Mader, Dynamic positioning using GPS carrier phase measurements, *Manuscripta Geodetica*, **11**, 1986, 272–277.

[3] R. Hatch, Instantaneous ambiguity resolution, Proceedings of IAG International Symposium No.107 on Kinematic Systems in Geodesy, Surveying and Remote Sensing. New York: Springer Verlag, 1991, pp 299–308.

[4] B. W. Remondi, Pseudo-kinematic GPS results using the ambiguity function method, navigation, *Journal of the Institute of Navigation*, **38**, (1), 1991, 17–36.

[5] P. J. G. Teunissen, A new method for fast carrier phase ambiguity estimation, Proceedings IEEE Position, Location and Navigation Symposium PLANS'94, Las Vegas, NV, 11–15 April, 1994, pp. 562–573.

[6] G Lachapelle, M. E. Cannon, and G. Lu, High-rrecision GPS navigation with emphasis on carrier-phase ambiguity resolution, *Marine Geodesy*, **15**, 1992, 253–269.

[7] H. Z. Abidin, D. E. Wells, and A. Kleusberg, Multi-monitor station 'On the fly' ambiguity resolution: Theory and preliminary results, Proceedings of DGPS'91, First International Symposium on Real Time Differential Applications of the Global Positioning System, 1, 16–20 Sept. 1991, Braunschweig, Federal Republic of Germany, pp 44–56.
[8] G. Seeber, *Satellite Geodesy*, Berlin/New York, Walter de Gruyter, 1993.
[9] N. C. Talbot, High-precision real-time GPS positioning concepts:Modeling and results, *Navigation, Journal of the Institute of Navigation*, **38**, (2), 1991, 147–161.
[10] P. J. G. Teunissen, GPS carrier phase ambiguity fixing concepts, in *GPS for Geodesy*, 2nd edition, P. J. G. Teunissen and A. Kleusberg (editors), Berlin, Springer-Verlag, 1998.
[11] P. J. de Jonge and C. C. J. M. Tiberius, *The LAMBDA Method for Integer Ambiguity Estimation: Implementation Aspects*, Delft Geodetic Computing Centre LGR series, No. 12, 1996.
[12] T. Tsujii, M. Harigae, T. Inagaki, and T. Kanai, Flight tests of GPS/GLONASS precise positioning versus dual frequency KGPS profile, *Earth Planets Space*, **52**, (10), 2000, 825–829.
[13] H. Landau, Precise kinematic GPS positioning, *Bull. Géodésique*, **63**, 1989, 85–96.
[14] C. E. Cohen, B. D. McNally, and B. W. Parkinson, Flight tests of attitude determination using GPS compared against an inertial navigation unit, navigation, *Journal of the Institute of Navigation*, **41**, (1), 1993, 83–97.
[15] M. E. Cannon and H. Sun, Assessment of a non-dedicated GPS receiver system for precise airborne attitude determination, Proceedings of ION-94, Salt Lake City, 20–23 Sept., 1994, pp 645–654.
[16] T. Tsujii, M. Murata, and M. Harigae, Airborne kinematic attitude determination using GPS phase interferometry, *Advances in the Astronautical Sciences*, **95**, 1997, 827–838.

10 Electromagnetic scintillation of GNSS signal

"Long Telescopes may cause Objects to appear brighter and larger than short ones can do, but they cannot be so formed as to take away that confusion of the Rays which arises from the Tremors of the Atmosphere. The only remedy is a most serene and quiet Air, such as may perhaps be found on the tops of the highest mountains above the grosser Clouds."

Sir Isaac Newton, Opticks (1730)

An effect of electromagnetic scintillation has been noted in the visible frequency range with the introduction of telescopes. It is a random rapid fluctuation of signal instant amplitude and phase. Newton identified this phenomenon with the atmosphere and recommended locating telescopes on the tops of the highest mountains. Further advance in scintillation research was achieved when astronomy moved to radio frequencies. Scintillation effects had been discovered by monitoring signals from other galaxies, in particular from Cassiopeia [1]. The source of scintillation had been established by making measurements with a set of receivers located at a distance from each other and operating on various frequencies. Correlations between scintillation effects in the receivers at different locations established that the source of scintillation is in the Earth's atmosphere. The dependence on frequency showed that the medium in which scintillation occurs is dispersive.

A GNSS signal in particular is subject to amplitude and phase scintillation caused by its propagation through the atmosphere, though the scintillation effects on GNSS from signal propagation in the troposphere are much smaller than those caused by signal propagation in the ionosphere. The effect of signal scintillation is important for navigation and geodetic applications, because it affects receiver performance to a point where it may lose a lock and stop tracking a signal. Amplitude scintillation results in signal quality degradation. Phase scintillation affects the carrier tracking loop in such a way that it requires a wider bandwidth due to higher dynamics of the carrier. In Chapter 12 we consider such effects and how they can be mitigated for some applications.

On the other hand, scintillation effects can serve as a valuable source of geophysical information. They can provide us with information about atmosphere internal structure and its dynamics. Scintillation effects may also be related to some as yet unexplored mechanisms in the atmosphere. In particular, by closely monitoring the ionosphere's fine structure and dynamics we may be able to enhance weather forecasting, predict climate changes, get early warnings about hurricanes and typhoons, and even predict earthquakes. We look at some of such applications in the next chapter. In Chapter 4 we looked at the main errors in GNSS signal propagation caused by the Earth's atmosphere. In this chapter we look at more subtle effects caused by the atmosphere. As before, we also

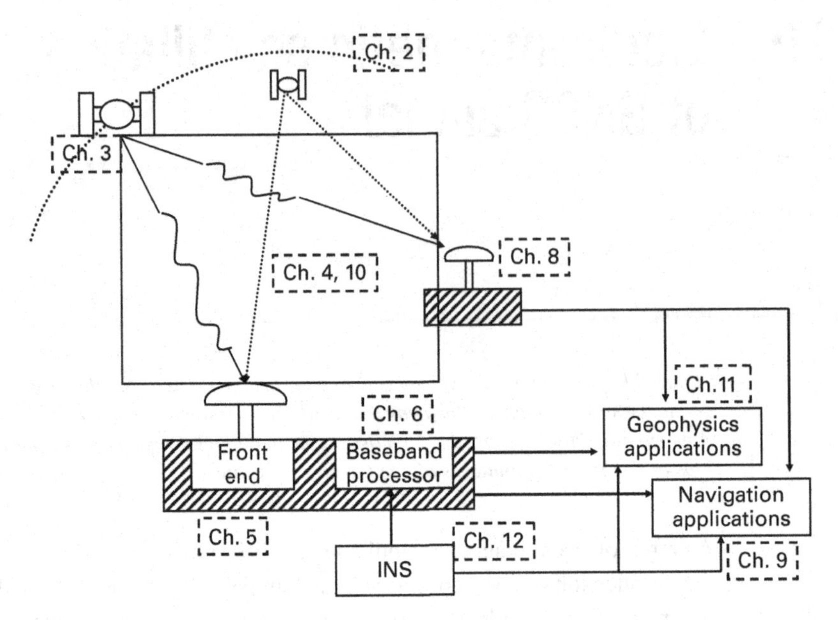

Figure 10.1 Subject of Chapter 10.

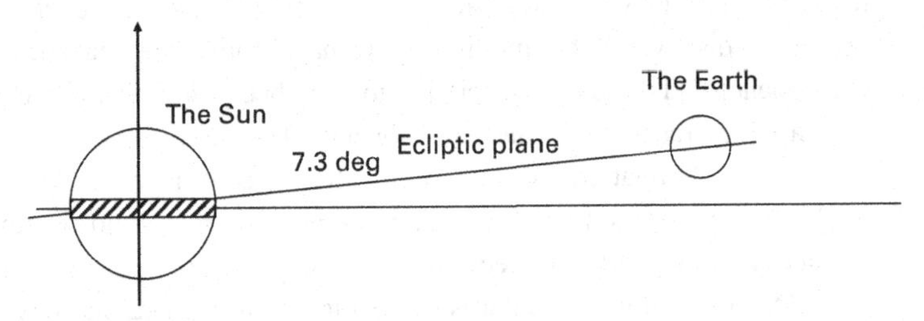

Figure 10.2 The Earth in relation to the Sun's equatorial belt (after [2]).

consider how to model and simulate such effects. Figure 10.1 shows the topic of this chapter in relation to other topics in the book.

10.1 Ionospheric irregularities

Solar radiation consists of ultraviolet and soft X-ray radiation. Sun-emitted radiation emerges in the direction normal to the Sun's surface. The ecliptic plane is titled by about 7.3° in relation to the Sun's equatorial plane. So as far as radiation is concerned, the Earth is affected only by the Sun's equatorial belt of 14.6° (see Figure 10.2) [2]. The ionosphere is created by this solar radiation, and also to a much lesser degree by space radiation. In fact the ionosphere shields everything on Earth from this radiation, which, when it encounters the atmosphere, ionizes the molecules within it. The level of ionization can be seen from the electron density profile (Figure 10.3).

In order to understand why the electron density profile exhibits such behavior, we need to consider the mechanism of its creation. The Earth's magnetic field interacts with

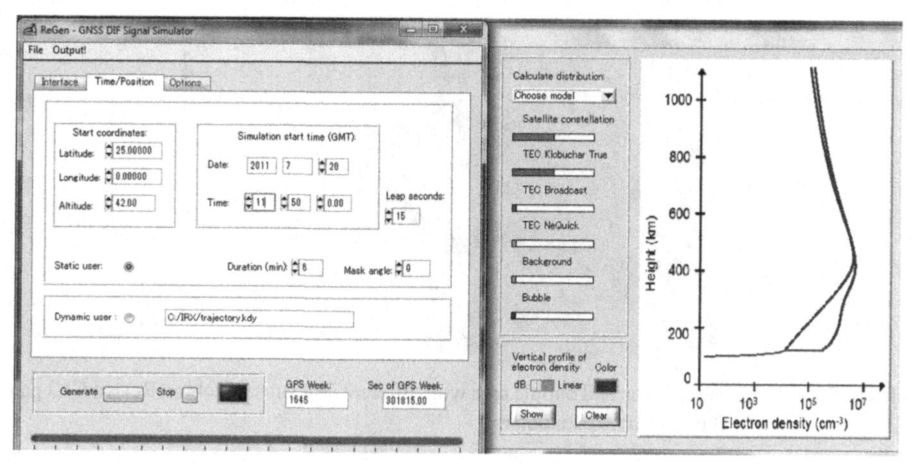

Figure 10.3 Electron density profiles calculated for two conditions by ReGen simulator.

ionized particles, the particles become aligned with the magnetic field, flow down the field towards the poles, recombine in the atmosphere, and create the aurora. Ionospheric plasmas on closed magnetic fields are co-rotating with the Earth (known as Ferraro's theorem [3]). The upper part of the plot in Figure 10.3 corresponds to the low density, high altitude atmosphere. The number of ionized electrons reaches its maximum at an altitude between 300 and 400 km (F layer), where the number of molecules available for ionization becomes large. From this point down to the Earth the number of ionized electrons rapidly decreases, because the radiation is losing its power [4].

It was discovered in the 1950s by analyzing radio signals from other galaxies that the source of scintillations is in the Earth's atmosphere. By comparison of the results at various frequencies, it has been established that the medium is dispersive and therefore ionized (see Chapter 4 for details of signal propagation in dispersive and non-dispersive media).

Further research on these extragalactic sources and later satellite data have established that scintillation is caused by spatial irregularities. GPS has allowed collection of a lot of information related to ionospheric irregularities. We are mostly interested here in the large irregularities which are responsible for scintillation. They can be classified in two major groups with enhanced and depleted ionization [1],[4],[5]. Irregularities with enhanced ionization are aligned with the Earth's magnetic field and highly elongated, with axial ratios of at least 20 or more. Irregularities with depleted ionization or ***plasma bubbles*** are also highly elongated along a South to North direction (Figure 10.4). Other areas of depleted ionization in the middle latitudes (about 60°) are called the ionospheric main trough. The spatial irregularities appear mostly in equatorial and auroral regions, but the scintillation which results from these irregularities has a global impact because, basically, for a receiver located at any place the line-of-sight (LOS) to some satellites may go through these structures.

Scintillation has a negative impact on GNSS measurements, but on the other hand it can be used to derive information about the ionosphere. We explore this subject in the next chapter. From that perspective it is interesting to look at the sources of these irregularities. They can be arranged into two categories: global large-scale irregularities

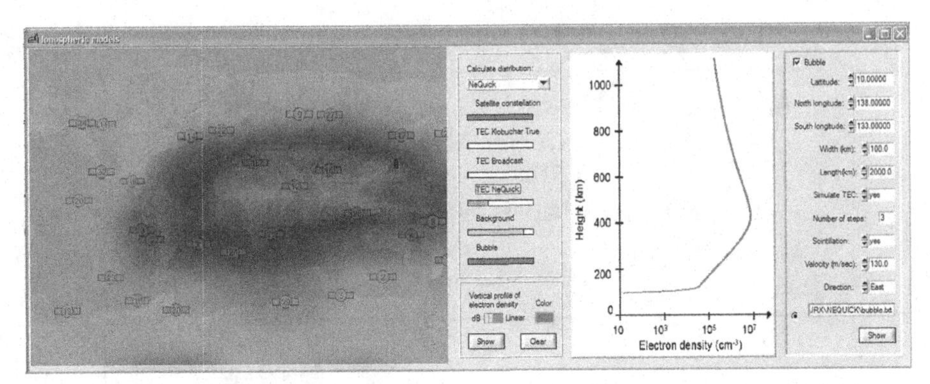

Figure 10.4 GPS constellation and bubble area with different ionization levels on ReGen GUI panel. Rightmost panel shows bubble parameters.

and local small-scale ones. Large-scale irregularities originate through the influence of the Sun and the geomagnetic field. Local-scale irregularities may appear as a result of volcanoes, earthquakes, and other disturbances [4]. The main characteristics of ionospheric scintillation can be summarized as follows [6]:

1. Number of scintillation events depends on the sun-spot number.
   1.1 There is no strong correlation with daily sun-spot number. The peak of daily scintillation is about one hour after sunset at the ionospheric height.
   1.2 There is a strong correlation with annual sun-spot number. Annual scintillation varies with the 11-year sun-spot cycle. The peak of annual scintillation is around the equinox.
2. The location of strong gigahertz scintillations is within ±30° of the magnetic equator.
3. Amplitude of scintillation shows a strong correlation with monthly sun-spot number.
4. A period of scintillation is generally less than 15 seconds.
5. The power spectra roll off as f^{-3}.

The main observations using satellites were made in the 1960s starting from extensive observation of scintillations in West Africa from satellites made by G. S. Kent in 1960. At that time all the main characteristics of ionospheric scintillations were defined. Investigators even determined the size of patches as 100 km across. They found the sun-spot cycle effect, with a maximum at the equinox [7]. At that time it was also established that patches can exist over night. The irregularities were attributed to fluctuations in TEC. The elipsoidal symmetry of the small patches (10 × 1 km) was also established. There is a seasonal and a sun-spot dependence. The maximum scintillation occurs near local midnight.

The scintillations are caused by TEC irregularities, mainly in the F-layer. During the solar maximum period, the scintillations are greater because of the increase in the background TEC. The irregularities' amplitude is not changed much because of the solar activity, but the background TEC can become at least 10 times higher.

In a quiet year with monthly sun-spot number less than 30, scintillation may not appear at all. On the other hand, scintillation during years with the highest sun activity may affect a GNSS signal severely. Signal fading can be up to 7dB for 10 % of the time [8], and up to 25 dB as isolated spikes in low- and mid-latitudes on an L1 frequency [9]. During the solar maximum worst case scenario, fading at the L-band in low latitude (±15°) regions in the plume areas can easily reach 20 dB [10].

The scintillation is caused by wave scattering. The mechanism depends on the size of the irregularities which cause the scattering in comparison to the Fresnel zone. ***Fresnel zone*** length is defined as the fundamental length scale when diffraction effects are important as follows:

$$l_F = \sqrt{\lambda H}, \tag{10.1}$$

where λ is wavelength, H is distance from a receiver to the Fresnel zone. In terms of scintillation, Fresnel length defines the scale at which a irregularities produce amplitude scintillation when a receiver is at distances farther than H. The structures that are larger than that contribute directly to phase, as their contribution to amplitude is suppressed by Fresnel filtering.

The physics behind equatorial scintillation can be explained as follows. Plume-like structures or funnels contain rising bubbles with low electron density. The size of the bubble exceeds the Fresnel zone as observed from the ground station. Therefore the signal is affected by refraction. Refractive effects may result in high S_4, up to 1 or higher [11]. Diffraction comes in to play when the size of irregularities is equal to or smaller than the Fresnel zone. Bubbles when they are moving generate small irregularities, with size down to centimeters. These irregularities generate scintillation by diffraction.

Bubble-type irregularities often have a double effect on a radio signal, through depletion of the ionization level, which affects code delay and phase advance, and through scintillation. The first effect is modeled in ReGen by ray tracing as shown in

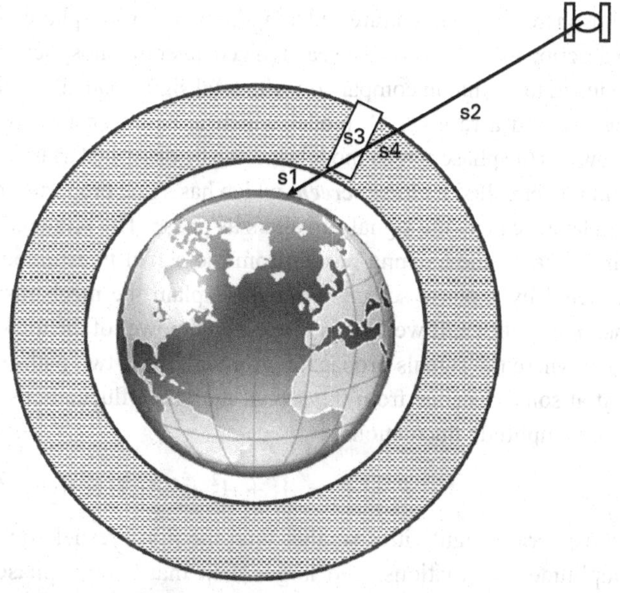

Figure 10.5 Slant TEC calculation for ray passing through a bubble.

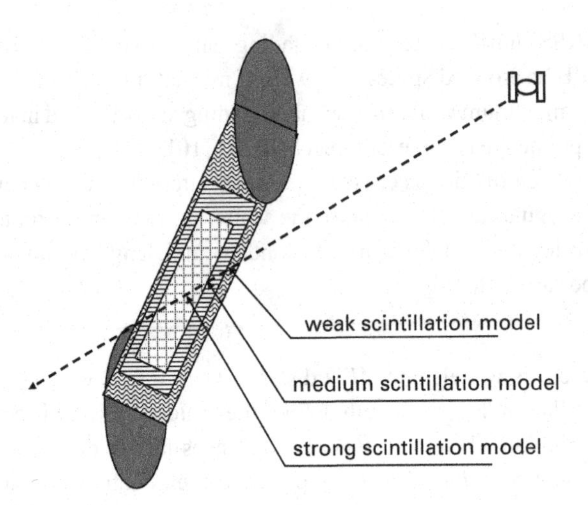

Figure 10.6 Stepwise scintillation model.

Figure 10.5. The bubble affect on ionospheric delay is then calculated as a difference: TEC (Ray "s1-s2") – TEC (Ray "s3-s4").

Bubble scintillation is modeled in ReGen using a stepwise model (Figure 10.6) and scintillation signals, either from observations or modeled using models as described below.

10.2 Rytov approximation for weak scattering

Scintillation is caused mostly by fluctuations of the refractive index. The fluctuations of the refractive index in their turn result from fluctuations of magnetic field and TEC in the ionosphere, and temperature and humidity in the troposphere. In order to provide a model to describe scintillation effects, an ionized layer of ionosphere affected by irregularities is assumed to be thin in comparison with total signal path along the line-of-sight between a satellite and a receiver. A model which is based on this assumption is consequently known as the phase screen model, and can be described as follows. A signal goes through a thin layer called a ***phase screen***, which has some properties disturbing the signal. This thin layer scatters the signal and alters its phase. The scattered signal propagates further. The signal at the antenna results from the signal transmitted by a GNSS satellite and scattered by a phase screen. Here we explain the mechanism of ***Fresnel filtering*** as mentioned above. If we look at the received power of the total signal (or intensity) at the point where the signals propagating through these two paths come together, we can see that at some distance from the screen the phase fluctuations along different paths will cause amplitude fluctuations:

$$P = A^2 = A_1^2 + A_2^2 + 2A_1A_2\cos(\varphi_1 - \varphi_2). \tag{10.2}$$

Ionospheric irregularities smaller than the first Fresnel zone produce both phase and amplitude scintillations, and larger than that – only phase scintillations. This is a basic model for phase and amplitude scintillations. We need to integrate this over

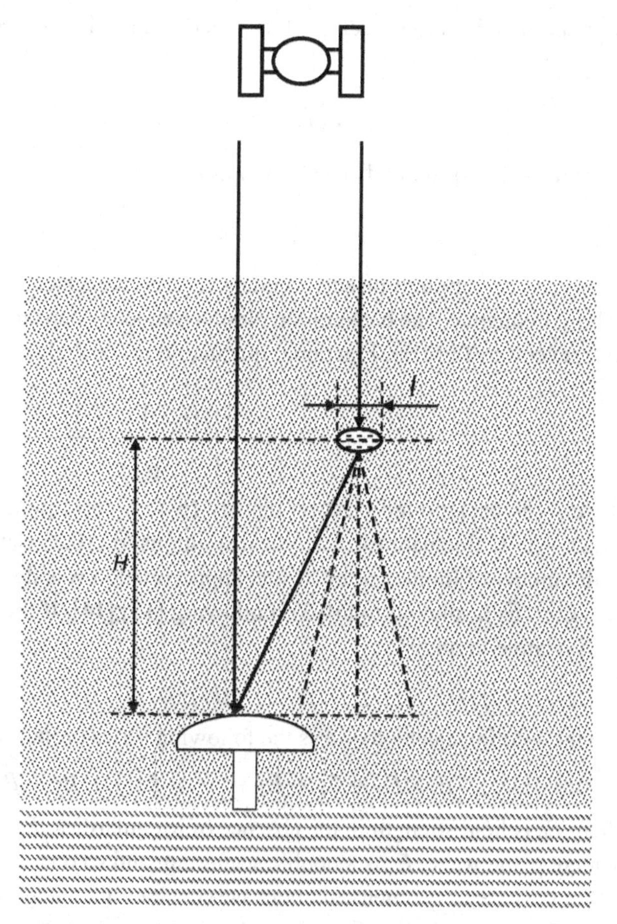

Figure 10.7 Receiver and ionospheric screen layer (after [1]).

all possible paths, and we can estimate the amplitude and phase of the received signal at that point.

We assume that ionospheric irregularities which cause scintillation are located at an altitude about 350 km above the receiver antenna (see Figure 10.7). We need to derive a solution for Maxwell's equations as the electromagnetic field parameters at the location of the receiver antenna.

As we saw in Chapter 4, a solution to Maxwell's equations can be shown in the form of a general ***scalar wave equation*** (4.10):

$$\nabla^2 E(r) + k^2 \varepsilon(r,t) E(r) = 0, \tag{10.3}$$

where $k = 2\pi\sqrt{\varepsilon\mu}f = 2\pi/\lambda$ is the electromagnetic wavenumber. By considering a scalar form of wave equation we neglect all issues related to polarization. We need to note that the changes of polarization caused by ionospheric irregularities can also manifest themselves as amplitude scintillation.

Let us consider as a model a slab of homogeneous ionosphere with TEC irregularities [12],

$$\varepsilon\left(\vec{r},t\right) = \langle\varepsilon\rangle\left[1+\delta\varepsilon\left(\vec{r},t\right)\right], \tag{10.4}$$

where $\langle\varepsilon\rangle$ is background dielectric permittivity,

$$\langle\varepsilon\rangle = \varepsilon_0\left(1-f_p^2/f^2\right), \tag{10.5}$$

where f_p is background plasma frequency, ε_0 is free space dielectric permittivity, f is a probing frequency. The background plasma frequency depends on TEC. Usually it varies from several to 18 MHz. Fractional fluctuation in the dielectric permittivity is

$$\delta\varepsilon\left(\vec{r},t\right) = -\frac{(f_p/f)^2}{1-(f_p/f)^2}\frac{N\left(\vec{r},t\right)}{N_0}, \tag{10.6}$$

where N_0 is background electron density (with a corresponding plasma frequency f_p), $N(r, t)$ is TEC fluctuations. Analytical solutions to (10.3) can be found only for a very small number of limited cases. A first approximation was developed by Nobel Laureate Max Born as a solution to (10.3). Following [13],[14], if the dielectric fluctuations are small relative to unity:

$$\sigma(\delta\varepsilon) \ll 1, \tag{10.7}$$

then the solution can be found as the following series proportional to the power of $\sigma(\delta\varepsilon)$:

$$E = E_0 + E_1 + E_2 + \ldots = E_0(1 + B_1 + B_2 + \ldots), \tag{10.8}$$

where E_0 is field strength in the absence of irregularities, E_1 is a single scattering of the incident plane wave, E_2 is a double scattering, and so on.

If in the case of geometrical optics we consider line integrals along a LOS, then in the case of the Born approximation volume integrations are considered. For the Born approximation the following condition is applied:

$$\langle\varphi^2\rangle + \langle\chi^2\rangle < 1, \tag{10.9}$$

where $\langle\varphi^2\rangle$ is phase variance and $\langle\chi^2\rangle$ is log amplitude variance. The other solution of Equation (10.3) is given by a Rytov approximation [13],[15]. The Rytov approximation is valid for broader conditions in comparison to (10.9),

$$\langle\chi^2\rangle < 1. \tag{10.10}$$

The Rytov approximation to (10.3) is given by the following series:

$$E = E_0 e^{(\psi_1+\psi_2+\psi_3\ldots)}, \tag{10.11}$$

and

$$E_0 = e^{\psi_0}. \tag{10.12}$$

The Rytov approach transforms Equation (10.3) to the following Riccati equation [15]:

$$\nabla^2\psi + (\nabla\psi)^2 + 2\nabla\psi_0\nabla\psi + k^2\delta\varepsilon = 0, \tag{10.13}$$

where

$$\psi = \psi_1 + \psi_2 + \psi_3 + \psi_4 + \ldots \tag{10.14}$$

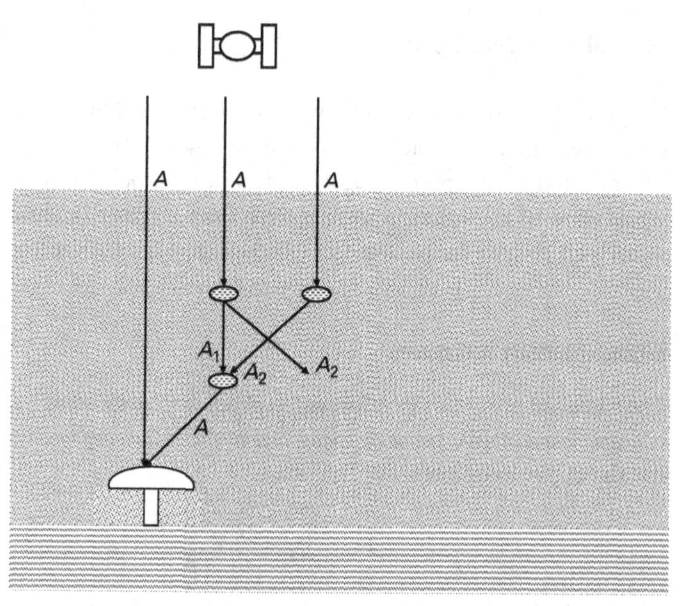

Figure 10.8 Double scattering model (after [1]).

The Riccati equation also plays a very important role in many areas from differential geometry to Bessel functions. We encounter the Riccati equation in the Kalman filter, where it describes covariance matrix propagation.

An unperturbed field is normally described by the ***Helmholtz equation***, which can be solved using a ***Green function***. The scattering terms of the Rytov approximation can then be expressed from (10.13) through unperturbed solution (10.12). The first two terms for double-scattering (Figure 10.8) are expressed as follows (see [15] for derivation):

$$\psi_1\left(\vec{R}\right) = -k^2 \int d^3r\, G\left(\vec{R},\vec{r}\right) \delta\varepsilon\left(\vec{r},t\right) \frac{E_0\left(\vec{r}\right)}{E_0\left(\vec{R}\right)} \tag{10.15}$$

and

$$\psi_2\left(\vec{R}\right) = -\int d^3r\, G\left(\vec{R},\vec{r}\right) \left[\nabla\psi_1\left(\vec{r}\right)\right]^2 \frac{E_0\left(\vec{r}\right)}{E_0\left(\vec{R}\right)}. \tag{10.16}$$

Thus the electric field for the double scattering model can be described as follows:

$$E\left(\vec{R}\right) = E_0\left(\vec{R}\right) \mathrm{e}^{\left(-\int d^3r\, G\left(\vec{R},\vec{r}\right) \frac{E_0\left(\vec{r}\right)}{E_0\left(\vec{R}\right)} \left[k^2\delta\varepsilon\left(\vec{r}\right) + \left(\nabla\psi_1\left(\vec{r}\right)\right)^2\right]\right)}. \tag{10.17}$$

Analytical solutions can be derived only for special cases.

A connection between members of Rytov and Born approximations can be easily established (see [13],[15]). Rytov and Born approximations describe a case of a plane incident wave. A complex phase method (CPM) provides an extension of Rytov's approximation for a more general case of a spherical incident wave [16].

10.3 Scintillation monitoring

Born and Rytov approximations provide a model for describing the electromagnetic field at the point where a receiver antenna is located for cases when a GNSS signal is disturbed when passing through ionospheric irregularities. Now we would like to have an instrument which would allow us to measure quantative parameters of signal disturbance. If we record a GNSS signal itself, then we can recover a scintillation signal and analyze it by autocorrelation analysis.

10.3.1 Signal intensity measures

The signal intensity carries information about amplitude scintillation (see (10.2)). We can estimate a quantative value by using one of the following two indices expressed through the signal power or intensity. The first is

$$S_3 = \frac{\left|E(P - \hat{P})\right|}{\hat{P}}, \tag{10.18}$$

where $\hat{P}$ is the mean power (intensity), $E(\cdot)$ is a mathematical expectation operator. The index S_3 gives the mean deviation of the power normalized by the mean power. The second index is

$$S_4 = \frac{\sqrt{E\left((P - \hat{P})^2\right)}}{\hat{P}}, \tag{10.19}$$

or more often expressed as follows:

$$S_4 = \sqrt{\frac{\langle P^2 \rangle - \langle P \rangle^2}{\langle P \rangle^2}}. \tag{10.20}$$

The index S_4 gives root mean square deviation of power, normalized by the mean power. There are also S_1 and S_2 indices similar to S_3 and S_4, but expressed in terms of amplitude rather than intensity [4]. The Equation (10.20) usually uses intensity notation I instead of power P. We use power notation in order to avoid confusion with in-phase accumulator measurements later in this chapter.

By definition, the scintillation parameter S_4 always satisfies the following condition:

$$S_4 \leq \sqrt{2}. \tag{10.21}$$

The weak scattering conditions for the Rytov approximation model validity (10.10) can be expressed in terms of intensity fluctuations [15]. The field strength is given by

$$E = E_0(1 + \chi), \tag{10.22}$$

hence received power

$$P = E_0^2(1 + \chi)^2 \approx P_0(1 + 2\chi) \tag{10.23}$$

and

$$\left\langle \left(\frac{\delta P}{P_0}\right)^2 \right\rangle = 4\langle \chi^2 \rangle, \tag{10.24}$$

where $\delta P = P - P_0$.

10.3.2 Scintillation measurements in baseband processor

The signal power can be measured as a difference between narrow-band power and wide-band power [17]:

$$P = P_{\mathrm{W}} - P_{\mathrm{N}}, \tag{10.25}$$

where wide-band power is defined as

$$P_{\mathrm{W}} = \sum_{j=1}^{M} \left(I_j^2 + Q_j^2\right) \tag{10.26}$$

and narrow-band power

$$P_{\mathrm{N}} = \left(\sum_{j=1}^{M} I_j\right)^2 + \left(\sum_{j=1}^{M} Q_j\right)^2, \tag{10.27}$$

both computed over M samples (see Section 6.6). We also need to remove a part of the signal power which comes from ambient noise. It can be calculated using signal-to-noise ratio [17]:

$$S_{4\mathrm{N}} = \sqrt{\frac{100}{S/N_0}\left[1 + \frac{500}{19S/N_0}\right]}. \tag{10.28}$$

Finally the S_4 index is calculated as follows:

$$S_4 = \sqrt{\frac{\langle P^2 \rangle - \langle P \rangle^2}{\langle P \rangle^2} - \frac{100}{S/N_0}\left[1 + \frac{500}{19S/N_0}\right]}. \tag{10.29}$$

For weak scintillations, for $2 < p < 6$, the S_4 index can also be expressed analytically through parameters of a phase screen model [18]:

$$S_4^2 = 4\sqrt{\pi} r_e^2 \lambda^2 (L \sec\theta) \sigma_{\mathrm{N}}^2 \kappa_0^{p-3} \Gamma\left(\frac{p/2+1}{2}\right) \Gamma\left(\frac{3-p/2}{2}\right) Z^{p/2-1} g/(p/2-1), \tag{10.30}$$

where θ is the incident angle of the wave on the slab, σ_{N} is the variance of the fluctuating electron density, g is a geometric factor which depends on the degree of anisotropy ($g = 1$ for isotropic irregularities), p is spectral index, r_e is classical electron radius,

$$r_e = \frac{e^2 \mu_0}{4\pi m}, \tag{10.31}$$

$$\kappa_0 = 2\pi / r_0, \tag{10.32}$$

where r_0 is the outer scale of the irregularities,

$$\mathbf{Z} = \lambda z_R \sec\left(\frac{\theta}{2\pi}\right), \tag{10.33}$$

$$z_R = z z_s / (z + z_s), \tag{10.34}$$

where z_s is the distance of the slab to the source, λ is a wavelength, .

From (10.30), for weak scintillations S_4 follows the frequency scaling law f^{-n}, where $n = (p + 2)/4$. The radius of the first Fresnel zone is $\sqrt{\lambda L}$, L is the slab thickness. S_4 can exceed unity in the focusing regions.

The power according to (10.29) and carrier phase measurements come from a baseband processor. These measurements are functions of satellite dynamics. We are interested in carrier phase fluctuations, which are in the phase measurements, but hidden by satellite movement, which is roughly 800 meters per second along the LOS for GPS satellites. The fluctuations are normally within 0.2 radian. In order to see them we need to remove satellite movement, which is done by de-trending using a high-pass filter. We also need to use a low-pass filter to remove high frequency variations in power measurements. In both cases a sixth-order Butterworth filter can be used.

The S_4 and σ_φ parameters for other than L1 frequency can be recalculated by the following [19]:

$$S_4(f) = S_4(L1)\left(\frac{f_{L1}}{f}\right)^{1.5}, \tag{10.35}$$

$$\sigma_\phi(f) = \sigma_\phi(L1)\frac{f_{L1}}{f}. \tag{10.36}$$

Phase fluctuations may also come from a front end clock, as we discussed in Chapter 6. For the purposes of scintillation monitoring it is necessary to use an OCXO in the front end.

We also need to measure code–carrier divergence (CCD). The CCD contains information about ionospheric delay/advance proportional to TEC, and information about multipath.

Raw measurements are normally delivered with a 50 Hz sampling rate. The iPRx software receiver allows output of data with a much higher sampling rate, up to 1000 Hz, though in real-time mode it depends also on the host PC specification. Figure 10.9 shows an ionospheric scintillation monitor (ISM) version of an iPRx receiver.

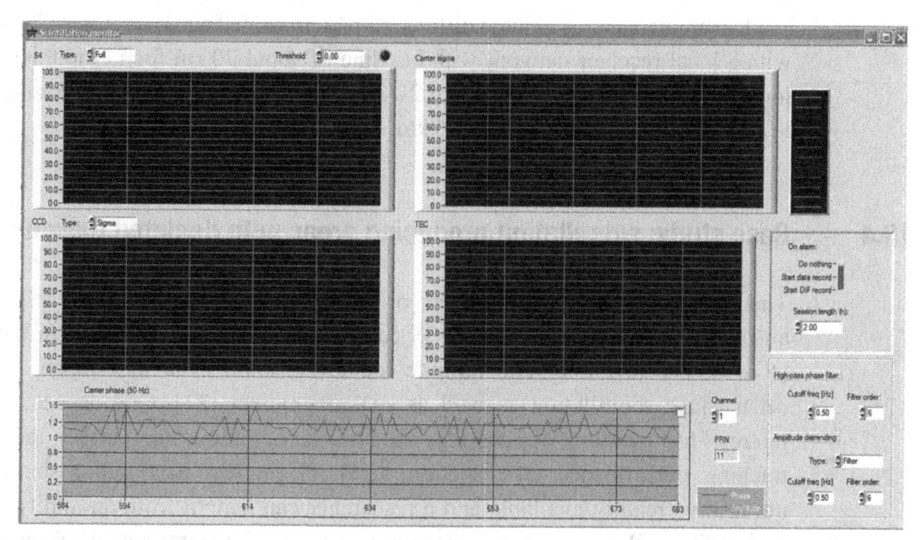

Figure 10.9 ISM scintillation monitoring panel. Bottom plot shows de-trending filtering for phase and amplitude.

10.3.3 Differential scintillation monitoring

Spaced scintillation monitor receivers allow us to establish a diffraction pattern of ionospheric irregularities. This allows us to deduce the size and shape of patches. Baseline size and direction can be optimized for the measurements. The thickness of the layer may be estimated based on the changes of form of the diffraction pattern. The first time such research was conducted in the 1960s it used Transit satellites [20]. If irregularities have a Gaussian distribution in height, with a standard deviation $\sqrt{2}\Delta$, then the thickness of the layer can be calculated as follows:

$$\Delta = \frac{V_c u H}{(u+V)^2}, \tag{10.37}$$

where u is satellite velocity, H is satellite altitude, V is drift pattern of the diffraction pattern, V_c is "intrinsic velocity" of the diffraction pattern equal to the ratio of the pattern size to the lifetime of a feature of the pattern.

Height of irregularities can be deduced from the velocity of the diffraction pattern. We should look for time displacement between similar fades in the neighboring receiver records. Scale of irregularities, from the transverse scale of the diffraction pattern, is given by

$$D_A = 2\left(\frac{z_B}{z_B - z}\right)d, \tag{10.38}$$

where Z_B is the altitude of the satellite, z is the height of irregularities, d is deduced from autocorrelation of a fading record. This is valid if the satellite beam is not parallel to the magnetic field.

Measurements of amplitude diffraction pattern from Transit satellites [20] over a year with a local receiver network with baseline of about 20 km found the thickness of the ionospheric irregularity to be about $2\sqrt{2}\Delta = 60$ [km], and height in the range from 300 to 600 km, with average about 400 km.

10.4 Case study: scintillation producing areas with depleted ionization

The areas with depleted ionization or ionospheric ***plasma bubbles*** are getting more and more attention from the GNSS community since it is common in the countries at relatively low magnetic latitude including south Asia and Japan. These areas often cause scintillation as well as spatial gradient [21]. One aspect of particular interest for aviation applications is how scintillation may affect the INS-aided tracking performance (see Section 12.3). Flight test data under scintillation are not always readily available and a researcher can benefit from an ability to simulate their effects. ReGen allows simulation of the digital intermediate frequency signal affected by scintillation. The scintillation data can be either simulated or come from measurements. The resulting DIF signal can be processed by a software receiver.

In this section we consider scintillation which is associated with a plasma bubble. The observation data were obtained by using GPS Silicon Valley's GPS Ionospheric Scintillation and TEC Monitor (GISTM) system GSV4004 at Naha (N26°13′ 43, E127°40′ 44), Japan, on 14 April 2004. The data were analyzed and the characteristics were extracted [21]. The GSV4004 is designed for monitoring scintillation and gives scintillation related parameters as well as GPS pseudorange and carrier phase. Figure 10.10 shows the trajectory of GPS satellites from 12:40 to 15:05 (GPS time) during which two series of scintillation were observed and indicated by lines with squares (first for PRN 15 from 12:45 to 13:05, second for PRN 22 from 14:20 to 15:05). Figure 10.11 shows TEC (total electron content) during the first period where the TEC for PRN 15 varied with a minimum at 12:59 (21:59, local time). The rapid decrement and increment of TEC are likely to be due to the plasma bubble.

Next, the index of amplitude scintillation (S_4) is shown in Figure 10.12. The index S_4 is defined by (10.29), where P denotes the GPS signal intensity. The S_4 for PRN15 increases and reaches up to 0.5. Large values of S_4 for PRN21 may be caused by multipath error because the satellite elevation is low (see Figure 10.10). In order to verify the source of S_4 variation, the difference between pseudorange and carrier phase (code carrier divergence, CCD) can be used, since multipath error on pseudorange is much larger than that on carrier phase [17]. The CCD (1 σ) values are plotted against the S_4 index for various satellites (Figure 10.13). If the CCD is larger than 0.25, then the intensity variation is considered to be caused by multipath. Therefore, we can conclude that large S_4 values for PRN21 result from multipath.

The statistics of the residuals of de-trended carrier phase (σ_ϕ) for PRN15 and PRN21 are shown in Figure 10.14. The σ_ϕ values are computed over 60 seconds and plotted at

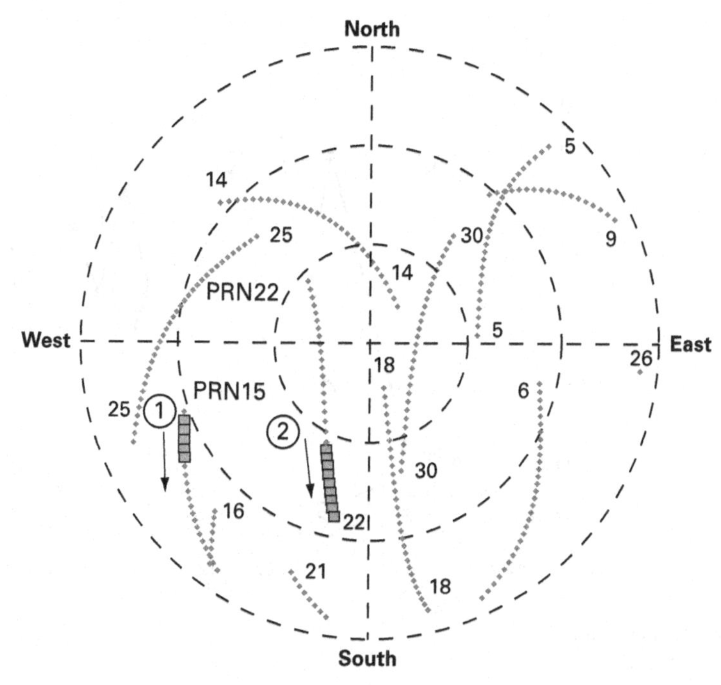

Figure 10.10 Trajectory of observed GPS satellites.

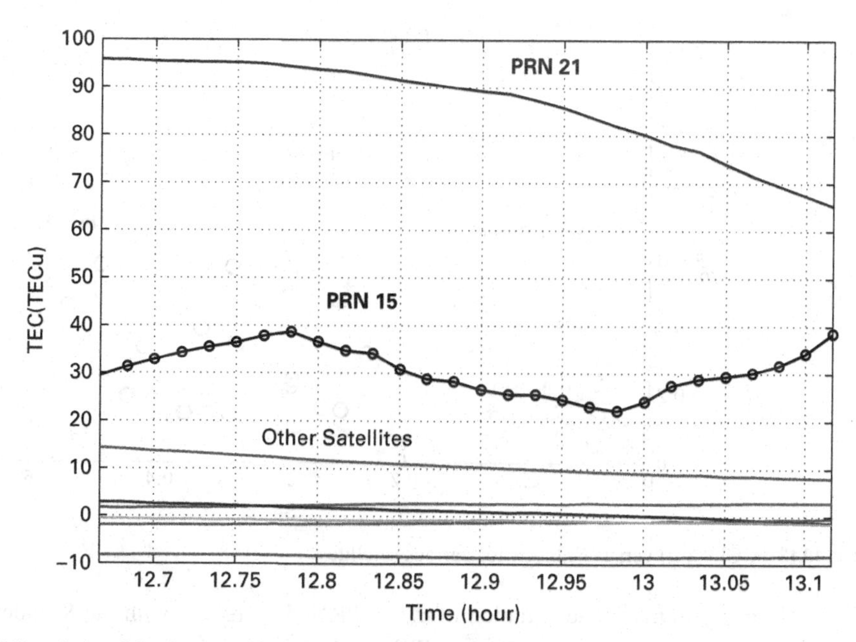

Figure 10.11 TEC variation (bias is not calibrated).

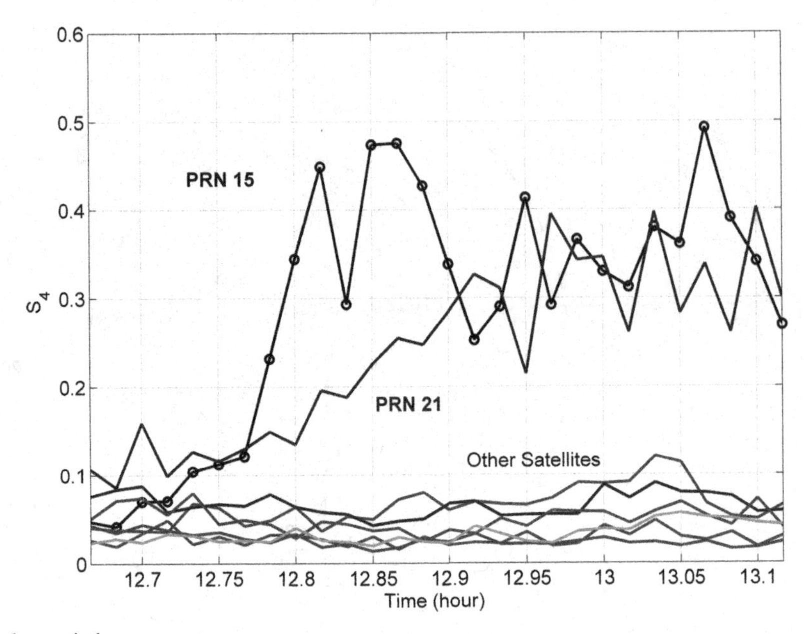

Figure 10.12 S_4 index variation.

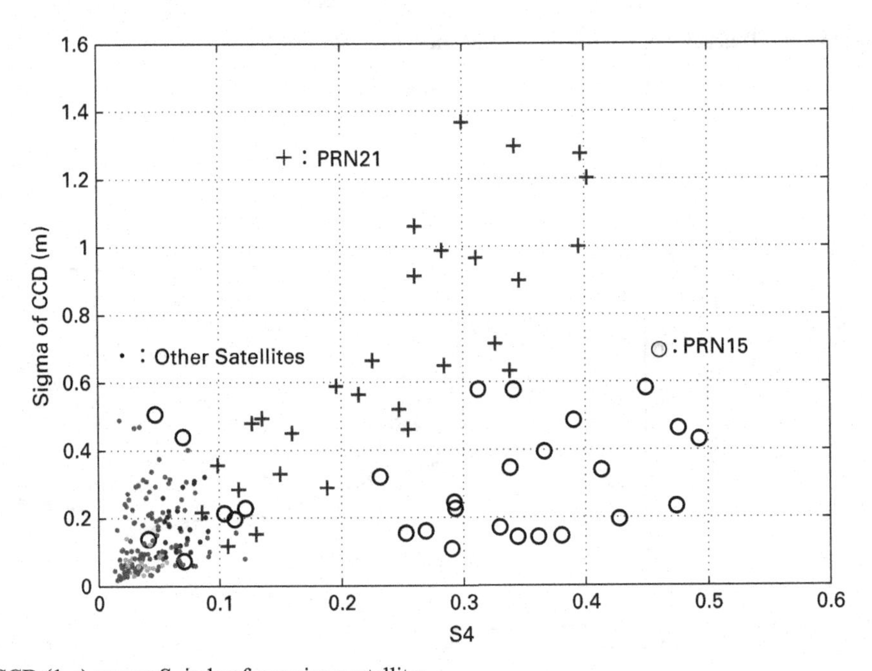

Figure 10.13 CCD (1 σ) versus S_4 index for various satellites.

every minute. The variation of σ_ϕ for PRN15 correlates with the S_4 index, while there seems no scintillation effect on PRN21.

The variations of normalized intensity and phase error from 12:48 to 12:49 are shown in Figure 10.15 for PRN15.

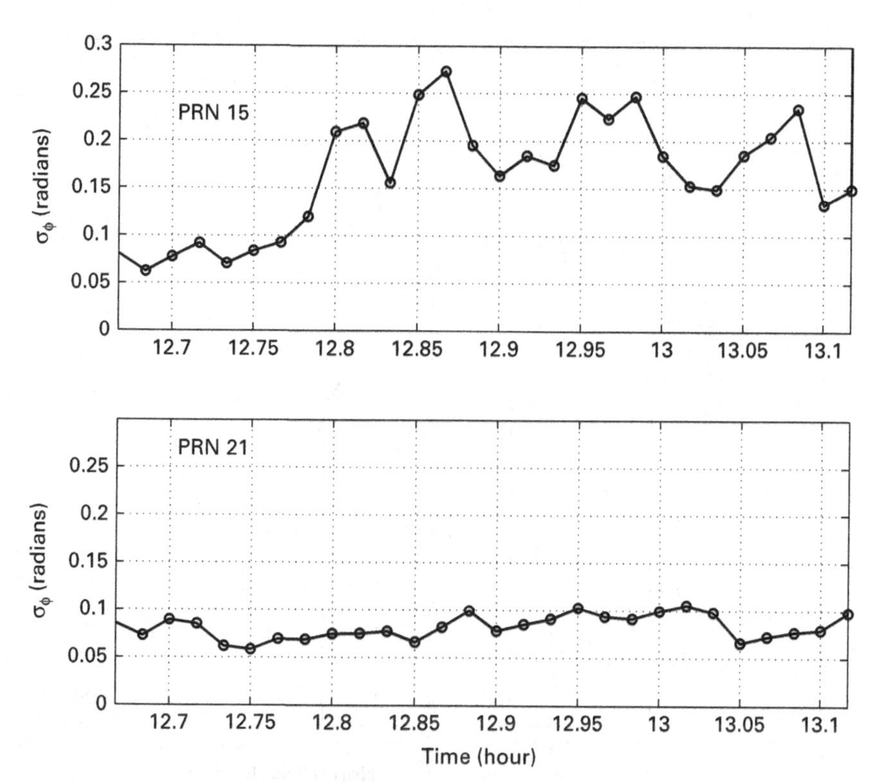

Figure 10.14 Variation of σ_ϕ for PRN15 (top) and PRN21 (bottom).

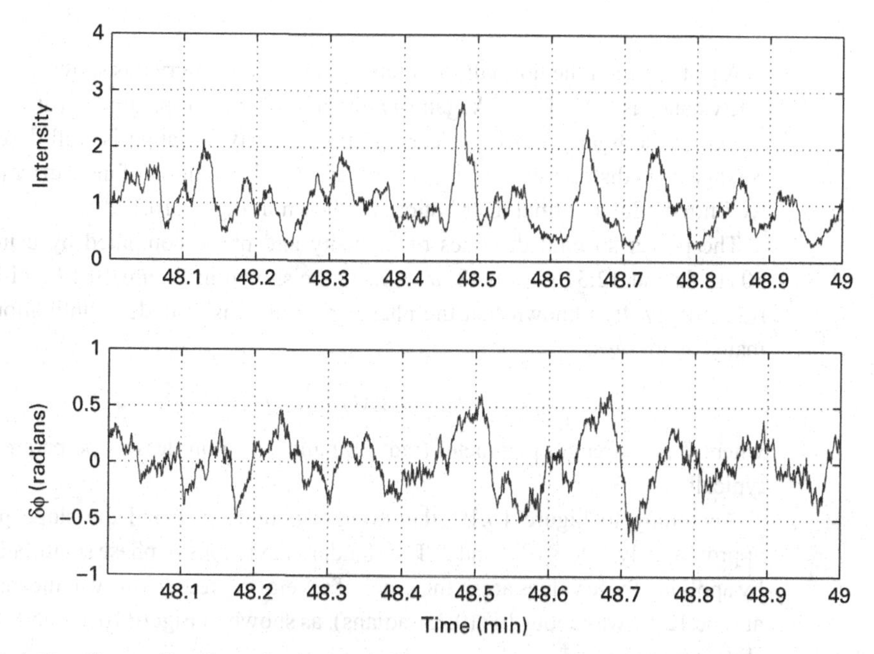

Figure 10.15 Variation of normalized intensity (top) and phase error (bottom) for PRN15 for one minute interval.

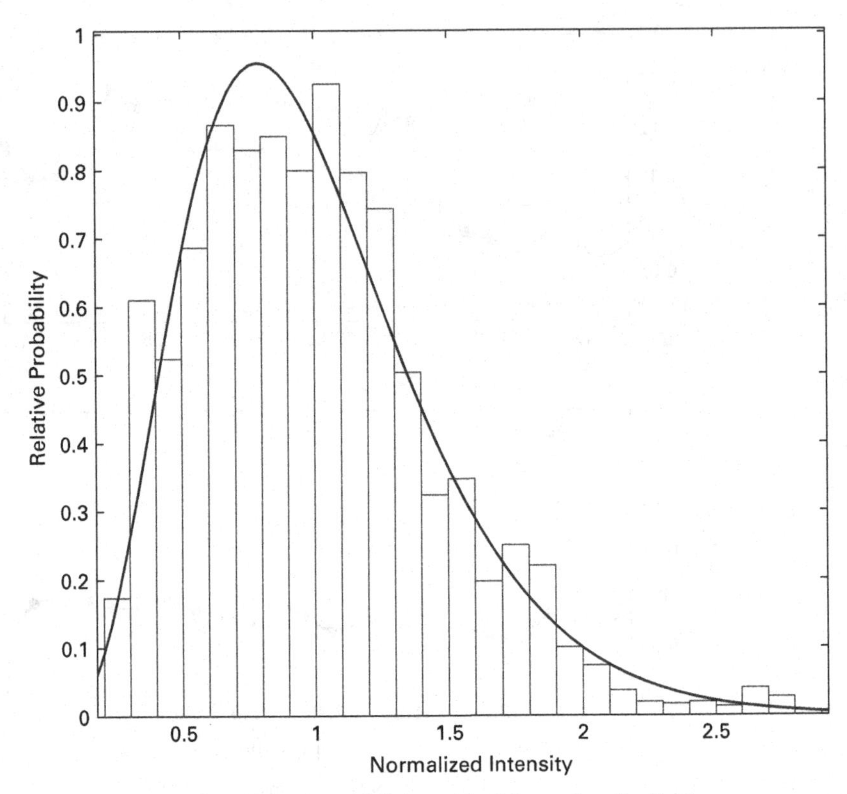

Figure 10.16 Histogram of the normalized intensity and fitting with Nakagami-*m* distribution.

A histogram of the normalized intensity during this interval is shown in Figure 10.16. The vaiance is 0.21. The Nakagami-*m* distribution with *m*-parameter ($m = 1/S_4^2$) is also shown in the figure, and it is clear that the intensity variation is well described by the Nakagami-*m* distribution. On the other hand, the distribution of the carrier phase error is well fitted with a normal distribution, as shown in Figure 10.17.

The power spectral densities of intensity and phase computed by using GSV4004 50 Hz data at 12:51 ($S_4 = 0.48$, $\sigma_\phi = 0.25$) are shown in Figure 10.18 and Figure 10.19 respectively. It is known that the phase spectral density under scintillation is approximated as follows:

$$P_{\delta\phi}(f) = T \cdot f^{-p}, \tag{10.39}$$

where T is a strength parameter (rad^2/Hz) and p is a unitless slope parameter which is typically 2.0–3.0.

According to Figure 10.19, the strength parameter T and the slope parameter are approximately $3.0e \times 10^{-3}$ and 2. The standard deviations of phase scintillation computed by applying these values are in the order of several degrees. However, the measured value at time 12:51 was about 14° (0.25 radians), as shown in Figure 10.14, and frequent cycle slips were observed.

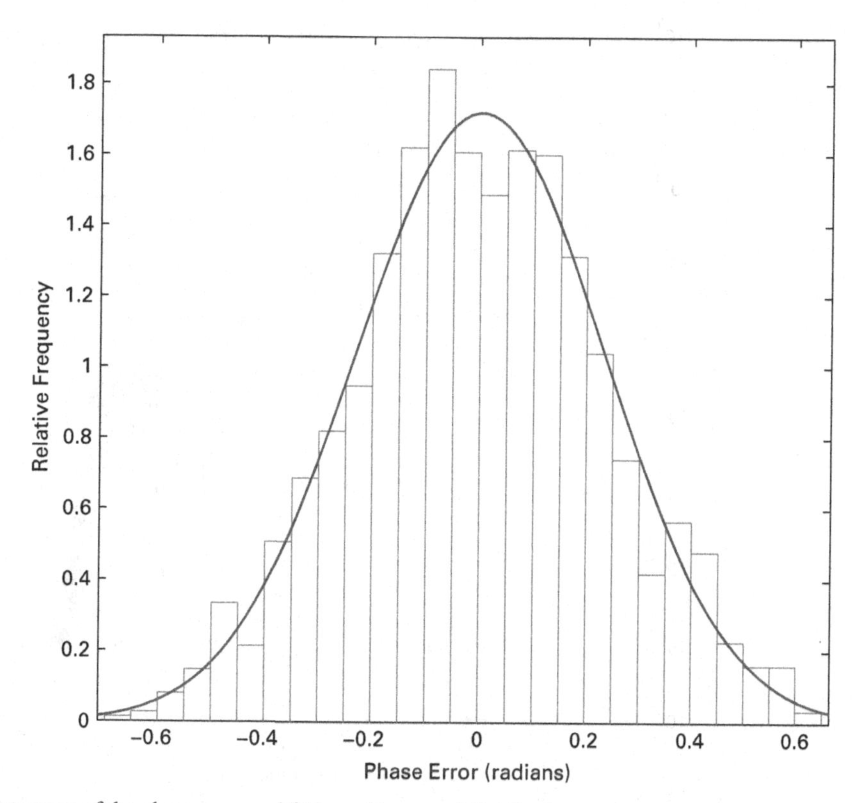

Figure 10.17 Histogram of the phase error and fitting with normal distribution.

Next, the parameters during the second period with rather weak scintillation from 14:20 to 15:05 are plotted in Figure 10.20. The top figure shows the TEC variation of PRN 22 and small fluctuations of TEC are observed rather than a deep depletion as seen in Figure 10.11.

Since the plasma bubble usually moves eastward, the second scintillation may be related to the same plasma bubble which caused the first scintillation. The distance between the ionospheric pierce point (IPP) of PRN 15 at 12:45 and PRN 22 at 14:20 is 534 km assuming that the height of the ionosphere is 400 km. If the plasma bubble moved eastward with a typical speed of 100 m/s, the travel distance in 95 minutes was 570 km and coincides well with the distance between IPPS. Therefore, it would be reasonable to consider that the second scintillation was associated with the plasma bubble as it moved from the west.

10.5 Characteristics of the experimentally derived scintillation data

The measurement results in the previous section as well as results from other tests [22], [15] show only a partial agreement with the theory given in Section 10.2, which can give an explanation of the phase scintillation and amplitude scintillation power spectrum. The

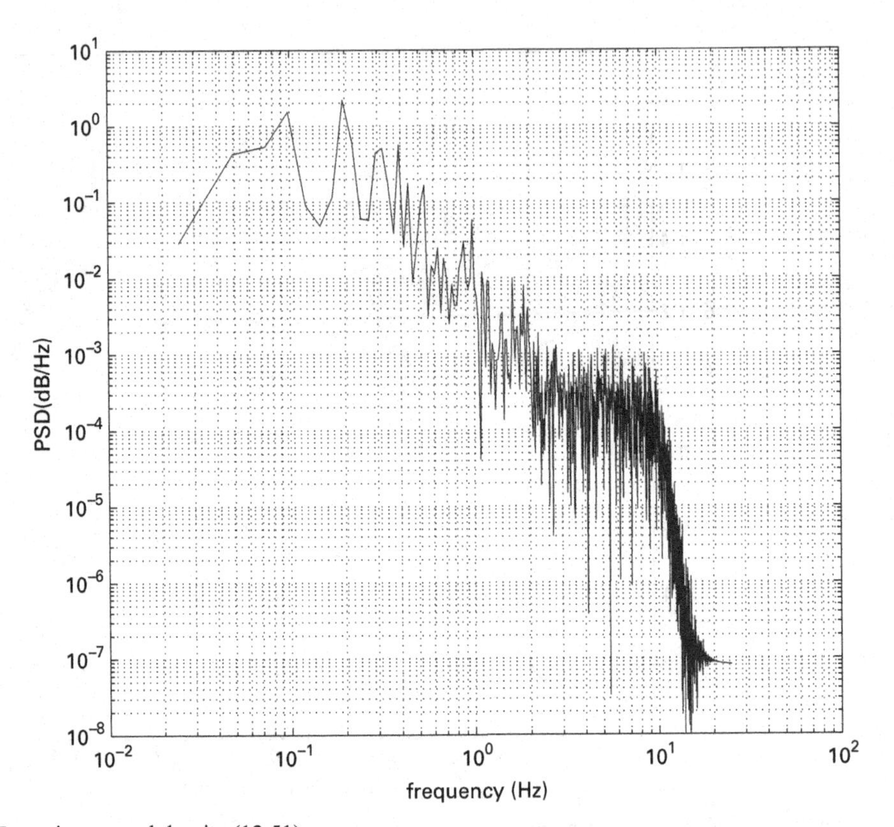

Figure 10.18 Intensity spectral density (12:51).

phase scintillation ***power spectral densities (PSD)*** can usually be expressed as power-law spectra [23]:

$$P_\phi(f) = T \cdot f^{-p}, \tag{10.40}$$

$$T = P_\phi(f = 1), \tag{10.41}$$

where f is frequency (Hz), T is a strength parameter about 0.02 (rad^2/Hz), and p is a unitless slope typically between 2.0 and 3.0. For amplitude scintillation, PSD also fits to a power-law with slope between 2.5 and 3, and for strong scintillation is 5.5 [24]. The experimental results in the previous section are in agreement with (10.40).

Stronger scintillation corresponds to a higher Fresnel minimum. The first Fresnel minimum may appear as the first minimum on a power spectrum plot. It can be found as an inverse value of the correlation interval, defined here as a time lag for which the level of correlation has decreased by 50% [25].

However, there are also some features that are not in good agreement with the theory.

The amplitude scintillation can be described by a gamma distribution [23],[26]. This specific type of gamma distribution is called Nakagami-m (see Figure 10.16).

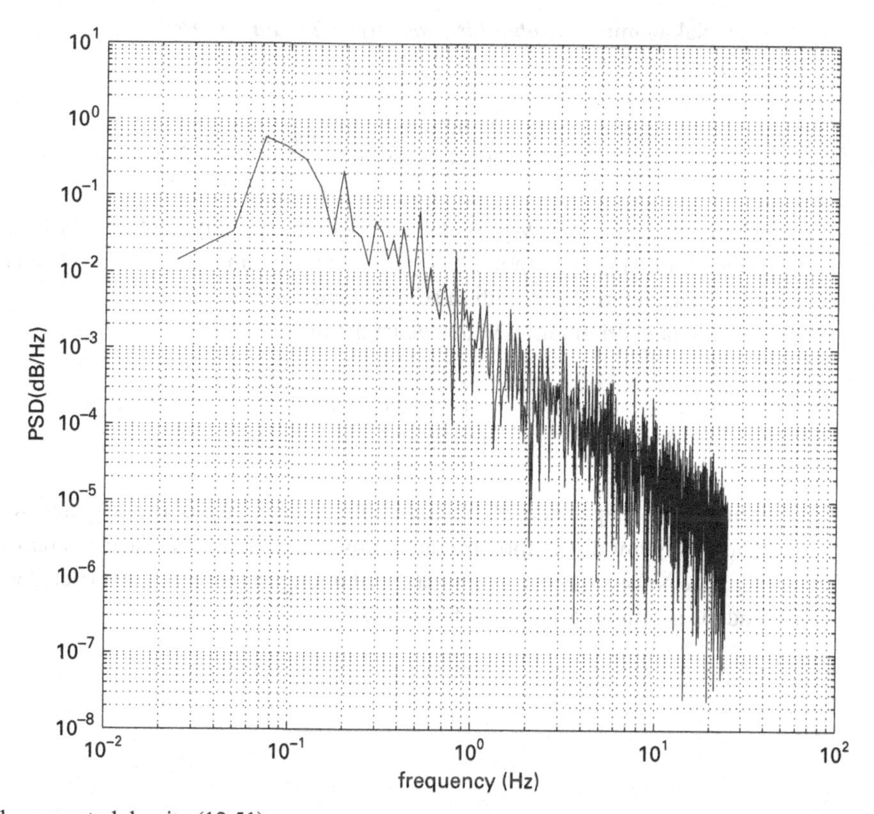

Figure 10.19 Phase spectral density (12:51).

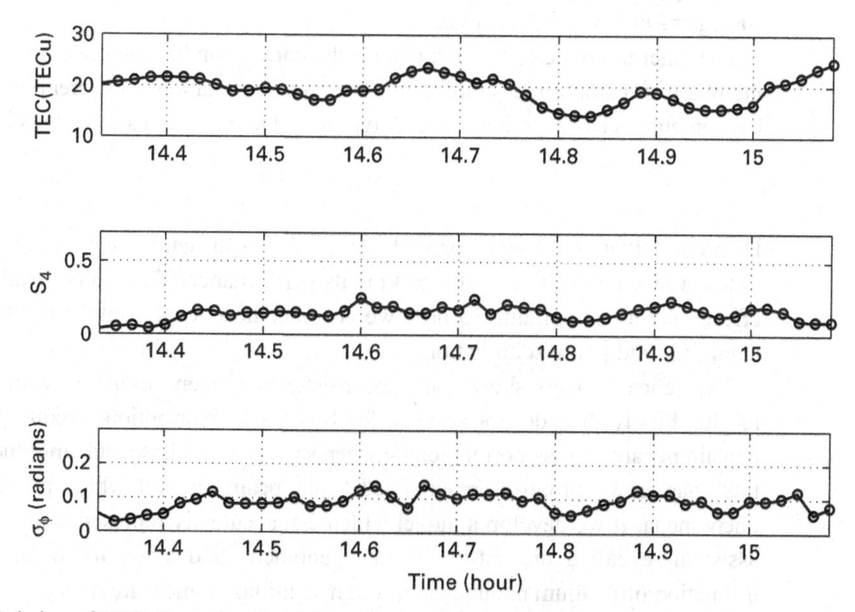

Figure 10.20 Variation of TEC, S_4, and σ_ϕ for PRN 22 in the second scintillation (14:20 to 15:05).

The Nakagami-m ***probability density function (PDF)*** is generally described as follows [27],[26]:

$$f(\delta P)=\frac{2\cdot m^m\cdot\delta P^{m-1}}{\Gamma(m)\cdot\langle\delta P\rangle^m}\mathrm{e}^{-m\cdot\delta I/\langle\delta I\rangle},\ \delta P=(\delta A)^2\geq 0,\tag{10.42}$$

where δP is the received signal scintillation intensity (power), $\langle\delta P\rangle$ is the average scintillation intensity (power), $\Gamma(\cdot)$ – is the gamma function, and m is a parameter which characterizes strength of scintillation.

Gamma distribution PDF in a general case is given by:

$$f_{\alpha,a,\beta}(x)=\begin{cases}\frac{1}{\Gamma(\alpha)\cdot\beta^\alpha}(x-a)^{\alpha-1}\mathrm{e}^{-(x-a)/\beta}, & x\geq a\\ 0, & x<a,\end{cases}\tag{10.43}$$

where α is a shape parameter, a is a displacement, and β is a scale parameter, $\alpha>0, \beta>0$. The Nakagami distribution becomes closer to normal when shape parameter is increasing. The Nakagami-m and gamma distribution parameters are connected with each other and the scintillation index as follows [26]:

$$\alpha=m=\frac{1}{S_4^2},\tag{10.44}$$

$$\beta=S_4^2.\tag{10.45}$$

The Nakagami-m distribution reduces to the Gaussian when scattering is weak, correspondingly m is large. The solutions we consider so far in this chapter do not account for a Nakagami-m type of distribution.

The other unpredicted phenomenon is the correlation between two processes describing phase and amplitude scintillation. The cross-correlation is different in different tests but sometimes has negative value. A typical value we use in simulations is [23]:

$$\rho_{I\phi}=-0.6.\tag{10.46}$$

Heavy scintillation is mostly unpredictable and it is difficult to get the data sets required to test a receiver, and analyze and tune its performance. Therefore, simulation of such effects is very important. So now we consider how to generate a GNSS signal with amplitude and phase scintillation.

Theoretical models show some inconsistencies when compared with experimental results. Firstly they do not account for Nakagami distribution, secondly they do not explain negative cross-correlation between scintillation phase and amplitude. Further in this chapter we propose an extended “split beam” model which accounts for both phenomena. If we develop a model which can accurately represent a signal, then it can assist in revealing the nature of the phenomena and allow us to tune methods of estimation of medium parameters through scintillation measurements.

10.6 Phenomenological model for weak scintillation

In this section we consider only a phenomenological model for scintillation, derived to fit experimental data following [24]–[26]. Both phase and amplitude scintillations can be described as random processes. We need to generate two negatively correlated random processes with Gaussian and gamma distribution functions describing phase and amplitude scintillation respectively. We described a GPS signal in Chapter 3 as follows:

$$A = A_0 \sin(\omega t + \varphi) \cdot D, \tag{10.47}$$

where A is an amplitude , φ is a phase and D is a C/A code. We omit a navigation message, because it is affected only through the spread code. Here it is more convenient to describe a GPS signal at the input of the receiver using a complex presentation as follows:

$$A = A_0 \mathrm{e}^{\mathrm{j}\phi} \cdot D. \tag{10.48}$$

When the signal is affected by scintillation, it can be represented by the following equation:

$$A = A_0 \cdot \delta A \cdot \mathrm{e}^{\mathrm{j}(\phi_0 + \delta\phi)} \cdot D, \tag{10.49}$$

where δA is amplitude scintillation, and $\delta\phi$ is phase scintillation.

The phase scintillation can be described as a zero-mean Gaussian random process with standard deviation $\sigma_{\delta\phi}$. A Gaussian random process has a normal distribution. We discussed normal distribution in Chapter 1. After the scintillation processes are generated we pass them through shaping filters in order to create a necessary autocorrelation in these processes. Otherwise, as we established in Chapter 3, the processes have unlimited power, and their neighboring values are completely independent from each other. We need to describe phase and amplitude scintillations as one two-dimensional random process with gamma distribution or bivariate gamma distribution. Its vector components should be correlated.

To create such stochastic processes we use a trivariate reduction method following [28]. It called trivariate, because it uses three independent random variables to construct two dependent random variables. Let us consider three independent gamma random variables G_1 , G_2, G_3 with parameters α_1, α_2, α_3. Now let us define a two-dimensional random vector as follows:

$$\{\delta A, \delta\phi'\} = \{G_1 + G_3, G_2 + G_3\}. \tag{10.50}$$

This is a bivariate gamma variable. The marginal gamma distributions of δA and $\delta\phi'$ have parameters $\alpha_A = \alpha_1 + \alpha_3$ and $\alpha_\varphi = \alpha_2 + \alpha_3$ respectively. The correlation between the components is now defined as follows:

$$\rho_{A\phi} = \frac{\alpha_3}{\sqrt{(\alpha_1 + \alpha_3)(\alpha_2 + \alpha_3)}}. \tag{10.51}$$

If $\rho_{A\phi}$ and the marginal gamma parameters α_i are specified beforehand, then there is exactly one solution, providing that the following condition is satisfied:

$$0 \leq \rho_{A\phi} \leq \frac{\min(\alpha_A, \alpha_\phi)}{\sqrt{\alpha_A \cdot \alpha_\phi}}. \tag{10.52}$$

In order to provide such a correlation, the parameters for the marginal gamma distributions should be defined as follows:

$$\alpha_1 = \alpha_A - \rho_{A\phi}\sqrt{\alpha_A \cdot \alpha_\phi}, \tag{10.53}$$

$$\alpha_2 = \alpha_\phi - \rho_{A\phi}\sqrt{\alpha_A \cdot \alpha_\phi}, \tag{10.54}$$

$$\alpha_3 = \rho_{A\phi}\sqrt{\alpha_A \cdot \alpha_\phi}. \tag{10.55}$$

We can in fact use

$$\{\delta A, \delta\phi'\} = \{G_1 + G_3, G_2 - G_3\} \tag{10.56}$$

in order to provide a negative correlation, as follows:

$$\rho'_{A\phi} = -\rho_{A\phi} \tag{10.57}$$

and Equation (10.54) becomes

$$\alpha_2 = \alpha_\phi + \rho_{A\phi}\sqrt{\alpha_A \cdot \alpha_\phi}. \tag{10.58}$$

Because the phase scintillation is in fact better represented by a random process with zero-mean Gaussian distribution, the parameters of the second component of the bivariate gamma distribution are adjusted to a special case when gamma distribution is close to Gaussian with large $\alpha = 100$ and $\beta = 1$. For large values of α, the gamma density is close to the normal density.

We need now to consider the scale parameter, which is a special kind of numerical parameter of a parametric family of probability distributions. The larger the scale parameter, the more spread out the distribution, while if it is small then the distribution is more concentrated. If we formally rewrite gamma PDF as follows:

$$f_{\alpha,a}(x) = \frac{1}{\beta} \cdot f_{\alpha,a}(x/\beta), \tag{10.59}$$

then the scale parameters should be the same within $\{G_1 + G_3\}$, $\{G_2 + G_3\}$ pairs and the resulting variable $\delta\phi'$ must be scaled back to $\delta\phi$ in order to have a correct standard deviation $\sigma_{\delta\phi}$. The generation of the independent G_i gamma random variable is trivial task and can be done by a standard function in many libraries.

After the phase and amplitude scintillation processes are generated as random values, we need to apply shaping filters to create a proper autocorrelation. We can use an ***infinite impulse response (IIR)*** filter as a shaping filter. These filters have a feedback loop and theoretically their response to an input effect is not limited in time. This is opposite to ***finite impulse response (FIR)*** filters, that have no feedback and therefore their response lasts only while an input effect is applied. The IIR filter can be implemented using the direct form filter equations:

$$x_i = \varphi_i - a_1 \cdot x_{i-1} - a_2 \cdot x_{i-2} - a_3 \cdot x_{i-3} \ldots - a_n \cdot x_{i-n}, \tag{10.60}$$

where n is the filter order and coefficients a_i are calculated to fit experimental spectra [29].

According to the Nyquist theorem, the sampling rate should be chosen to be at least twice the value of maximum frequency we need to consider. From experimental results, the maximum scintillation frequency in a GPS signal is 10 Hz , therefore we can choose sampling about 20 Hz. The sampling interval for a scintillation signal is significantly larger than for scenario sampling; therefore in a simulator we have to generate the scintillation process in an external loop in relation to a one millisecond signal loop, interpolating the scintillation for each sample.

10.7 Split-beam scintillation model

Though a conventional model is created to fit to experimental data, it provides no insight into the mechanism of scintillation generation. Besides, it cannot cover the whole range of S_4 and cross-correlation. In this section we present an alternative model, which is based on a hypothesis of physical mechanism of signal scintillation. In the proposed model the signal, when it encounters an eddy, is split into two beams. One beam is scattered and the related field can be described by a Rytov approximation. The second beam comes through the eddy and is attenuated. The phase scintillation is caused by the scattered beam. The amplitude scintillation in this model comes not only from phase scintillation after Fresnel filtering, but also from attenuation of the direct signal. This model involves attenuation and therefore requires consideration of a complex value of the refractive index. The scattered signal can be described by (10.17):

$$E_\mathrm{S} = E_{\mathrm{S}0}\mathrm{e}^{\left(-\int d^3r\, G\left(\vec{R},\vec{r}\right)\frac{E_0\left(\vec{r}\right)}{E_0\left(\vec{R}\right)}\left[k^2\delta\varepsilon\left(\vec{r}\right)+\left(\nabla\psi_1\left(\vec{r}\right)\right)^2\right]\right)}. \tag{10.61}$$

The attenuated direct signal can be described by considering the complex component of the refractive index ς. The direct signal can be expressed as

$$E_\mathrm{D} = E_{\mathrm{D}0}\mathrm{e}^{-Y}, \tag{10.62}$$

where Y is attenuation index, expressed as

$$Y = 20\,\mathrm{lg}\frac{E_{\mathrm{D}0}}{E_\mathrm{D}} = 20\frac{\omega}{c_0}\int \varsigma\, dl\ \mathrm{lg\,e}. \tag{10.63}$$

The total signal therefore is a sum of two components. The fluctuations due to source movement should satisfy conservation of energy:

$$P_\mathrm{S} + P_\mathrm{D} = P, \tag{10.64}$$

where P_S is scattered signal and P_D is direct. We should also apply double scattering to conserve energy (Figure 10.8). When the beams are combined the phase scintillation is defined by the first beam and the amplitude scintillation is defined by both beams.

Thus we can see that the scintillation amplitude spectrum is built from two signals. The amplitude scintillation is now caused by two different mechanisms and should not fall under the normal distribution law. Some amplitude scintillation is cause by attenuation in the eddy with changing path geometry. The other part of the amplitude scintillation is caused by interferences of scattered signals. The Nakagami-m distribution allows us to see effects of both beams. The phase scintillation is caused only by the scattered beam. Therefore the phase scintillation spectrum has a normal distribution. The Nakagami-m distribution was actually applied initially to a similar model to simulate attenuation of wireless signals traversing multiple paths [30].

The total energy of the signal is divided into two beams. If the larger part of the signal is scattered, then the phase scintillation become higher. It also affects the shape of Nakagami-m distribution for the amplitude scintillation. The part of the amplitude scintillation coming from the direct beam is diminishing. If the amplitude scintillation comes from the direct path only, then the cross-correlation between amplitude and phase would be -1. The cross correlation -0.6 means that about 60% of the amplitude scintillation is caused by attenuation in the direct path and 40% by weak scattering.

The maxima in amplitude scintillation PSD correspond to the Fresnel frequency. Large-scale irregularities cannot generate scintillation within the given distance (see (10.2). The steepness of the slope before the Fresnel frequency according to the proposed model should correlate with amplitude-phase cross-correlation value. If the parameter m is large, then the amplitude distribution is closer to Gaussian, and according to the present model it means that the scintillation is due to one channel only. The value of cross-correlation should then be less negative and the slope before the Fresnel frequency should be correspondingly flatter. A higher value of negative cross-correlation between amplitude and phase scintillation should correspond to a steeper slope before the Fresnel frequency. The falling part of the spectrum corresponds to irregularities in the range below that defined by (10.1), which is about 5 Hz for GPS.

10.8 Project: ionospheric scintillations monitoring with an iPRx receiver

A scintillation monitor version of an iPRx receiver can measure and record scintillation data. We have described the theory behind scintillation monitoring with a single frequency receiver. The scintillation monitoring was conducted in order to improve reliability and accuracy of airborne INS/GPS coupled navigation systems, especially at the approach and landing stages. Such measurements are also important for many geophysical applications, as described in the next chapter. The scintillation monitor receiver also gives an insight into general receiver operation and the physics of GNSS. In this section we look at scintillation monitor operation. If you have such a scintillation monitor receiver in your lab, this section helps you to get hands on experience with it. A scintillation monitor receiver is available as an upgrade to the bundled iPRx receiver.

There are also other scintillation monitor receivers including software [31] with similar functionality. We have used such an off-the-shelf scintillation monitor receiver for scintillation measurements along with iPRx. The advantage of a software receiver such as an iPRx for scintillation monitoring is that one can use a number of inexpensive RF recorders to collect scintillation network data to measure, for example, the size of ionospheric irregularities (see Section 10.3.3), and process all data with a single receiver in post-processing mode, rather than use a few scintillation monitor receivers. The iPRx receiver can also process the scintillated signal generated by ReGen software. Below we sum up some receiver features regarding software implementation for scintillation monitoring.

1. The receiver clock requirement is OCXO, otherwise the receiver oscillator's phase noise can interfere with phase scintillation measurements.
2. A software receiver can output data with sampling up to 1000 Hz.
3. Simultaneous TEC measurements can be provided from IGS maps in the case of post processing the data. One can also use a dual frequency software receiver.
4. For scintillation measurements, a dual frequency receiver does not actually provide any advantages [18].
5. A software receiver can process the signal with multi-correlators.
6. The antenna environment should be checked for multipath.
7. Automatic gain control (AGC) should be disabled in order not to affect the measured signal intensity [18].
8. A software receiver allow optimization of variable tracking loop parameters such as code loop bandwidth and carrier loop bandwidth to improve noise rejection. This is possible because one can process recorded signal repeatedly with different parameters in post-processing mode.
9. The ability of a software receiver to work with recorded signals has special advantages in the case of working with a network of receivers. We have already mentioned that a user can process records from many stations with one set of software. Besides, if for example we know that a DIF signal record from one station shows scintillation of a specific satellite signal, then we can re-process other recorded signal sets for the same satellite, using parametric methods. It is possible to find and measure scintillation, if we know it is there and have a model. The specialized hardware receiver will show no information in this case.
10. Software receivers in general can provide complete access to interior features including signal processing and scintillation data measurement algorithms.

References

[1] A. D. Wheelon, *Electromagnetic Scintillation.* Vol. I, *Geometrical Optics*, Cambridge, Cambridge University Press, 2001.

[2] J. E. Allnut, *Satellite-to-Ground Radiowave Propagation*, London, UK, Peter Peregrinus Ltd., 1989.

[3] A. Brekke, *Physics of the Upper Polar Atmosphere*, Worthing, UK, Praxis Publishing Ltd., 1997.
[4] R. D. Hunsucker and J. K. Hargreaves, *The High-Latitude Ionosphere and its Effects on Radio Propagation*, Cambridge, Cambridge University Press, 2003.
[5] S. Saito, T. Yoshihara, and N. Fujii, Study of effects of the plasma bubble on GBAS by a three-dimensional ionospheric delay model, 22nd International Technical Meeting of the Satellite Division of the Institute of Navigation, 22–25 Sept. 2009, Savannah, Georgia, US, pp. 1141–1148.
[6] D. Fang, 4/6 GHz ionospheric scintillation measurements, AGARD Conference Proceedings 284, *Propagation Effects in Space/Earth Paths*, NATO, Advisory Group for Aerospace Research and Development, Neuilly-sur-Seine CEDEX, France, 1980.
[7] G. S. Kent and R. W. H. Wright, Movements of ionospheric irregularities and atmospheric winds, *J. Atm. Terr. Phys.*, **30**, (5), 1968, 657–691.
[8] J. P. Mullen, E. Mackenzie, S. Basu, and H. Whitney, UHF/GHz scintillation observed at Ascension Island from 1980 through 1982, *Radio Science*, **20**, (3), 1985, 357–365.
[9] Y. Karasawa, K. Yasukawa, and M. Yamada, Ionospheric scintillation measurements at 1.5 GHz in mid-latitude regions, *Radio Science*, **20**, 1985.
[10] S. Basu, *et.al.*, 250 MHz/GHz scintillation parameters in the equatorial, polar, and auroral environments, *IEEE Journal of Selected Areas in Communications*, SAC-5, (**2**), 1987, 102–115.
[11] T. Ogawa, K. Sinno, M. Fujita, and J. Awaka, Severe disturbances of UHF and GHz waves from geostationary satellites during a magnetic storm, *J.Atmos.Terr.Phys.*, **42**, (7), 1980, 637–644.
[12] K. C. Yeh and A. W. Wernik, On ionospheric scintillation in *Wave Propagation in Random Media (Scintillation)*, V. I. Tatarskii, A. Ishimaru, and V. U. Zavorotny (editors), Bellingham, WA, The Society of Photo-Optical Instrumentation Engineers and Institute Of Physics Publishing Ltd., USA, 1993.
[13] M. Born and E. Wolf, *Principles of Optics. Electromagnetic Theory of Propagation, Interference and Diffraction of light*, seventh expanded edition, Cambridge, Cambridge University Press, 1999.
[14] G. Gbur, *Mathematical Methods for Optical Physics and Engineering*, Cambridge, Cambridge University Press, 2010.
[15] A. D. Wheelon, *Electromagnetic Scintillation.* Vol. II, *Weak Scattering*, Cambridge, Cambridge University Press, 2003.
[16] N. Zemov, V. Gherm, and H. Strangeways, On the effects of scintillation of low-latitude bubbles on transionospheric paths of propagation, *Radio Science*, **44**, 2009, RSOA 14. (http://www.agu.org/journals/ABS/2009/2008RS004074.shtml)
[17] A. J. Van Dierendonck, *et al.*, *Measuring Ionospheric Scintillation in the Equatorial Region Over Africa, Including Measurement From SBAS Geostationary Satellite Signals*, Long Beach, CA, ION GNSS-2004, 21–24 September 2004.
[18] C. L. Rino , A power law phase screen model for ionospheric scintillation 2. Strong scatter, *Radio Science*, **14**, (6), 1979, 1135–1145.
[19] A. J. Van Dierendonck, J. Klobuchar, and Q. Hua, *Ionospheric Scintillation Monitoring Using Commercial Single Frequency C/A Code Receivers*, Salt Lake City, USA, ION GPS-93, 1993.
[20] J. P. Debarber and W. J. Ross, The diffraction of HF radio waves by ionospheric irregularities, in *Spread F and its Effects upon Radiowave Propagation and Communications*, P. Newman (editor), Agardograph 95, NATO, Advisory Groups for Aerospace Research and Development, Neuilly-sur-Seine CEDEX, France, 1966.

[21] T. Yoshihara, N. Fujii, K. Matsunaga, *et al.*, *Preliminary Analysis of Ionospheric Delay Variation Effect on GBAS due to Plasma Bubble at the Southern Region in Japan*, Proceedings of ION NTM 2007, Jan. 2007, San Diego, US, pp. 1065–1072.

[22] E. J. Fremouw, R L. Leadabrand, R C. Livingston, *et al.*, Early results from the DNA wideband satellite experiment – Complex-signal scintillation, *Radio Science*, **13**, (1), January–February 1978, 167–187.

[23] E. Fremouw, Signal statistics of transionospheric scintillation, COMSAT Symposium: *Space Weather Effects on Propagation of Navigation & Communication Signals*, Bethesda, MD., 22–25 Oct. 1997.

[24] C. J. Hegarty, M. B. El-Arini, T. Kim, and S. Ericson , Scintillation modeling for GPS-wide area augmentation system receivers, *Radio Science*, **36**, (5), 2001, 1221–1231.

[25] R. S. Conker, M. B. El-Arini, C. J. Hegarty, and T. Hsiao, Modeling the effects of ionospheric scintillation on GPS/SBAS availability, Bedford, MA, MITRE product MP 00W0000179, 2000.

[26] S. Pullen, G. Opshaug, A. Hansen, *et al.*, *A Preliminary Study of the Effect of Ionospheric Scintillation on WAAS User Availabilty in Equatorial Regions*, in Proceedings of ION GPS-98, Alexandria, VA, Institute of Navigation, 1998.

[27] M. Nakagami, *The m-distribution, a general formula of intensity of rapid fading*, In William C. Hoffman (editor) *Statistical Methods in Radio Wave Propagation*: Proceedings of a Symposium held 18–20 June 1958, pp. 3–36, London/New York, Pergamon Press, 1960.

[28] L. Devroye, *Non-Uniform Random Variate Generation*, New York, Springer-Verlag, 1986.

[29] *Streamlining Digital Signal Processing. Tricks of the Trade.*, Richard G. Lyons (editor), Hoboken, NJ, John Wiley & Sons, Inc, 2007.

[30] J. D. Parsons, *The Mobile Radio Propagation Channel*, New York, Wiley, 1992.

[31] J. Seo, T. Walter, E. Marks, T.-Y. Chiou, and P. Enge, Ionospheric scintillation effects on GPS receivers during solar minimum and maximum, Stanford University, *International Beacon Satellite Symposium 2007*, 11–15 June 2007, Boston, MA.

Exercises

Exercise 10.1. Use ReGen to calculate the electron density profile using a NeQuick model, as described in Section 4.4.4, to analyze dependency of TEC on time of day, time of year, solar activity, and latitude. Figure 10.3 shows an example of such plots. The TEC profiles are calculated for the same conditions for midnight and midday. One can see the dynamics of day–night changes in ionospheric layers (shown by Figure 4.6), in particular a disappearance of the D-layer at night.

Exercise 10.2. See how plumes on the NeQuick or IGS TEC plots in ReGen move with time in accordance with Ferraro's theorem.

Exercise 10.3. Use a ReGen simulator to simulate ionospheric irregularities with ionization level different from the background. The background model can be calculated with NeQuick or Klobuchar models (Figure 10.4). Analyze the impact of such irregularities on GNSS receiver behavior.

11 Geophysical measurements using GNSS signals

"And whilst all these nations have magnified their Antiquities so exceedingly, we need not wonder that the Greeks and Latines have made their first Kings a little older then the truth."
Sir Isaac Newton, The Chronology of Ancient Kingdoms, Chap. VI.

Though GNSS satellites are just some among many other satellites which are implemented for geophysical measurements, GNSS play a very significant and increasing role in collecting geophysical data. They have become an indispensable tool for many geophysical applications. Apart from this, GNSS satellites have also allowed spread spectrum radiowave technology to mature. This technology can be used on its own for atmosphere probing in other parts of the electromagnetic spectrum than GNSS's. Together with other atmosphere sensing instruments, GNSS provides information for weather prediction and monitoring climate change. GNSS has also become an essential tool for estimation of the Earth's rotation parameters. There are also emerging areas of paramount importance, which are yet under development, such as using GNSS signals for earthquake studies. Figure 11.1 depicts the subject of this chapter in relation to other chapters in this book.

Data from global reference networks are routinely used to monitor movement of the Earth's tectonic plates relative to each other and to an inertial frame. In this task the same type of GNSS observables as described in Chapter 8 are used. Such observables are created using measurements from globally distributed networks of GNSS receivers, supplying various data in raw data formats, mostly in RINEX. The most prominent among such networks is the IGS network. There are also various regional and local networks, which can provide large volumes of geophysical information. Among such networks, for example, is the Geographical Survey Institute (GSI) network in Japan, which consists of more than 1200 high-end geodetic GNSS receivers. The data from these networks include code and carrier phase measurements, Doppler measurements, and signal-to-noise ratio. All measurements are delivered from dual frequency receivers using two frequencies (L1, L2). We consider implementation of data from such networks in Sections 11.1, 11.3, and 11.4.

It is difficult to collect such data from the parts of the Earth covered by ocean. The oceans cover most of the Earth's surface and it is essential for many geophysical applications to access such data. Section 11.5 describes a system which makes this possible. The case study in this section covers a system which allows seafloor monitoring.

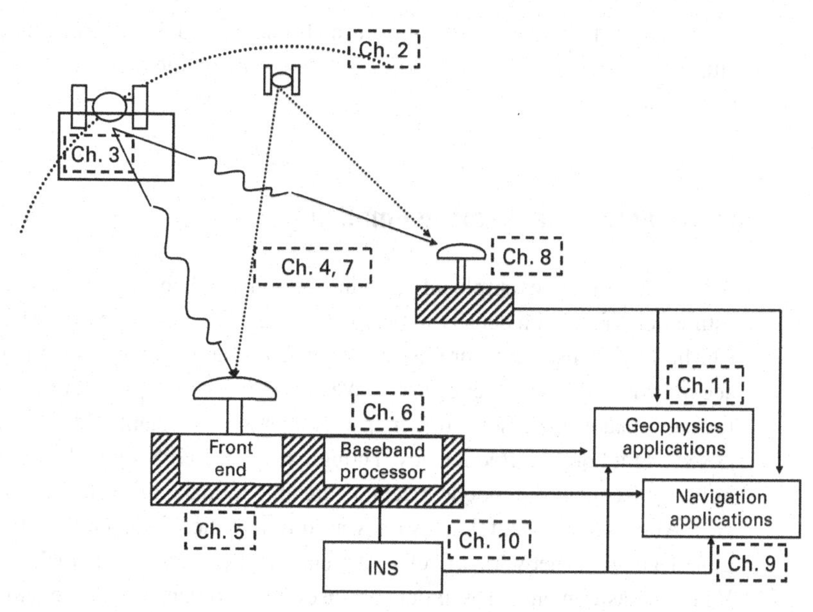

Figure 11.1 Subject of Chapter 11.

Section 11.1 describes how the concentration of free electrons in the ionosphere (TEC) is estimated. Information about evolution of TEC distribution may contain, apart from other valuable information, various earthquake precursors, which we consider in more detail in Section 11.6.

It is possible to increase the information derived from GNSS greatly if one can use measurements from receivers located on LEO satellites. In particular, it is possible to derive information on the vertical distribution of various atmospheric parameters. In Section 11.2 we consider radio occultation methods, which are implemented by using GNSS receivers on LEO satellites. This section describes how some information can be derived from the properties of the radio signal itself, in particular its amplitude and phase statistical parameters.

This type of information from the signal itself is not available in general from the networks. If a geodetic network provides data, usually with sampling once every 30 seconds, then in order to analyze the signal at least 50 Hz sampling is required (see Chapter 11). In the previous chapter we described a scintillation monitoring receiver and what kind of data it can provide. For signal analysis purposes such data may include de-trended raw phase and amplitude measurements, their statistics, narrow and wide-band noise estimations, signal-to-noise measurements, and code-carrier divergence. The recovery of information about the medium from scintillation data is based on the principles of ***diffraction tomography*** [1]. In medicine X-ray tomography is based on the same principles, and can determine a distribution of absorption coefficient in the medium. Diffraction tomography in addition can determine distribution of the refractive index. The power spectrum of amplitude and phase scintillation process (see the previous chapter) contains information about spatial distribution of irregularities in the medium. This spatial distribution can be calculated from frequency properties of the amplitude and phase scintillation. A similar method is also used for seafloor monitoring.

11.1 Ionospheric parameters estimation

A very important example of deriving geophysical information from GNSS data is estimation of electron density in the ionosphere or TEC (see Chapter 4). This information can be further used for various purposes, but mainly for improvement of positioning accuracy for single frequency users. We also consider an application of such information for earthquake precursors further in this chapter. The slant TEC is calculated using a GNSS signal on two frequencies. Though so far the GPS signal has a civilian signal on only one frequency, geodetic receivers are able to make measurements on both frequencies. After that the slant TEC is converted to a vertical TEC (see Section 4.4.1). With a global geodetic network of reference stations one has a potentially large number of VTEC measurements. These data are used in accordance with the adopted TEC distribution model. The parameters of such a model are estimated using accumulated TEC measurements. In Chapter 4 a few such models have been described, in particular the Klobuchar model. If we compare the Klobuchar model with the NeQuick model we see that the Klobuchar model gives a rather simplified representation of TEC distribution, and does not account, for example, for two bulges near the Earth equator. The reason is that this model has been designed to be broadcast in a navigation message and has therefore been optimized to be both compact and accurate enough for stand alone positioning. The NeQuick model described in Chapter 4, however, requires data on vertical electron distribution, and therefore its parameters cannot be estimated in full using ground based networks. Therefore, we have to look at other models which allow more precise mapping without vertical profiling.

We follow [2] to define such a TEC distribution model. This model is implemented, for example, by CODE (Center for Orbit Determination in Europe). Ionospheric maps using this model are available from the CODE website. They are calculated using the IGS network on a regular basis. We can describe TEC distribution over the globe using spherical harmonic expansions:

$$\mathrm{TEC}(\beta, s) = \sum_{n=0}^{n_{\max}} \sum_{m=0}^{n} \tilde{P}_{nm}(\sin\beta)(a_{nm}\cos ms + b_{nm}\sin ms), \tag{11.1}$$

where $n_{\max}$ is the maximum degree of the spherical harmonic expansion, β is geographic or geomagnetic latitude, s is sun-fixed longitude of the ionospheric pierce point, $\tilde{P}_{nm}$ is normalized polynomial associated Legendre functions, a_{nm}, b_{nm} are TEC coefficients of the spherical harmonics, i. e., the global ionosphere model parameters. Ths sun-fixed longitude is calculated as follows:

$$s = \lambda - \lambda_0, \tag{11.2}$$

where λ is the longitude of the ionospheric pierce point, λ_0 is the longitude of the sun. The associated normalized Legendre function can be expressed as follows:

$$\tilde{P}_{nm} = \Lambda(n, m)P_{n,m}, \tag{11.3}$$

where $P_{n,m}$ is a classical non-normalized Legendre function, and $\Lambda(n, m)$ is a normalization function of degree n and order m and can be expressed as

$$\Lambda(n, m) = \sqrt{\frac{(n - m)!(2n + 1)(2 - \delta_{om})}{(n + m)!}}, \tag{11.4}$$

where δ_{om} denotes the Kronecker delta.

This model is a single-layer model as a Klobuchar model, but with higher accuracy. However, this model requires a much greater quantity of data and therefore may not be well suited for broadcast purposes, at least from navigational satellites.

Summarizing, the observables formed from dual frequency code and phase measurements from the global network are used as measurements of slant TEC. Vertical TEC data calculated from slant TEC are used to find all a_{nm}, b_{nm} parameters, which best fit the model of (11.1) in the least squares sense.

If we add a network of GNSS receivers on LEO satellites to a global ground-based network, then we can estimate a three-dimensional electron distribution, providing that we have good a-priori model, similar to the NeQuick model described earlier. Such an approach is implemented also by the parameterised ionospheric model (PIM) using data from the IGS network and CHAMP LEO satellites [3]. The measurements are accumulated over time in order to compensate for the effects of satellite to receiver geometry.

The TEC information we can derive in such a way is important for two reasons. Firstly, in geodetic and navigation tasks it allows a GNSS user to compensate for ionospheric error in single frequency observables. In fact, for geodetic tasks it is recommended that even a dual frequency user uses single frequency observables and an IGS ionospheric map rather than dual frequency observables, if the baseline is less than 10 km [4].

Secondly, TEC distribution and its evolution with time can provide valuable information important for various geophysical tasks and even serve as an earthquake precursor database.

11.2 Radio occultation technique

In the previous section we looked at a method for ionospheric TEC estimation using a ground-based network of reference stations. This method does not allow us to create a vertical profile of the electron distribution, one for example similar to the NeQuick model considered in Chapter 4. If we consider a network of GNSS receivers located on LEO satellites, then we should be able to add much more information to those we have in the case of a ground network.

Such information has proved itself especially valuable when it comes from a network of spaceborne GNSS receivers located on LEO satellites. Because these satellites are constantly moving and the GNSS signal traverses the Earth's atmosphere by various paths, information on distribution of various parameters with altitude can also be derived from these data. This technique is mostly using eclipsing or ***occultating*** satellites.

In principle a data processing mechanism is not that different from the case of a ground network. The coordinates of the spaceborne receiver antenna should be estimated for each epoch as a part of the total solution along with the parameters of interest. The main problem in position estimation of LEO satellites is related to models of forces applied to the satellite. For LEO satellites we should include drag force in the consideration, because air resistance cannot be disregarded for low Earth orbits. Drag force is a most difficult force to model due to the many factors which may affect it. Luckily these forces can be safely neglected for GNSS MEO satellites.

The methods which work with signal paths from a GNSS satellite to a LEO satellite are referred to as ***GNSS radio occultation***. These methods were originally proposed for inter-planetary missions and then further developed by the Jet Propulsion Laboratory, USA. The first application of radio occultation was in 1964, using data from the US Mars mission spacecraft Mariner-4 [5]. Figure 11.2a shows the principle of radio occultation, as developed for interplanetary missions. The radio occultation method works in a similar way for studying the Earth's atmosphere by implementing GNSS – LEO satellites (Figure 11.2b). These methods make it possible to obtain data which are not available from any other remote sensing, including data on solar wind, and on the atmospheres of Venus, Titan, and Uranus, and the ionospheres of Jupiter, Saturn, and Venus. The methods have provided a lot of new information on temperature inhomogeneities,

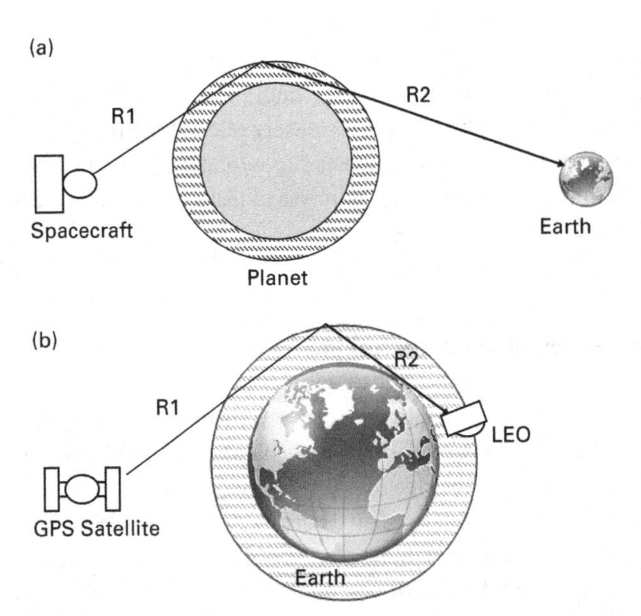

Figure 11.2 Principle of radio occultation for interplanetary missions, (a) (after [6]), and (b) for Earth atmosphere study with GNSS and LEO satellites.

turbulence, waves in planetary atmospheres, TEC irregularities and magnetic fields [7]. What we learn from these data about interactions of the solar wind with planetary ionospheres can help us to understand similar mechanisms in the Earth's atmosphere which affect weather, climate, and the very life of our planet.

Scintillation has been observed during every extra-planetary mission for more than four decades. Amplitude scintillation allows us to determine the structure of inhomogeneities and their orientations (because of Fresnel filtering features shown in the spectrum (all the minimums)). Phase scintillation provides information on scales larger than the radius of the first Fresnel zone. The radius of the first Fresnel zone in the case of planetary radio occultation measurements (see Figure 11.2) is defined [7] as

$$R_{\mathrm{F}} = \sqrt{\lambda R},$$
$$R = \frac{R_1 R_2}{R_1 + R_2}. \tag{11.5}$$

Similar occultation measurements are conducted in the Earth's atmosphere.

Apart from the dual frequency observables, we can derive much more information from the signal itself when it passes from a GNSS satellite to a LEO satellite. GNSS radio occultation methods provide information on temperature, water vapor, and electron density profiles in the atmosphere. Today GNSS radio occultation provides the best available precision for temperature profiling [8]. The main error sources from GNSS observations may be considered as quasi-random. By averaging multiple profiles one can derive temperature profiles with an accuracy better than 0.1 K. In long-term climate studies, a GNSS based method provides ten times better accuracy than any other method today [9].

In order to process all the large amount of data collected from LEO satellites and derive necessary information from it, it is necessary to use elaborate data assimilation algorithms. Figure 11.3 shows how information about various atmospheric parameters is derived from GNSS signals [10]. Using dual frequency observables and precise orbits, one can derive ray bending angles and consequently a refractive index. This information serves as a basis for deriving all essential atmospheric parameters. The number of methods for data assimilation includes use of a Kalman filter [11]. The rigorous theory of radio occultation methods is given in [7].

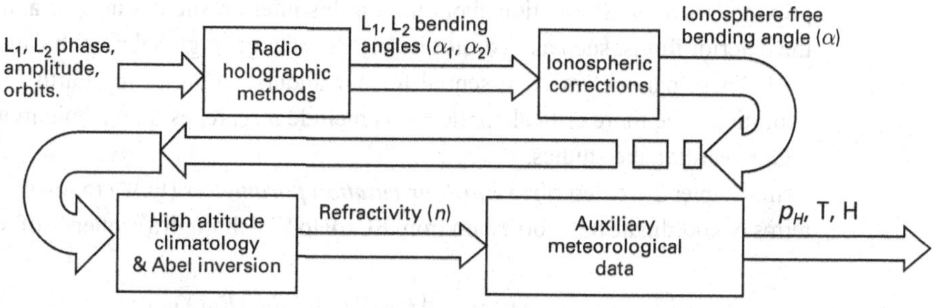

Figure 11.3 Flowchart of GNSS data assimilation method (after [10]).

Atmospheric monitoring using GPS receivers on LEO satellites has been in progress for years. The first successful implementation of a GNSS occultation method was the GPS/Meteorology (GPS/MET) experiment launched in 1995 on board MicroLab-1. The LEO satellites which were at different times used to derive such information are Challenging Minisatellite Payload (CHAMP), Gravity Recovery and Climate Experiment (GRACE), FORMOSAT-3/COSMIC, and others. Some data from LEO satellites are available to the public. For example, the Information System and Data Center for Geoscientific Data provides data from their satellite CHAMP [12].

As an example of weather prediction by occultation methods we note FORMOSAT-3. It provides approximately 2500 globally and uniformly distributed vertical ionospheric electron density profiles per day. The data have been successfully used for the prediction of tropical cyclogenesis. For example, these data allowed prediction of Hurricane Ernesto over the western Atlantic in August 2006. Data other than GNSS data can be particularly affected by clouds and precipitation. The GNSS data also provide much more information on low-level moisture. This type of information was especially important for predicting the evolution of the hurricane [13].

Another example of application of GNSS occultation methods is development of a global warming model and monitoring of its evolution. Global warming scenario monitoring requires access to vertically profiled data. The predicted scenario is associated with a warming troposphere and cooling stratosphere. The other key feature of the scenario is water vapor evolution in various regions [6]. These data can be delivered on such a scale only by GNSS occultation methods.

The next mission of LEO satellites for GNSS occultation should include receivers capable of tracking GLONASS and GALILEO satellites. It would allow construction of profiles with much better spatial and temporal resolution.

11.3 Earth rotation parameters estimation

Earth rotation parameters can be estimated with very long baseline interferometry (VLBI), satellite laser ranging (SLR), and lunar laser ranging (LLR) methods. Recently GNSS started to make a significant contribution to Earth rotation parameters estimation. Earth rotation parameters are essential for all geodetic tasks, when high-accuracy coordinate estimation is required. Apart from this, Earth rotation parameters provide essential information about the Earth's interior, such as its core and mantle, and the exterior fluids, such as hydrological cycle, change in global mass balance, and so on [14]. These parameters are essential for our understanding of the Earth's structure and evolution. The more critical applications include researches on tectonic movements and related earthquake studies.

In Chapter 2 we described ***Earth orientation parameters*** (EOP) following GPS ICD in terms of coordinate transformation from ECEF to ECI through a sequence of rotations [15]:

$$\vec{X}_{\mathrm{ECEF}} = [R_{\mathrm{PM}}][R_{\mathrm{ER}}][R_{\mathrm{N}}][R_{\mathrm{P}}]\vec{X}_{\mathrm{ECI}}, \tag{11.6}$$

where $[R_{PM}]$, $[R_{ER}]$, $[R_N]$, $[R_P]$ are rotation matrices for polar motion, earth rotation, nutation, and precession, respectively. We refine the definitions here following [16]. Earth rotation parameters describe the following:

1. The position of the non-rigid Earth axes in ECI. These axes are also called Tissérand axes after Félix Tissérand (1845–1896). This position is defined by nutation and precession.
2. The position of the Earth's rotation axis in relation to the figure axis and the angular velocity of the Earth. Polar motion and length of day (LOD) parameters are used to define a position of the Earth's rotation axis and angular velocity. Length of day is defined as follows:

$$\mathrm{LOD} = \frac{2\pi}{\omega_E}, \tag{11.7}$$

where ω_E is Earth angular velocity. Excess LOD (ΔLOD) is defined as follows:

$$\Delta\mathrm{LOD} = \mathrm{LOD} - 86\,400\ [\mathrm{sec}]. \tag{11.8}$$

3. Greenwich sidereal time, which is defined by an angle describing the position of a conventional meridian in ECI. It is described by the integral of the angular velocity over time. It gives Earth angular position around the figure axis at the particular epoch.

In particular, GNSS allows estimation of polar motion and length of day parameters. The estimates of these parameters using GNSS are among the best available today with respect to all other methods [17]. These estimates are freely available through the IGS. Reference [17] describes a method of LOD estimation and provides free tools to assist in making an estimate.

In the first section of this chapter we have seen that we use some assumptions about an underlying model. Then we find parameters that fit to this model best. This is a general approach. There are also identification methods, which can use a black box approach without any a-priori knowledge about a model. Such approaches may be required when information about the model is insufficient. In geophysical tasks an interdisciplinary approach is often required. The pitfall of this approach is that information from outside the field is often taken as axiomatic. It sometimes leads to discrepancies that cannot be resolved without relaxing constraints in the predefined models.

Information about changes in the LOD during the thousands of years of humankind's history can be derived from historical records of eclipses [18]. Such records are available starting from ancient times and in many countries [19]. We owe all the oldest eclipse records to Ptolemy. The oldest one is Babylonian, dated from 721 BC. However, such information is assimilated using underlying chronological models and constraints inherited from neighboring disciplines. These models are attributed mostly to the seventeenth century. Analyses of historical eclipses based on the planets' ephemerides and all forces laid over the chronological models leads to a significant discrepancy between physical and chronological models.

Taking chronological models as axiomatic leads to the necessity to revise the physical models. Analyses of historical eclipse observations from 500 BC to 1990 AD showed

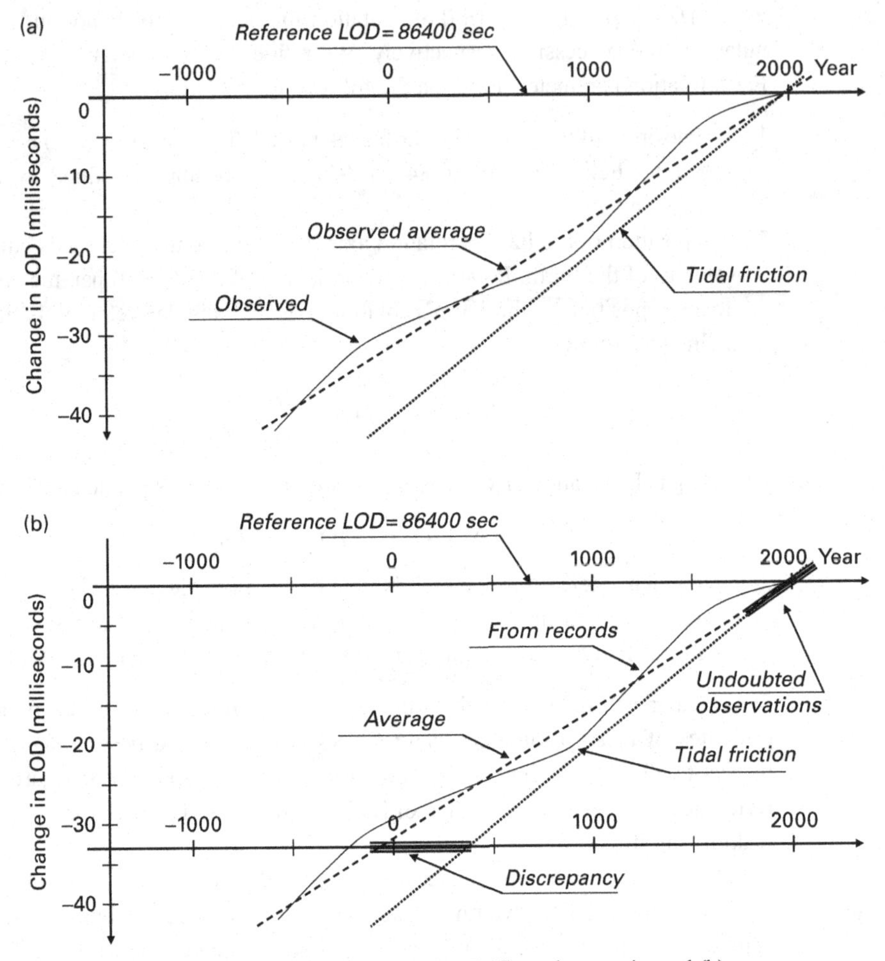

Figure 11.4 Changes in the LOD from historical data (a) (after [18]), and approximated (b).

that changes in long-term rate of the Earth's rotation cannot fit into the physical model (see Figure 11.4a). It was found that there is a significant and changing non-tidal component which should be added to tidal friction [18]. However, we can point out another way to resolve this inconsistency by relaxing constraints in chronological models. From Figure 11.4b we can see that we have about 500 years of discrepancy around Ptolemy's time.

Here we also need to mention Ptolemy's critics. It has been noted that the data in Almagest fit to the model developed by Ptolemy rather too well [20]. It has also been noted that the observational data in Almagest do not correspond to data which come from backward interpolation of planetary movement. Therefore one may conclude that some of the data were falsified. Regarding the accuracy of data fit, Ptolemy's model was constructed by fitting the available data. We mentioned before that Ptolemy's model, as almost any other series expansion, can be developed basically to any accuracy. The discrepancy in backward extrapolated data is, however, a much more interesting point.

Also it cannot be resolved within the constraints of current chronological models [21]. If, however, we assume that chronological models can be revised, then we may drop our suspicions of Ptolemy's dishonesty as a scientist and regain our trust in the physical models, which embody our understanding of the physical world.

Indeed, if we try to recalculate the time of Ptolemy's life using his observations from Almagest and fit them to the backward extrapolated planetary data, we find that these data bring Ptolemy's lifetime about 500 years forward [21]. This would remove discrepancies in Ptolemy's observations and the discrepancy between observed average change in LOD (1.7 ms/cy) and tidal friction (2.3 ms/cy). That also gives the best solution in the Ockham's razor sense. It is interesting that the correctness of the accepted chronological models has been doubted since Isaac Newton's work [22]. His work covered chronological models AD and also suggested shortening the time between ancient events and the present.

11.4 GNSS in earthquake study

We consider here earthquakes, which are caused by relative movement of tectonic plates and mostly appear in the places where the tectonic plates come together. Deformation of the Earth's crust, which results from relative movement of tectonic plates, causes an accumulation of large strain energy. The slow movement is very small. Only GPS has made it possible for the first time to measure these small movements routinely at the level of millimeters. This has provided a unique opportunity to constantly monitor the Earth's crustal deformation. The GSI GPS network has over 1200 high-end geodetic receivers providing a wealth of observational data. Figure 11.5 shows the results of crustal stress monitoring with GPS in the Japan Izu peninsula area, where at least three tectonic plates come together. Accumulation of GPS observations allows various analyses of the data. Figure 11.6 shows maximum shear strain rates in central Japan, based on two year GPS data [23]. It is interesting to note that most of evidence of plate tectonics comes from seafloor measurements [43], which we consider in the next section.

The strain energy accumulated over time is released catastrophically from time to time in the form of earthquakes. The longer the time elapsed since an earthquake, the higher the probability of a new one. This method of forecasting is called the ***seismic gap method*** and it is based on elapsed time since the previous stress release in a plate boundary area. White circles in Figure 11.6 indicate epicenters of earthquakes with magnitude greater than 3.0. The parameters of these releases can also be monitored by GNSS.

An earthquake generates body waves, which propagate underground from the epicenter, and surface waves, which propagate above ground. The body waves propagate faster than surface waves, but the surface waves can cause more damage than body wave. There are two types of body waves, compressional waves, also called primary waves or P-waves, and secondary waves, called S-waves. The P-waves move with a velocity about 6 km/sec and cause volume and density changes in the material. The S-wave moves much slower with a velocity of about 3.5 km/sec and cause changes in material shape but not its volume. Their waveforms can be very complicated and they are most dangerous for

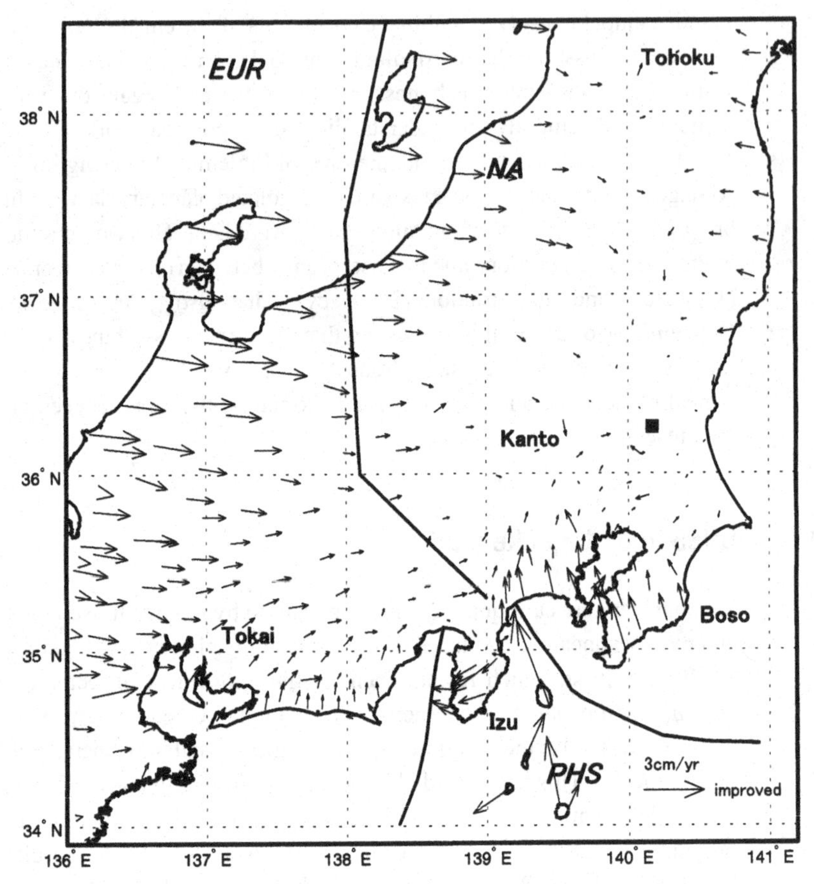

Figure 11.5 Monitoring of crustal stress with GPS (from [23], reproduced with permission).

buildings. The location of the epicenter is determined using measurements of time difference between the first arrivals of P and S waves from specific locations around the earthquake in a similar manner to position determination from GNSS satellites (see Figure 11.7). A warning system triggered by registration of the body waves in the stations closest to an epicenter can give a warning time within some tens of seconds for people to leave a building before the surface waves arrive [24].

11.5 Case study: seafloor monitoring system

The oceans cover two thirds of the Earth's surface. Earthquake monitoring and warning systems could benefit greatly if the GNSS network could cover these surfaces as well. We can also gain a lot of knowledge if the GNSS network is expanded towards the oceans. Seafloor monitoring systems considered in this chapter allows monitoring of earthquake dynamics. Because such systems can achieve high accuracy, they can also provide data for a better understanding of tectonic plate geometry and motions.

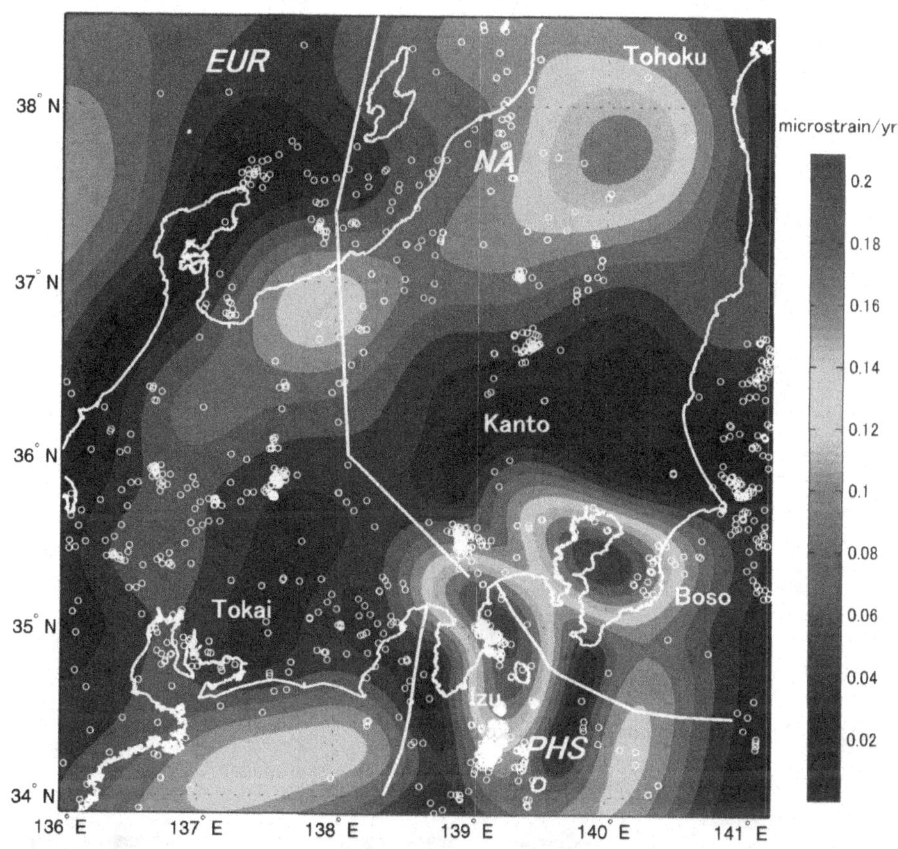

Figure 11.6 Maximum shear strain rates in central Japan, based on two year GPS data. White circles indicate epicenters of earthquakes with magnitude greater than 3.0. (from [23], reproduced with permission).

Land-based GPS observations enable the detection of co-seismic and post-seismic displacement with millimeter levels of accuracy. However, seafloor motion cannot be measured by GPS directly though many earthquakes occur in the ocean, for example, the earthquake in Japan on 11 March 2011 with magnitude 9. A GPS and acoustic combined methodology to detect seafloor crustal movement was first proposed by the Scripps Institute of Oceanography [25]. The system allowed successful detection of the movement of the Juan de Fuca plate [26]. Since then, similar systems have been developed in Japan [27],[28] and in Taiwan [29]. A configuration of the GPS/acoustic seafloor positioning system is shown in Figure 11.8.

The system comprises onboard GPS receivers, acoustic transducers, and mirror transponders on the seafloor. The onboard GPS receivers provide for real-time kinematic positioning and attitude determination of the vessel. If the seafloor stations are located within a hundred kilometers offshore, then reference GPS receivers are installed near the coast and medium range kinematic positioning is implemented. If the seafloor stations are located further away from the coast, then the precise point positioning (PPP) method is

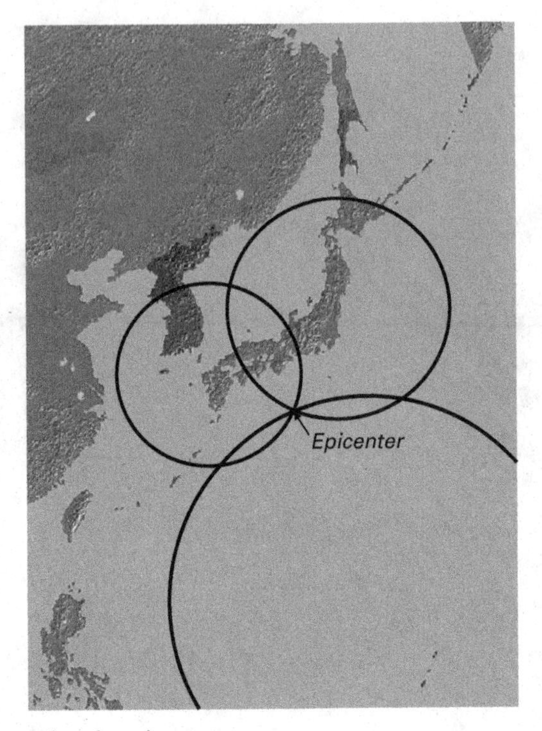

Figure 11.7 Determination of an earthquake epicenter.

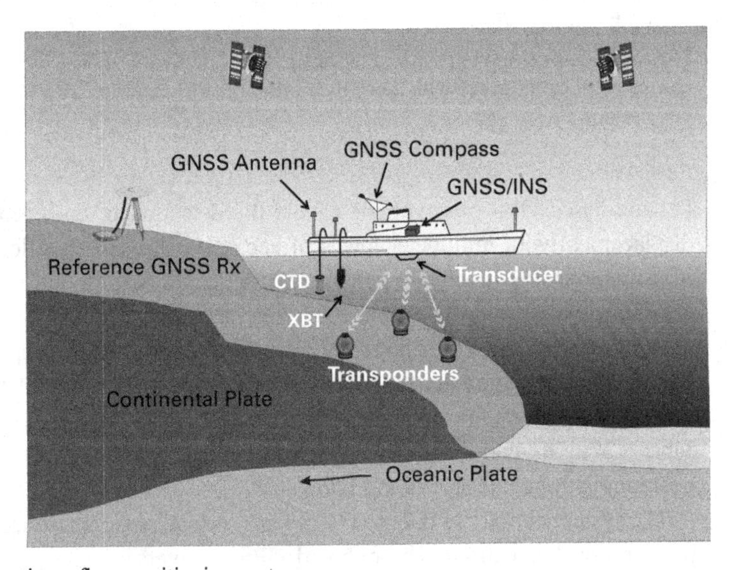

Figure 11.8 GPS/acoustic seafloor positioning system.

applied. Off-the-shelf GPS/INS is implemented for the vessel attitude determination. A longer distance between on-board antennas is used in order to improve the attitude accuracy [28].

The acoustic transducer and transponders constitute the range measurement system. The transducer transmits a ranging signal, the transponder receives it and transmits it back. The distance between transducer and transponder is determined by measuring the two way travel time of the acoustic signal. Usually three or four transponders are installed on a seafloor observation site. Since the range is computed from the acoustic wave travel time, the accuracy of acoustic wave speed determination is a limiting factor. The measurements of acoustic wave velocity profile are conducted using a conductivity, temperature, and depth profiler sensor (CTD) and an eXpendable bathy thermograph (XBT) sensor. This system allows measurement of the horizontal seafloor velocity with an accuracy of 1 cm/year. Figure 11.9 shows a transponder which is used for measurement of the seafloor off the eastern coast of Taiwan [29]. The transducer is installed underneath a buoy, which is dragged by a vessel as shown in Figure 11.10. Three GPS antennas are mounted on the buoy for attitude determination purposes.

Now we look at an example of measurements of seafloor displacement caused by an earthquake. In this example we consider two large earthquakes, which occurred successively off the coast southeast of the Kii Peninsula, Japan on 5 September 2004 within five hours. The moment magnitudes (M_w) were 7.2 and 7.5 correspondently. The epicenters of the earthquakes are shown in Figure 11.11 by crosses. The seafloor observation site with three transponders was located 60 km offshore on the Eurasian plate. It is shown in the figure by a circle. The locations of three ground reference GPS receivers are shown in the figure by the solid circles. The epicenters of the earthquakes were 60–70 km from the seafloor observation site and close to the Nankai Trough, which is at the border of the Philippine Sea plate and the continental plate. As a result of comparing the position estimates made before and after the earthquakes, a 21.5 cm southward displacement has been detected.

Figure 11.9 Transponder.

Figure 11.10 Buoy with GPS antennas for attitude determination dragged by a vessel.

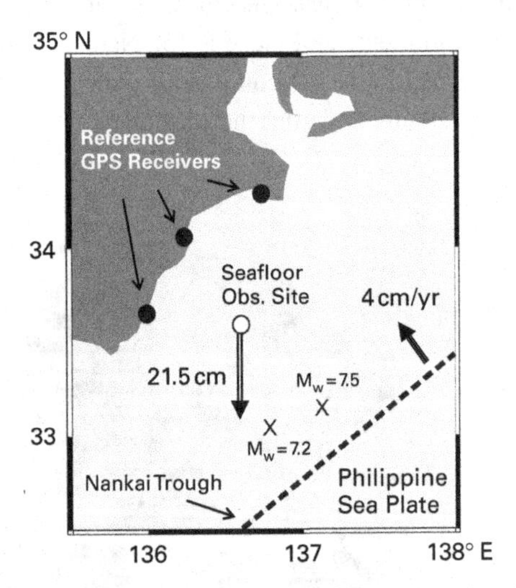

Figure 11.11 Results of observations of seafloor displacement caused by earthquakes (after [28]).

Although seafloor positioning with centimeter-level accuracy is possible, long time observation sessions for several tens of hours are necessary in order to average effects of acoustic speed variation in seawater. Several efforts have been made to improve the accuracy of estimating acoustic speed. For example, using five transponders could improve accuracy since two horizontal gradient parameters of acoustic speed could be estimated in addition to the usual three parameters such as longitude, latitude, and average acoustic speed [30].

11.6 Earthquake precursors in the ionosphere

There are many researches devoted to investigation of correlations between earthquakes and ionospheric anomalies which preceded them. GNSS today is the best tool for measuring and studying these anomalies. The TEC mapping method described in the Section 11.1 allows us to find and monitor such anomalies. Radio occultation methods described in Section 11.2 can also be used to find geographical and temporal distributions of the ionospheric disturbances from scintillation data [31].

11.6.1 Earthquake forecast and prediction using ionospheric precursors

Some specific anomalies in the ionosphere can be used as earthquake precursors. First of all we need to define what we can expect from observing precursors. ***Earthquake prediction*** can be defined as an ability to calculate the specific magnitude, place, and time for a given future earthquake. ***Earthquake forecasting*** is defined as estimation of probabilities that an earthquake of large magnitude and faulting type will occur within a specified period of time. It is a common view that earthquake prediction is not feasible in the near future. The forecasting is, however, based on existing models of strain energy accumulation and release (see Section 11.4).

Ionospheric precursors are observable variations in the state of the ionosphere before earthquakes. Earthquake precursors are somewhere closer to prediction. Precursors cannot give information about the earthquake location to an accuracy better than thousands of kilometers. Precursors cannot give information about the earthquake more accurately better than a day or two. Finally, precursors cannot give information about the earthquake magnitude with probability better than a certain value. In many cases such information may commonly be disregarded, even if available. The potential cost of acting on such uncertain information may exceed the cost of damage from the earthquake itself. However, studying such precursors may, on the one hand, allow future development of prediction methods, and, on the other hand may allow better understanding of earthquake mechanisms.

The first correlations between earthquakes and the ionospheric anomalies which preceded them were described in 1964 in relation to a big Alaska earthquake [32].

11.6.2 Mechanisms responsible for ionospheric precursors

We look at the events related to and associated with earthquakes, which may be connected to the appearance of precursors. Ionospheric precursors manifest themselves as anomalies in the ionosphere, which can be observed as disturbances in TEC distribution and through scintillations. There are many observations of a local increase in TEC in the two to three days prior to strong mid-latitude earthquakes, and a decrease up to 30% near the epicenter in the one to two days before the event. The disturbed region maximum in these cases was located in close proximity to the earthquake epicenter [33]. A number of researches show that TEC decreases within three–four days prior to an earthquake and this decrease in TEC is accompanied by anomalous crests with specific dynamics [34], [35]. The large number of such observations allows us to state that existence of ionospheric precursors within a few days interval prior to the earthquake is statistically proven [36]. There is a number of theories to explain the ionospheric precursors [37],[36].

In general we associate ionospheric anomalies with two types of process, global-scale and local-scale:

1. *Global-scale* mechanisms of precursor ionospheric anomalies. Variations in ionospheric TEC distribution are correlated with variations in other geophysical parameters. These TEC variations can be explained by variations of these correlated parameters, including electromagnetic parameters and temperature, or by the fact that they are caused by the same source. Radiogenic heat, tidal dispatch in the solid Earth, and frictional dissipation in the connecting mantle ensure strong correlations between the Geoid and seismic velocity anomalies in the lower mantle [38]. Consequently, complex global telluric currents may be partly responsible for ionospheric precursors, i.e. correlations between seismic activity, TEC distribution, and distribution of other physical parameters.
2. *Local-scale* mechanisms of precursor ionospheric anomalies. Earthquakes are connected to faults in the Earth's crust and upper mantle. Generally earthquakes are recorded at plate boundaries, though a small number can be observed to originate within plates. When plates are moving relative to each other, stress is built up in the material until it exceeds the material's strength [39]. The local process associated with the accumulation of stress may be responsible for TEC anomalies. Such local-scale mechanisms are also shown in physical changes in rock properties in the vicinity of faults, release of trace gas, and foreshocks. These have already provided successful short-term predictions. For example, an earthquake near Haicheng in Northeast China in February 1975 was predicted five hours before it occurred, based on such precursors.

We look below at further anomalies in physical parameters which are correlated with ionospheric precursors.

11.6.3 Earthquake lights

Sometimes some precursory anomalies are even visible to the naked eye. We noted the importance of the visible part of the electromagnetic spectrum in Chapter 4. The first time earthquake lights were noted was in 1965 during the Matsushito earthquake in Japan. The lights were photographed [40]. There have been many other occurrences of earthquake lights. Some of such lights are reported as a red and white glow directly prior to an earthquake. A more recent example was observed in Tianshui, Gansu province about 30 minutes before the Sichuan earthquake on 12 May, 2008 at a distance of about 450 km from the epicenter. The luminous clouds resemble some characteristics of auroras. Earthquake lights can be explained by both global and local mechanisms. The global mechanism is similar to the one which causes auroras. The local mechanisms are of an electrochemical nature. Some theories regarding the nature of the lights are set out below:

1. *Piezolelectric effect.* Electric fields are formed in quartz rich crustal rocks and are subject to high mechanical pressure. However, in this case the lights would be accompanied by extreme heat, which has never been observed and reported [41].
2. *Oxidation.* Oxidation causes ionization of atoms. Oxidation is described as a common weathering reaction when a mineral originally formed in an oxygen-free environment becomes exposed to the air due to mechanical processes. Oxidation also diminishes the strength of the mineral, which in turn makes it more prone to further mechanical deformation [42].
3. *Aurora-type lights.* Earthquake lights have a similar nature to auroras and are caused basically by a current passing between the atmosphere and the Earth. The light comes from atoms that have been ionized or excited. When a molecule emits radiation, it releases energy and returns to its normal state. The emitted radiation covers a spectrum also visible to the human eye. Different gases emit radiation on different frequencies, which results in various colors. The process is similar to an aurora light phenomenon. In the case of aurora lights near the poles, electrons move along magnetic field lines down to the Earth. On their way, at between 100 and 200 km altitude they collide with ions. The result of the collisions is visible as aurora lights.

In our view, TEC anomalies are consistent with aurora theory, because they show the areas where the electron content is lower due to the current down to the Earth's surface.

11.6.4 Temperature variations

Thermal convection is considered as the driving mechanism of plate tectonics. The heat source hypothesis explains the concentration of radioactive elements near the surface, produced by a more than 70 km thick layer of rock [43]. The temperature variations can cause variations of electron content in the atmosphere. It is well known now that irregularities in the F layer may be artificially induced by high power, ***high frequency (HF)*** waves, called heaters [44]. The artificial irregularities are caused by thermal forces. They reach maximum amplitude 30–40 sec after the heater is turned on and disappear one–two minutes after the heater is off. These heaters also induce scintillation. For example, an HF heater at 5.1 MHz with effective radiated power of 200 MW can induce an intensity of scintillation with $S_4 = 0.12$ [45]. The Fresnel minima in the intensity spectra coincide with Fresnel maxima in the phase scintillation spectra. This indicates that heater generated irregularities are in the layer with thickness less than 50 km.

11.6.5 Radon emission

Ionospheric precursors can also be explained by escape of radon [41]. The Earth's lithosphere also contains uranium, sometimes with high concentration. It produces mineral radium as a daughter product. The radium is radioactive with a decay half-life of 1620 years. It is responsible for forming the heavy radioactive gas radon as a daughter product of its decay. Radon may accumulate underground. Sometimes it escapes and can accumulate in a building, causing a rather high health risk. It decays with a half-life of 3.8 days, producing daughter products of polonium, bismuth, and lead.

Earthquakes may cause a massive escape of radon starting within a few days prior to the event. Each radon atom can produce 1 million ion pairs. Earthquake lights then can originate from ionizing collisions between alpha particles and atoms. In relation to TEC anomalies, it was noted that the relative distribution of radon atoms and excess of positive ions as a function of altitude are similar [46]. Radon observations as precursors are given particular emphasis in China [47]. These precursors may be caused by premonitory land deformation.

11.6.6 Electric variations

Apart from earthquake lights, changes in the Earth's electric current are among electrical phenomena most often reported in relation to earthquake precursors. They have been reported in many researches. It is interesting that these changes have never been associated with aftershock activity [48].

The electromagnetic structure of the Earth is tremendously complex and our models are probably not yet sufficient to describe the earthquake precursory mechanisms. There is a combination of local and global mechanisms working in the Earth's lithosphere, atmosphere, low atmosphere, and local surface areas. Generally when considering atmospheric electricity we can accept a global model of the ionosphere as equivalent to a metallic conductor, with potential difference of the order 300 000 V above the potential of the Earth's surface. Earth-ionosphere current is about ~1800 A on average and total effective resistance ~200 ohms. To add to the complexity of this system, the Earth's liquid outer core consists of iron–nickel alloy, which is an electrical conductor. Furthermore, we need to account for the electrical currents in the lower atmosphere layers and between atmosphere and the Earth's surface. The electron density of the upper atmosphere is controlled by the electrical structure of the lower ionosphere [49]. This is partly defined by the fair-weather electric field. Fair-weather conditions are pictured as a uniform leak between the highly conducting ionosphere and the Earth spherical condenser plates. The fair-weather electric field was discovered by Lemonnier in 1752. It can be modeled by an excess of positive ions in the lower atmosphere [46]. Lemonnier detected electrical charges without clouds in the atmosphere and noticed that they are different at day and night. In this model thunderstorms are generators which provide return current. The currents associated with earthquakes have very different timescales and are associated with different atmospheric layers. The model should account for the current in the atmosphere as well for the earth current.

The earth current is measured as a difference in electric potential by electrodes buried in the ground, It is not possible to measure earth current in Japan, which otherwise provides a large bulk of research material in relation to earthquakes, because of stray electric currents leaking from DC (direct current) operated railways [48]. Many such measurements, however, have been made in Russia and China, where man-made noise is low. For example, earth potential dropped from 100 mV to 0 within 2 days at a distance of a few tens of kilometers from the Haicheng earthquake epicenter (M = 7.3, 1975). The distance between electrodes was 60 m. Similar measurements have been conducted in the USA and Russia. The first time such phenomena were reported was in 1972 [50].

Measurements related to earth resistivity have three different time scales, which can be attributed to different mechanisms:

1. *Long-term changes*. These changes have been detected using methods of electromagnetic sounding of the Earth's crust. These methods are used for monitoring the state of the crust in seismo-active regions. They were developed to facilitate offshore prospecting to search for areas rich with oil or gas. A progressive increase in conductivity of rocks in the two–three months before an earthquake has been reported. An increase in conductivity is explained by deformation, which causes micro cracks. These micro cracks fill with water, which causes an increase in conductivity. Subsurface sedimentary structures, such as faults, folds, and domes that trap oil can then be mapped [51].
2. *Medium-term changes*. Sometimes changes in resistivity were measured from 10 days before and disappeared about 5 days before an earthquake.
3. *Short-term changes*. There are also short-term precursors, in which case the time for a ground-based resistivity change is a few hours. For example, such changes in resistivity were measured four hours prior to the Izu Hanto-Oki earthquake (M = 6.9,1974). The epicentral distance was about 100 km. No such effects were reported, however, for the Izu Peninsula earthquake in 1978 (M = 7.0). It is interesting that it was in the same place with an even shorter distance to the epicenter [47].

In this case the physics behind development of TEC depletion areas prior to the earthquakes can be explained by the same mechanism which is used to explain ionospheric bubbles. These bubbles (Chapter 10) can be explained by Rayleigh–Taylor instability, which is coupled with an electric current [52].

11.6.7 Magnetic variations

The Earth's magnetosphere is also affected by earthquakes. There are also changes in geomagnetic field, but they are difficult to distinguish from noise and short-term variations. Some relatively small changes in the magnetograms are related to stress variations in the rocks. These changes can also appear prior to an earthquake. They are rather difficult to spot, because they are masked by other slow and long-term and especially diurnal variations [53].

References

[1] M. Born and E. Wolf, *Principles of Optics. Electromagnetic Theory of Propagation, Interference and Diffraction of Light*, seventh expanded edition, Cambridge, Cambridge University Press, 1999.

[2] S. Schaer, *Mapping and Predicting the Earth's Ionosphere using the Global Positioning System*, Volume 59 of Geodätisch-geophysikalische Arbeiten in der Schweiz, Schweizerische Geodätische Kommission, Institut für Geodäsie und Photogrammetrie, Eidg. Technische Hochschule Zürich, Zürich, Switzerland, 1999.

[3] M. Angling, An assessment of an ionospheric GPS data assimilation process, in *Earth Observation with CHAMP, Results from Three Years in Orbit*, C. Reigber, H. Lühr, P. Schwintzer, and J. Wickert (editors), Berlin/Heidelberg, Springer-Verlag, 2005.

[4] R. Dach, U. Hugentobler, P. Fridez, and M. Meindl (editors), *User Manual of the Bernese GPS Software Version 5.0*, Astronomical Institute, University of Bern, 2007.

[5] G. Fjeldbo, W. Fjeldbo, and R. Eshleman, Models for the atmosphere of Mars based on the Mariner-4 occultation experiments, *J. Geophys. Res.*, **71**(9), 1966, 2307.

[6] R. Woo, Spacecraft radio scintillation and solar system exploration, in *Wave Propagation in Random Media (Scintillation)*, V. I. Tatarskii, A. Ishimaru, and V. U. Zavorotny (editors), Bellingham, WA, The Society of Photo-Optical Instrumentation Engineers and Institute Of Physics Publishing Ltd., USA, 1993.

[7] W. G. Melbourne, *Radio Occultations Using Earth Satellites, A Wave Theory Treatment*, JPL Deep Space Communications and Navigation Series, Hoboken, NJ, John Wiley & Sons, Inc., 2005.

[8] E. R. Kursinski, *et al.*, Observing Earth's atmosphere with radio occultation measurements, *J. Geophys. Res.* **102**(D19), 1997: 23429–23465.

[9] T. P. Yunck and G. A. Hajj, Atmospheric and ocean sensing with GNSS, in *Earth Observation with CHAMP, Results from Three Years in Orbit*, C. Reigber, H. Lühr, P. Schwintzer, and J. Wickert (editors), Berlin/Heidelberg, Springer-Verlag, 2005.

[10] S. Syndergaard, Y.-H. Kuo, and M. S. Lohmann, Observation operators for the assimilation of occultation data into atmospheric models: A review, in *Atmosphere and Climate Studies by Occultation Methods*, U. Foelsche, G. Kirchengast, and A. Steiner (editors), Berlin/ Heidelberg, Springer-Verlag, 2006.

[11] G. Evensen, *Data Assimilation*, Berlin/Heidelberg, Springer-Verlag, 2007.

[12] N. Jakowski, K. Tsybulya, S. M. Stankov, and A. Wehrenpfennig, About the potential of GPS radio occultation measurements for exploring the ionosphere, in *Earth Observation with CHAMP, Results from Three Years in Orbit*, C. Reigber, H. Lühr, P. Schwintzer, and J. Wickert (editors), Berlin/Heidelberg, Springer-Verlag, 2005.

[13] C.-J. Fong, S.-K. Yang, C.-H. Chu, *et al.*, FORMOSAT-3 / COSMIC constellation spacecraft system performance: After one year in orbit, *IEEE Transactions on Geoscience and Remote Sensing*, **46**, (11), 2008, 3380–3394.

[14] R. Rummel, G. Beutler, V. Dehant, *et al.*, Understanding a dynamic planet: Earth science requirements for geodesy, in *Global Geodetic Observing System Meeting the Requirements of a Global Society on a Changing Planet in 2020*, H.-P. Plag, and M. Pearlman (editors), Berlin/ Heidelberg, Springer-Verlag, 2009.

[15] GPS IS, *Navstar GPS Space Segment/Navigation User Interfaces*, GPS Interface Specification IS-GPS-200, Rev D, Fountain Valley, CA, GPS Joint Program Office, and ARINC Engineering Services, March 2006.

[16] G. Beutler, *Methods of Celestial Mechanics, Volume II: Application to Planetary System, Geodynamics and Satellite Geodesy*, Berlin/Heidelberg, Springer-Verlag, 2005.

[17] G. Beutler, *Methods of Celestial Mechanics, Volume I: Physical, Mathematical,and Numerical Principles*, Berlin/Heidelberg, Springer-Verlag, 2005.

[18] F. R. Stephenson, *Historical Eclipses and Earth's Rotation*, Cambridge, Cambridge University Press, 1997.

[19] B. Hetherington, *A Chronicle of Pre-telescopic Astronomy*, Hoboken, NJ, John Wiley & Sons Ltd., 1996.

[20] Robert R. Newton, *The Crime of Claudius Ptolemy*, Baltimore, MD, The Johns Hopkins University Press, 1977.

[21] A. T. Fomenko, V. V. Kalashnikov, and G. V. Nosovsky, *Geometrical and Statistical Methods of Analysis of Star Configurations: Dating Ptolemy's Almagest*, London, UK, CRC Press, Inc., 1993.

[22] Isaac Newton, *The Chronology of Ancient Kingdoms Amended To which is Prefix'd, A Short Chronicle from the First Memory of Things in Europe, to the Conquest of Persia by Alexander the Great*, London, Printed for J. Tonson in the Strand, and J. Osborn and T.Longman in Paternoster Row, MDCCXXVIII.

[23] J. Li, K. Miyashita, T. Kato, and S. Miyazaki, GPS time series modeling by autoregressive moving average method: Application to the crustal deformation in central Japan, *Earth Planets Space*, **52**, 2000, 155–162.

[24] W. D. Mooney and S. M. White, Recent developments in earthquake hazards studies, in *New Frontiers in Integrated Solid Earth Sciences*, S. Cloetingh, and J. Negendank (editors), Dordrecht, Springer Science+Business Media B.V., 2010.

[25] F. N. Spiess, Suboceanic geodetic measurements, *IEEE Trans. Geosci. Remote Sens.*, **23**, 1985, 502–510.

[26] F. N. Spiess, C. D. Chadwell, J. A. Hildebrand, *et al.*, Precise GPS/acoustic positioning of seafloor reference points for tectonic studies, *Phys. Earth Planet. Inter.*, **108**, 1998, 101–112.

[27] M. Fujita, Y. Matsumoto, T. Ishikawa, *et al.*, Combined GPS/acoustic seafloor geodetic observation system for monitoring offshore active seismic regions near Japan, Proc. ION GNSS-2006, Fort Worth, Texas, 2006.

[28] R. Ikuta, K. Tadokoro, M. Ando, *et al.*, A new GPS-acoustic method for measuring ocean floor crustal deformation: Application to the Nankai Trough, *J. Geophys. Res.*, **113**, 2008, doi:10.1029/2006JB004875.

[29] H. Y. Chen, S. B. Yu, H. Tung, T. Tsujii, and M. Ando, GPS medium-range kinematic positioning for the seafloor geodesy off Eastern Taiwan, *Engineering Journal of the Chulalongkorn University*, **15**, (1), 2011, 17–24.

[30] M. Kido, Y. Osada, and H. Fujimoto, Temporal variation of sound speed in ocean: a comparison between GPS/acoustic and in situ measurements, *Earth Planets and Space*, **60**, 2008, 229–234.

[31] A. G. Pavelyev, J. Wickert, Y. A. Liou, A. A. Pavelyev, and C. Jacobi, Analysis of atmospheric and ionospheric wave structures using the CHAMP and GPS/METRadio occultation database, in *Atmosphere and Climate Studies by Occultation Methods*, U. Foelsche, G. Kirchengast, and A. Steiner (editors), Berlin/Heidelberg, Springer-Verlag, 2006.

[32] K. Davies and D. M. Baker, Ionospheric effects observed around the time of the Alaskan earthquake of March 28, 1964, *J. Geophys. Res.* **70**, 1965, 2251–2253.

[33] A. A. Namgaladze, M. V. Klimenko, V. V. Klimenko, and I. E. Zakharenkova, Physical mechanism and mathematical modeling of earthquake ionospheric precursors registered in total electron content, ISSN 0016□7932, *Geomagnetism and Aeronomy*, **49**, (2), 2009, 252–262, Pleiades Publishing Ltd.

[34] J. Y. Liu, Y. J. Chuo, S. J. Shan, *et al.*, Pre-earthquake anomalies registered by continuous GPS TEC measurements, *Annales Geophysicae*, **22**, 2004, 1585–1593.

[35] J. Y. Liu, Y. J. Chuo, S. A. Pulinets, H. F. Tsai, and X. Zeng, A study on the TEC perturbations prior to the Rei-Li, Chi-Chi and Chia-Yi earthquakes, in *Seismo-Electromagnetics: Lithosphere-Atmosphere-Ionosphere Coupling*, M. Hayakawa and O. Molchanov (editors), Tokyo, TERRAPUB, 2002, pp. 297–301.

[36] S. Pulinets and K. Boyarchuk, *Ionospheric Precursors of Earthquakes*, Berlin/Heidelberg, Springer-Verlag, 2004.

[37] V. A. Liperovsky, O. A. Pokhotelov, C. V. Meister, and E. V. Liperovskaya, Physical models of coupling in the lithosphere–atmosphere–ionosphere system before earthquakes, *Geomagnetism and Aeronomy*, **48**, (6), 2008, 795–806, Pleiades Publishing, Ltd. (Original Russian Text published in *Geomagnetizm i Aeronomiya*, **48**, (6), 2008, 831–843.)

[38] J. P. Poirier, *Introduction to the Physics of the Earth's Interior*, 2nd edition, Cambridge, Cambridge University Press, 2000.

[39] F. Press and R. Siever, *Earth*, 4th edition, New York, W. H. Freeman and Company, 1986.

[40] Y. Yasui, A summary of studies on luminous phenomena accompanied with earthquakes, *Proc. Kakioka Magnetic Observ.*, **15**, (1), 1973.

[41] H. Mitzutani, T. Ishido, T. Yokokura, and S. Ohnishi, Electrokinetic phenomena associated with earthquakes, *Geophys.Res.Lett.*, **3**, 1976, 365–368.

[42] *Encyclopedia of Geomorphology*, A. Goudie (editor), Abingdon, UK, Routledge, 2003.

[43] J. C. de Bremaecker, *Geophysics: The Earth's Interior*, Hoboken, NJ, John Wiley & Sons Inc., 1985.

[44] L. Erukhimov, V. Kovalev, A. Lerner, *et al.*, Spectrum of large-scale artificial inhomogeneities in the F layer of the ionosphere, *Radio Phys. & Quantum Electron.*, **22**, (10), 1979, 1278–1281.

[45] K. Rawer, *Wave Propagation in the Ionosphere*, Dordrecht, Kluwer Academic Publishers, 1993.

[46] L. Wåhlin, *Atmospheric Electrostatics*, Baldock, UK, Research Studies Press Ltd., 1986.

[47] T. Rikitake, *Earthquakes Forecasting and Warning*, Japan/Tokyo, Center for Academic Publications, 1982.

[48] *Current Research in Earthquake Prediction I*, T. Rikitake (editor), Japan/Tokyo, Center for Academic Publications, 1981.

[49] W. Webb, Electrical structure of the lower atmosphere, in *Developments in Atmospheric Science I, Structure and Dynamics of the Upper Atmosphere*, F. Verniani (editor), Burlington, MA, Elsevier Scientific Publishing Co., 1974.

[50] O. Barsukov, Variations in electric resistivity of mountain rocks connected with tectonic causes, *Tectonophysics*, **14**, 1972, 273–277.

[51] A. V. Guglielmi and O. A. Pokhotelov, *Geoelectromagnetic Waves*, Bristol, UK, IOP (Institute Of Physics) Publishing Ltd., 1996

[52] R. Schunk and A. Nagy, *Ionospheres: Physics, Plasma Physics, and Chemistry*, 2nd edition, Cambridge, Cambridge University Press, 2009.

[53] R. Lanza and A. Meloni, *The Earth's Magnetism: An Introduction for Geologists*, Berlin/Heidelberg, Springer-Verlag, 2006.

12 Aiding baseband and navigation processors using INS

In this chapter we consider how external aiding data can be used in acquisition, tracking, and positioning. We use aiding from Inertial Navigation Systems (INS). In general any system which provides measurements of vehicle dynamics such as velocity and acceleration can be used. In particular, we can use relatively cheap inertial measurement unit (IMU) sensors, such as MEMS accelerometers and gyros, to provide suitable information in a wide range of applications, including those less technically demanding. However, we base our examples on a rather more demanding INS for an airborne navigation complex. The theory, approach, and algorithms are applicable to low-end consumer applications in general, however, rather than only to aviation and space applications, which would have been the case just a decade ago at the time of writing.

12.1 Principles of GNSS and INS integration

The GNSS gives stable navigation solutions as long as the receiver maintains a signal lock on four or more satellites. We discussed in Chapter 1 that information about height allows us to discard one variable and reduce the number of satellites required to three. Furthermore, a stable enough clock (usually not a part of a conventional receiver) allows reduction of this number to two. In the case of aircraft, attitude information is essential. Clock stability in general is not good enough to assume a constant receiver clock error. Therefore, if the number of satellites is reduced to fewer than four due to signal blockage, intentional/unintentional interference, etc., the position and timing solution will be lost. Vulnerability of GNSS radiowave propagation is a critical issue for aviation safety.

Inertial navigation systems (INS) are widely used for aircraft navigation. They are independent, and immune to interference and radiowave conditions. An INS operates using dead reckoning. The accelerometer/gyro errors are integrated and the position/attitude errors grow cumulatively. For example, the position error of a navigation grade INS increases to a few nautical miles after some hours' cruise. Since the characteristics of GNSS and INS are complementary, an integrated system can have the advantages of both. The methods of GNSS and INS integration are classified as loosely-coupled, tightly-coupled, and ultra tightly-coupled, and are depicted in Figure 12.1.

The simplest method of integration is to use the position and velocity output of a GPS receiver to calibrate the sensor error of an IMU. This is the loosely-coupled method.

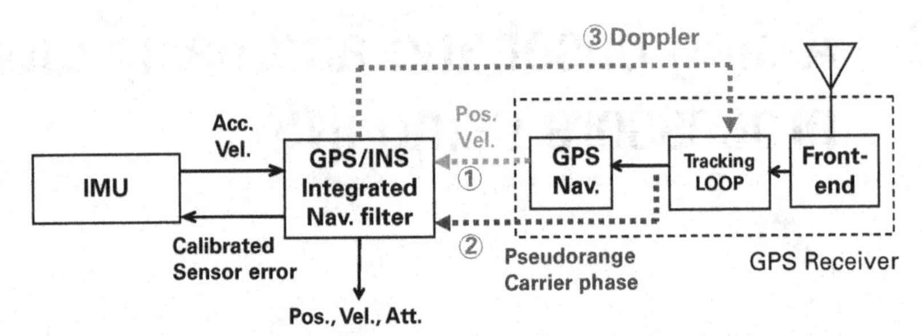

Figure 12.1 GPS/INS integration methods (1: loosely-coupled, 2: tightly-coupled, and 3: ultra tightly-coupled).

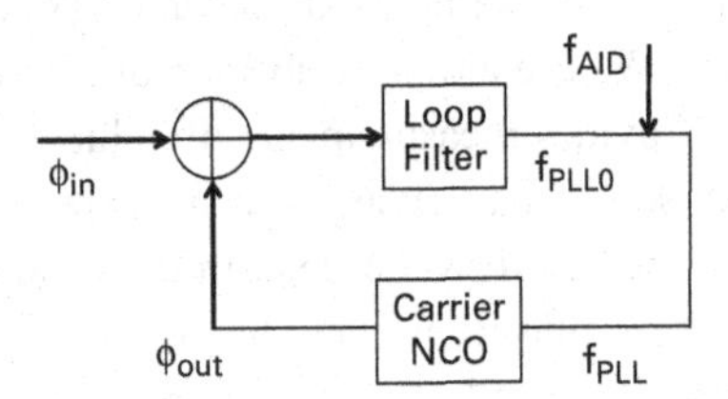

Figure. 12.2 Doppler-aided PLL model.

The second, tightly-coupled method, uses the measurements from a receiver such as pseudorange and carrier phase to calibrate IMU error. Since this method uses measurements from each satellite independently, faulty or poor quality measurements can be weighted or rejected, and a higher performance than that of a loosely-coupled system can generally be achieved. In addition to the aiding from a GPS receiver, the third method uses information from IMU to improve the tracking performance of the GPS signal. This is called the ultra tightly-coupled method, and it enables robust tracking in a weak signal environment, the details of which are described in the next section.

12.2 Case study: INS aiding for robust tracking

12.2.1 Doppler-aided PLL

Robust tracking of a GNSS signal in a harsh environment such as severe ionospheric scintillation and intentional/unintentional interference is a challenge for the safety of transportation such as civil aviation. To retain carrier tracking is important for a precision approach using GBAS, since the carrier phase is used for smoothing pseudorange measurements. If cycle slips have occurred in several channels, the corresponding smoothing procedures have to be restarted, and would cause a missed approach. Implementation of an inertial sensor improves the tracking performance if the Doppler frequency caused by aircraft dynamics is compensated by the inertial measurements [1]–[3].

The Doppler-aided PLL model is shown in Figure 12.2. Since the DLL is much more robust than PLL, we focus on the tracking performance of PLL in this section. Although third-order loop filters are more robust than second-order filters in high dynamics environments, they may be less stable and the transient response may be larger. Therefore, a second-order loop filter is used as an example loop filter for all analyses below.

Phase tracking is mainly affected by the clock dynamics, the receiver dynamics and the thermal noise. The phase noise due to clock dynamics for the second-order loop is given in radians^2 [2],[3]:

$$\sigma^2_{\delta\phi,\text{clock}} = \left(\frac{2\pi}{1.8856B_n}\right)^3 \frac{h_4\pi}{2\sqrt{2}} + \left(\frac{2\pi}{1.8856B_n}\right)^2 \frac{h_3\pi}{4} + \left(\frac{2\pi}{1.8856B_n}\right) \frac{h_2\pi}{2\sqrt{2}}, \tag{12.1}$$

where h_2, h_3, h_4 are clock coefficients. Also, the maximum phase error due to the receiver dynamics (acceleration) for second-order PLL in *radians* is expressed as follows [3]:

$$\delta\phi_{\text{dynamics}} = \frac{2\pi \cdot 2.7599a_{\max}}{\lambda B_n^2}, \tag{12.2}$$

where $a_{\max}$ is the maximum acceleration (g) of the receiver in the line-of-sight direction. The total 1-sigma phase error neglecting the vibration induced clock jitter in *radians* is then given as follows [4]:

$$\sigma_{\delta\phi} = \sqrt{\sigma^2_{\delta\phi,\text{thermal}} + \sigma^2_{\delta\phi,\text{clock}}} + \frac{\delta\phi_{\text{dynamics}}}{3}. \tag{12.3}$$

The total phase error (without dynamics) versus noise bandwidth for TCXO and OCXO are depicted in Figure 12.3 where C/N_0 is 40 dB and the clock coefficients are taken from [2]. Since the effect of receiver dynamics is not included, this plot indicates a typical performance for a static receiver. For the code/phase tracking loop, the noise bandwidth is an important parameter The narrower the noise bandwidth, the smaller the effect of thermal noise. Since there is no information on thermal noise, we want to reduce the noise by using a narrower bandwidth. However, with a narrower bandwidth, the effect of clock instability is large, and the loop may lose tracking. As seen in Figure 12.3, the phase error due to clock error is dominant for narrower noise bandwidth. Therefore, the use of a more stable clock (e.g. OCXO) enables application of a narrower bandwidth, which results in reducing the phase noise. For geodetic receivers, a noise bandwidth of a few hertz is usually used to achieve high positioning accuracy.

On the other hand, the total phase errors including dynamic stress are depicted in Figure 12.4 where the maximum acceleration of 0.1(g) is assumed. Since the phase error due to dynamic stress is dominant for narrower bandwidth, there seems no advantage to using OCXO if the dynamics is not compensated. Therefore, the noise bandwidth of a few tens of hertz is usually used for dynamic users. If the vehicle dynamics is almost compensated for by INS aiding, the phase error becomes as shown in Figure 12.3, and it makes sense to use OCXO to reduce the phase noise.

There is a number of methods for applying INS aid [1]–[3],[5]. We introduce here three basic approaches to INS aid [6].

The frequency of PLL can be expressed as follows:

$$f_{\text{PLL}} = f_{\text{D}} + f_{\text{clk}} + f_{\text{noise}}, \tag{12.4}$$

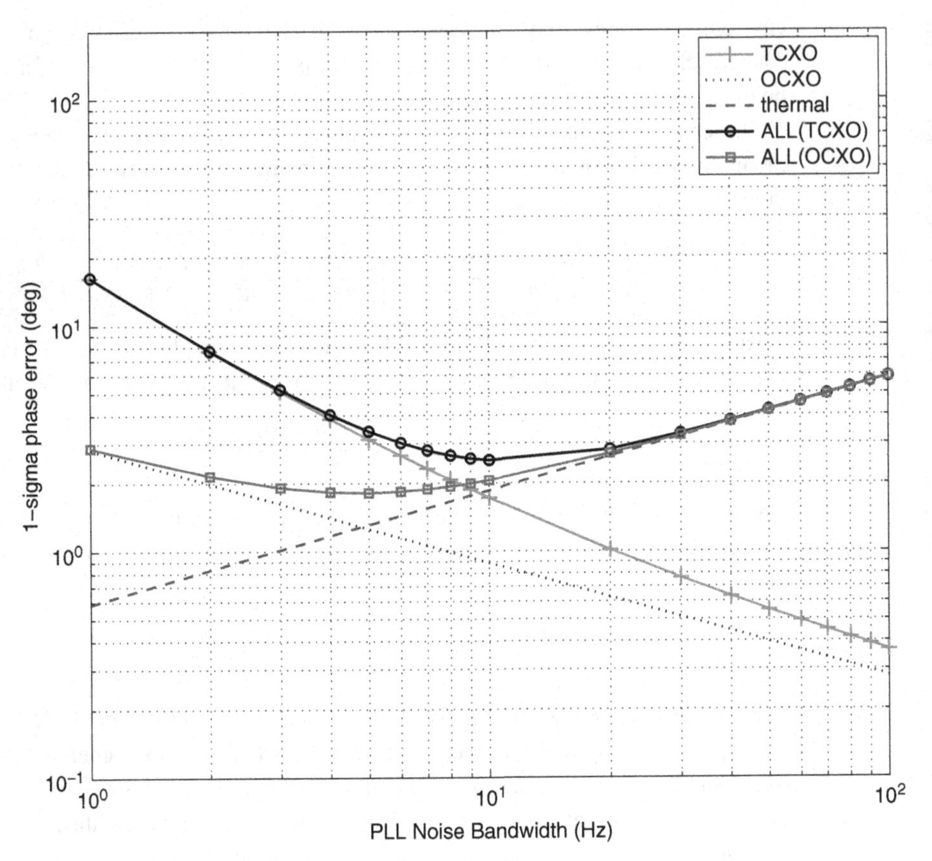

Figure 12.3 Phase error (without dynamics) versus noise bandwidth.

where f_D and f_{clk} are Doppler and clock frequency [2]. If the loop is aided, the frequency of PLL can be rewritten as follows:

$$f_{PLL} = f_{PLL0} + f_{AID}. \tag{12.5}$$

Three types of aid frequency such as delta Doppler (Δf_D), Doppler (f_D), and Doppler and clock frequency ($f_D + f_{clk}$) are considered here. The Doppler frequency is computed as follows, if the acceleration is negligible:

$$f_D = \frac{\vec{e} \cdot (\vec{v}_S - \vec{v}_R)}{\lambda}, \tag{12.6}$$

where $\vec{v}_S$, $\vec{v}_R$, $\vec{e}$, and λ are satellite velocity, receiver velocity, line-of-sight unit vector, and L1 wavelength, respectively. Before aiding Doppler information, we assume that the carrier is tracked by the usual loop. Therefore, it is reasonable to add the delta Doppler during the coherent integration time to the loop. When the coherent integration time is denoted as T_I, the delta Doppler is expressed as follows.

$$\Delta f_D = \frac{-\vec{e} \cdot \vec{a}_R \cdot T_I}{\lambda}, \tag{12.7}$$

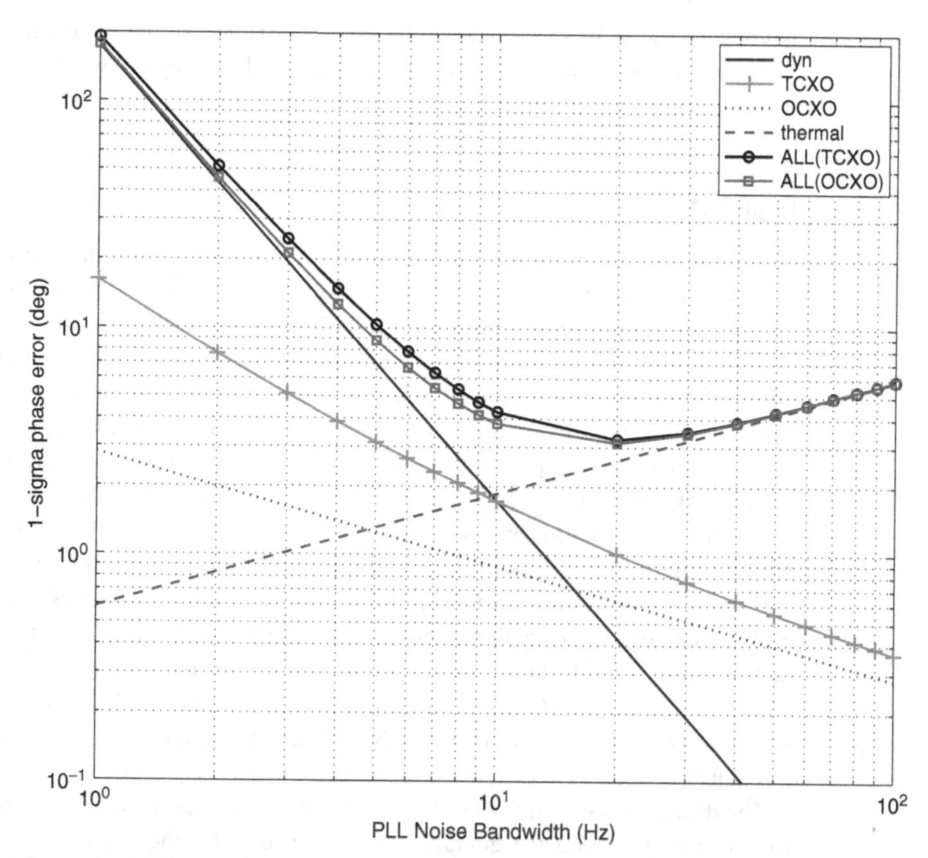

Figure 12.4 Phase error (with dynamics) versus noise bandwidth.

where $\vec{a}_R$ is receiver acceleration. Since the acceleration of a satellite during T_I is generally very small compared to the vehicle acceleration, the effect of satellite motion is neglected in (12.7).

In the second method, the Doppler frequency rather than Doppler increment is added. The receiver velocity ($\vec{v}_R$) is obtained from a GPS/INS integrated navigation filter. A better tracking performance would be expected because velocity is normally not noisier than acceleration. However, the initialization procedure of aiding should be considered carefully since f_{PLLO} in (12.5) is abruptly changed while f_{PLL} remains unchanged. Using a software-defined receiver can make it easier to treat such a transition.

In the third method, the clock frequency is aided in addition to the Doppler. It can be computed simply as follows:

$$f_{clk} = \frac{1}{N}\left(\sum_{i=1}^{N}(f_{PLL}^{i} - f_{D}^{i})\right), \tag{12.8}$$

where the superscript i indicates the ith channel and N is the number of tracked channels. The clock frequency is not aided by INS in a precise sense since the clock frequency is not obtained from INS and needs to be estimated. However, the information on clock frequency can be useful if it is used for acquisition and tracking of a weak signal because it is common for all channels [5],[7].

12.2.2 Flight test

As shown in the previous section, the noise bandwidth of the dynamic receiver can be narrowed if the dynamic stress is compensated for by INS aiding. The effect of aiding is demonstrated in this section by using real flight data at take-off.

The configuration of the onboard equipment is shown in Figure 12.5. Two types of GPS/INS navigation system were used for the flight experiments. The first one is a tightly-coupled GPS/INS which consists of a Kearfott T-24 inertial measurement unit (IMU) with ring laser gyro and servo accelerometer, and an Ashtech G12 single-frequency GPS receiver [8]. The second is a miniaturized GPS/INS navigation system which consists of MEMS gyros and accelerometers, a U-blox LEA-4T GPS receiver, and triaxial magnetometers [9]. A 15-state loosely-coupled GPS/INS Kalman filter is adopted to suppress growth of the position error caused by the MEMS inertial sensor errors. Also, two GPS front-end units with different clocks (TCXO and OCXO) were installed and GPS IF data were recorded. The IF frequency and sampling rate are 4130 400 Hz and 16 367 600 Hz, respectively. A low-cost GPS/INS and two front-end units are shown in Figure 12.6, and JAXA's research aircraft used for the test is shown in Figure 12.7.

The data samples from the low-cost GPS/INSs as well as the DIF data from the two front-end units during the aircraft take-off are used for the analyses hereafter, and included in this book (D_INS_mems, D_FLT_tcxo, D_FLT_ocxo). Figure 12.8 shows the flight profile at take-off, while Figures 12.9 and 12.10 show the velocity and acceleration in a navigation frame (ENU).

Now, let us consider the results of flight data processing. Figure 12.11 depicts the carrier error (in cycles) for two satellites (PRN 9 and 12) when the DIF data from the front-end with TCXO [D_FLT_tcxo] were processed and no aiding was applied. The carrier error is the output of a discriminator, atan(Q/I), where I and Q are in-phase and the quadrature phase signal is integrated for one millisecond. The standard deviations (s.t.d.) of the carrier error are also shown in the figure. The standard deviation calculated

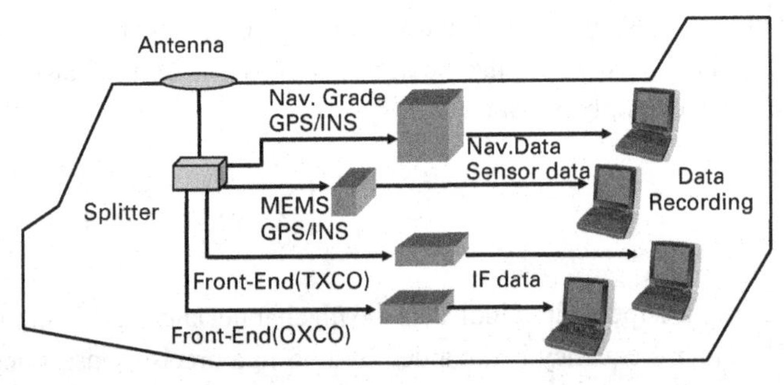

Figure 12.5 Configuration of onboard equipment.

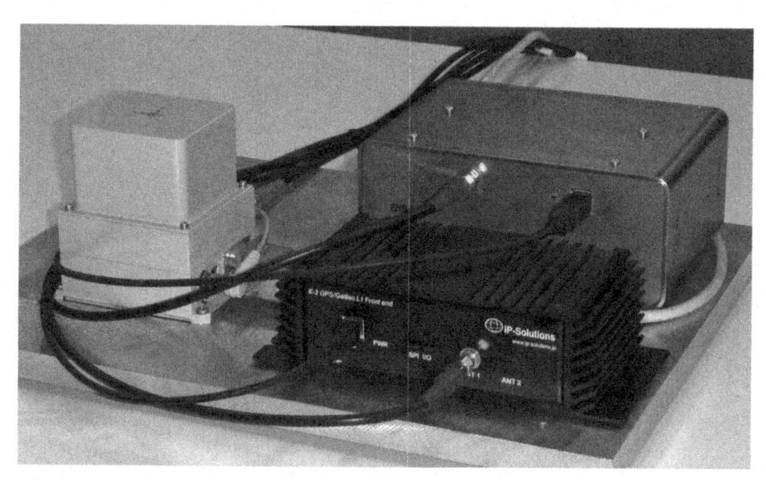

Figure 12.6 Low-cost GPS/INS and GPS front-end units.

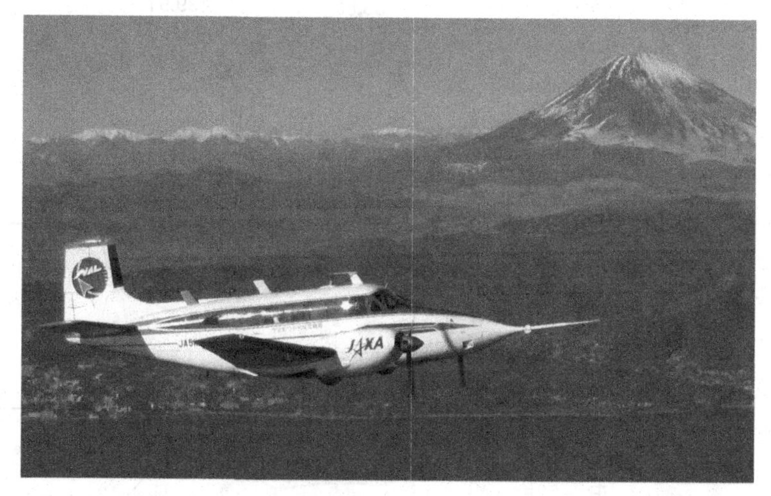

Figure 12.7 Research aircraft used for the flight demonstration.

from two channels was 7.4 mm. In order to see the aid effect, the noise bandwidth of the PLL loop filter was reduced to three Hz from the usual value of 25 Hz.

The delta Doppler (in hertz) frequencies computed by using the data from the low-cost MEMS INS (D_INS_mems) are shown in Figure 12.12. The trends in delta Doppler are different between satellites since the line-of-sight vectors for the satellites are different. Although the magnitude of delta Doppler seems very small, these values are added at every coherent integration time (1 msec) and accumulated in the loop frequency as long as the aiding is executed. When delta Doppler frequencies are added in the loop, the carrier errors are largely removed, as shown in Figure 12.13, with corresponding reduction of the standard deviations (3.8 mm). When the navigation grade INS was used for aiding, the standard deviation of carrier phase error was very similar. When

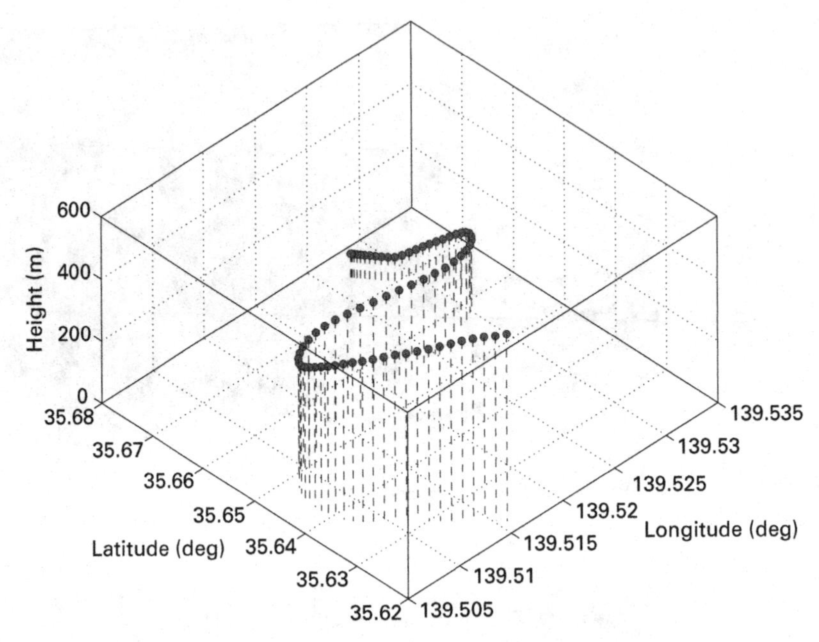

Figure 12.8 Flight profile at take-off.

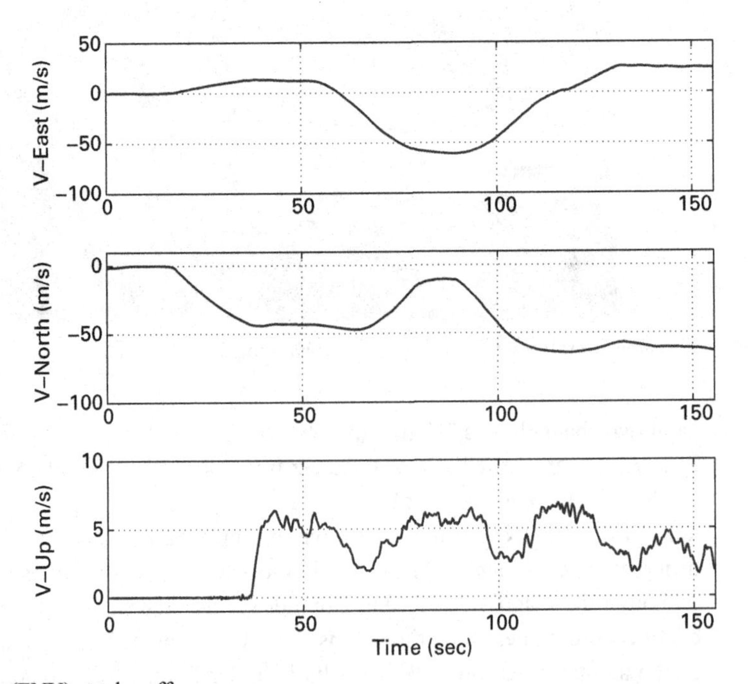

Figure 12.9 Velocity (ENU) at take-off.

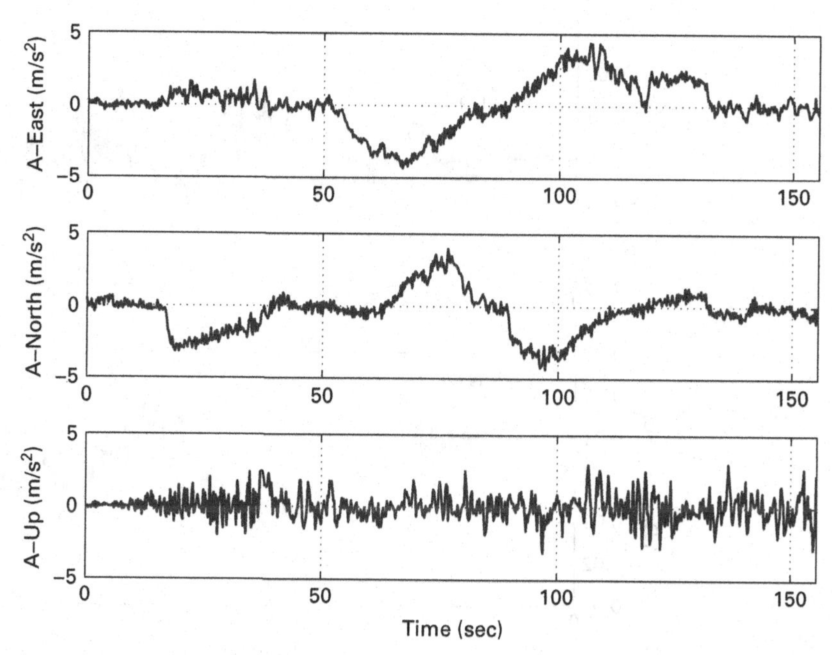

Figure 12.10 Acceleration (ENU) at take-off.

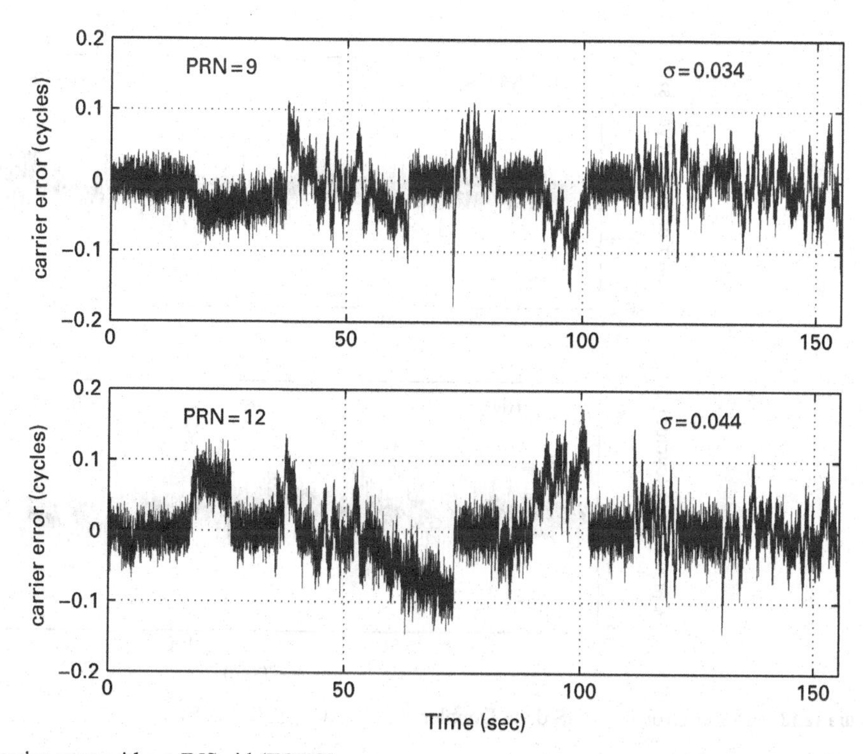

Figure 12.11 Carrier error without INS aid (TCXO).

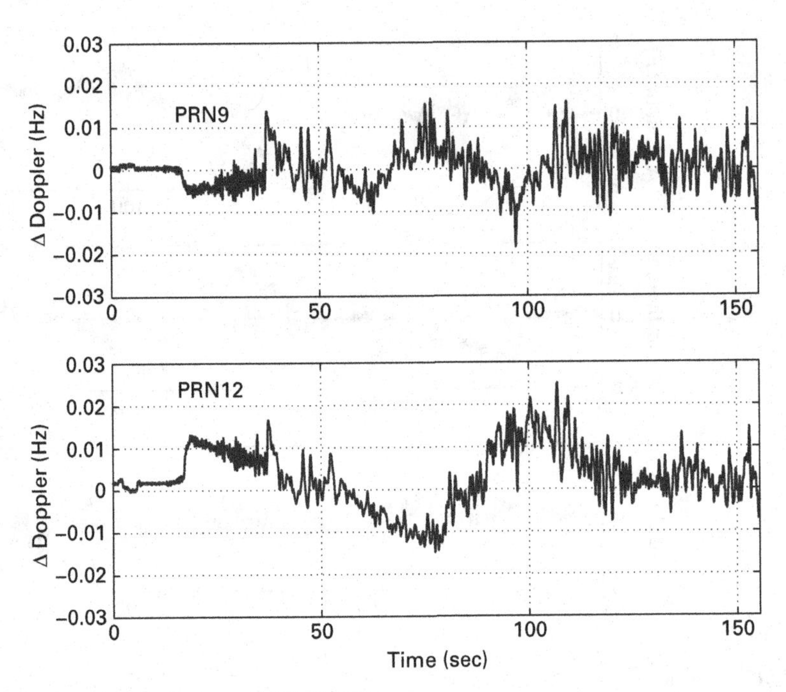

Figure 12.12 Delta Doppler added in the loop (MEMS INS).

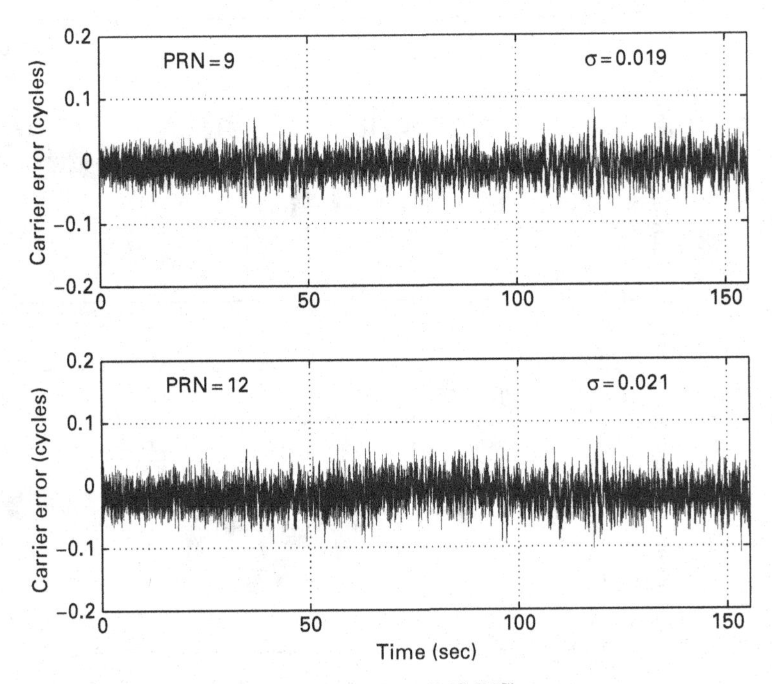

Figure 12.13 Carrier error with INS delta Doppler aid (TCXO, MEMS INS).

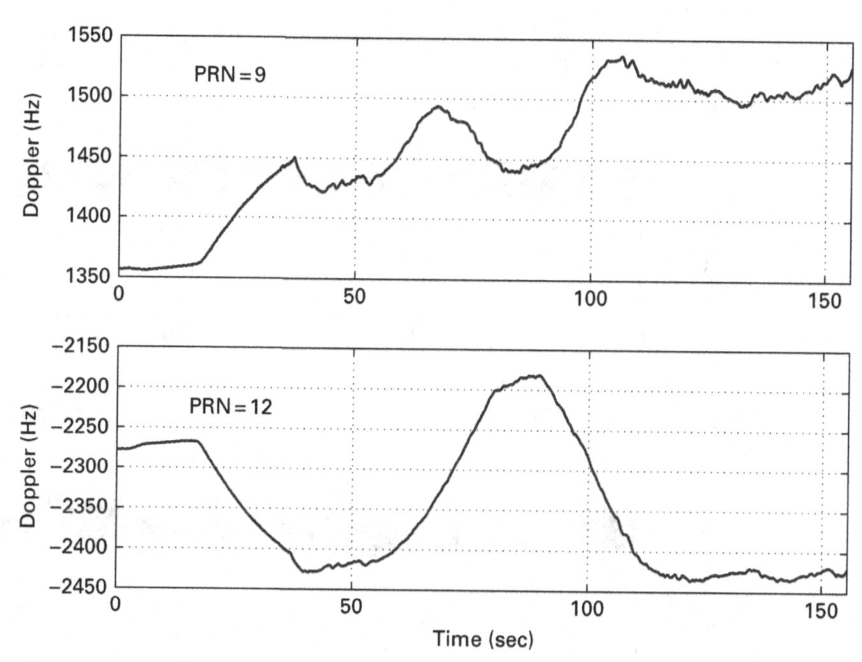

Figure 12.14 Doppler frequency added in the loop.

the data from the front-end with OCXO (D_FLT_ocxo) are used, the standard deviations are slightly reduced.

Next, we verify the effect of Doppler aiding. Figure 12.14 represents the Doppler frequency (in Hertz) added into the tracking loop by using the output of tightly-coupled GPS/INS. Comparing with Figure 12.12, it is seen that the Doppler frequency is the integration of delta Doppler though the sign is opposite. Since it is based on the velocity output rather than acceleration, it is very smooth compared with the delta-Doppler shown in Figure 12.12, and therefore a greater reduction of phase error is expected.

Figure 12.15 shows the carrier error (in Hertz) when the Doppler frequencies are added into the tracking loop by using the output of a tightly-coupled GPS/INS and the front-end with OCXO(FLT_ocxo). Compared with Figure 12.13, the standard deviation of carrier error was reduced from 3.8 mm to 3.3 mm as expected.

In order to apply the third integration method, the clock frequency has to be estimated. Figure 12.16 shows the estimated clock frequencies for the front-end with TCXO (top) and OCXO (bottom), respectively, where the data of tightly-coupled GPS/INS are used for aiding. The clock drift of TCXO is more significant than OCXO, as expected from the clock frequency characteristics shown in Figure 12.16. The estimated frequency is noisy since it is calculated based on (12.8) by using a loop frequency not by using an external sensor. The effect of applying the estimated clock frequency is slight, and the resultant standard deviation of phase error is similar when the aiding data of tightly-coupled GPS/INS are applied for the DIF data of the front-end with OCXO (FLT_ocxo). A more sophisticated algorithm to estimate clock frequency may be necessary to make use of this aiding method.

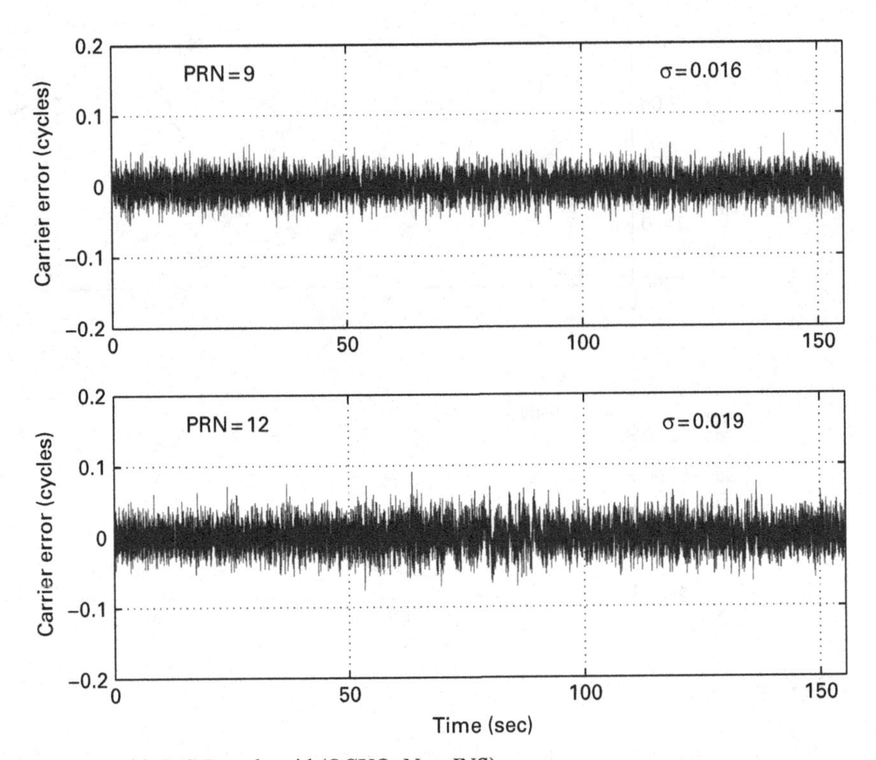

Figure 12.15 Carrier error with INS Doppler aid (OCXO, Nav. INS).

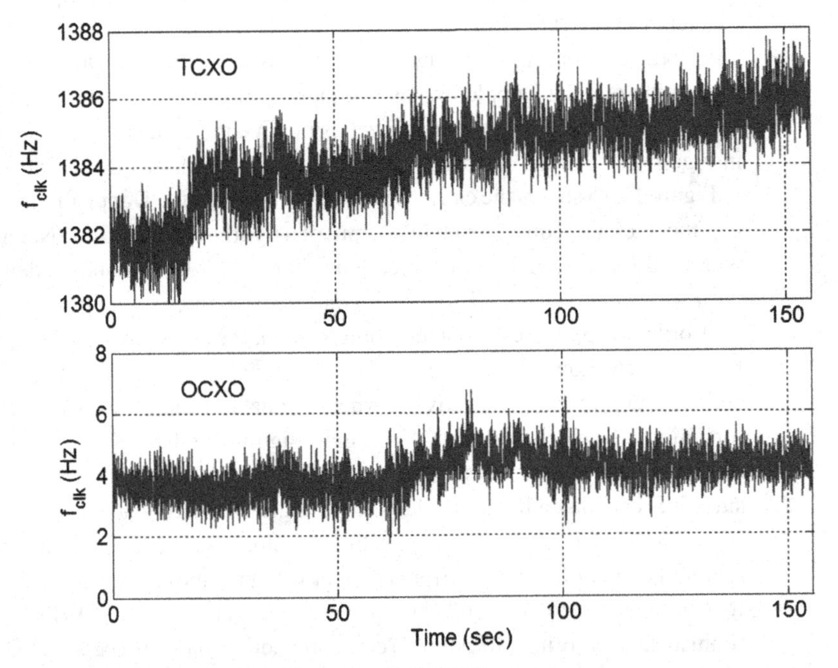

Figure 12.16 Estimated clock frequency of TCXO (top) and OCXO (bottom).

The effect of INS aiding is well demonstrated in this section by narrowing the noise bandwidth of PLL. However, the performance difference between TCXO and OCXO is not clearly seen. This could be because the residual Doppler error was more significant than the clock instability. Also, the effect of vibration-induced clock jitter may have to be considered [3],[4]. On the other hand, the Doppler aiding by a low-cost MEMS INS shows a similar performance to aiding by a navigation grade INS. Therefore, use of low-cost INS seems sufficient for unmanned aerial vehicles and general aviation. However, the integration with navigation grade INS guarantees superior continuity during GPS outage, and is suited for a precision approach.

12.3 Case study: INS-aided tracking under scintillation conditions

12.3.1 Theoretical performance of Doppler aiding

The theoretical performance of Doppler aiding under scintillation is described in this section. Figure 12.17 shows the carrier phase error (1 σ) vs. noise bandwidth of the second order PLL where scintillation is not included. The aircraft acceleration (0.25 g) is included and this effect is assumed to be removed by the Doppler aiding. The empirical tracking threshold (15 degrees) is shown by a dashed line while the tracking margins with/without aiding are shown by arrows. It can be seen that the phase error is generally reduced by the Doppler aiding. However, a rather wider bandwidth (e.g. 20 Hz) can guarantee stable tracking even without aiding. Next, the carrier error under a fairly strong scintillation ($S_4 = 0.48$, $T = 0.003(\text{rad}^2/\text{Hz})$, $p = 2.0$) is shown in Figure 12.18, where S_4 is an index for amplitude scintillation while T, and p are parameters for phase scintillation (see Chapter 10). The phase errors caused by amplitude/phase scintillation are computed by equations as follows [10]:

$$\sigma^2_{\delta\phi,\text{scin\_therm}} = \frac{B_n}{C/N_0} \cdot \frac{1}{1 - S_4^2}\left(1 + \frac{1}{2T_I C/N_0(1 - 2 \cdot S_4^2)}\right), \tag{12.9}$$

$$\sigma^2_{\delta\phi,\text{scin\_phase}} = \frac{\pi T}{2f_n^{p-1} \sin([p-1]\pi/4)}, \tag{12.10}$$

where $B_n(= 3\pi f_n/2\sqrt{2})$ is the noise bandwidth of the PLL and T_I is the pre-detection integration time (=0.001 sec in this chapter). Comparing Figure 12.17 and Figure 12.18, the phase error is greatly increased due to the scintillation. Figure 12.18 indicates that the Doppler aiding with a narrow bandwidth reduces phase error under scintillation, although a wide bandwidth (e.g. 20–30 Hz) without aiding can still have some margin for tracking. Note that the INS aiding does not directly improve the tracking performance under scintillation since the INS does not sense the GPS signal. The INS aiding gives a margin for tracking by removing the effect of dynamics. Even if the signal tracking were maintained by the INS aiding, the range measurement would be affected since the signal

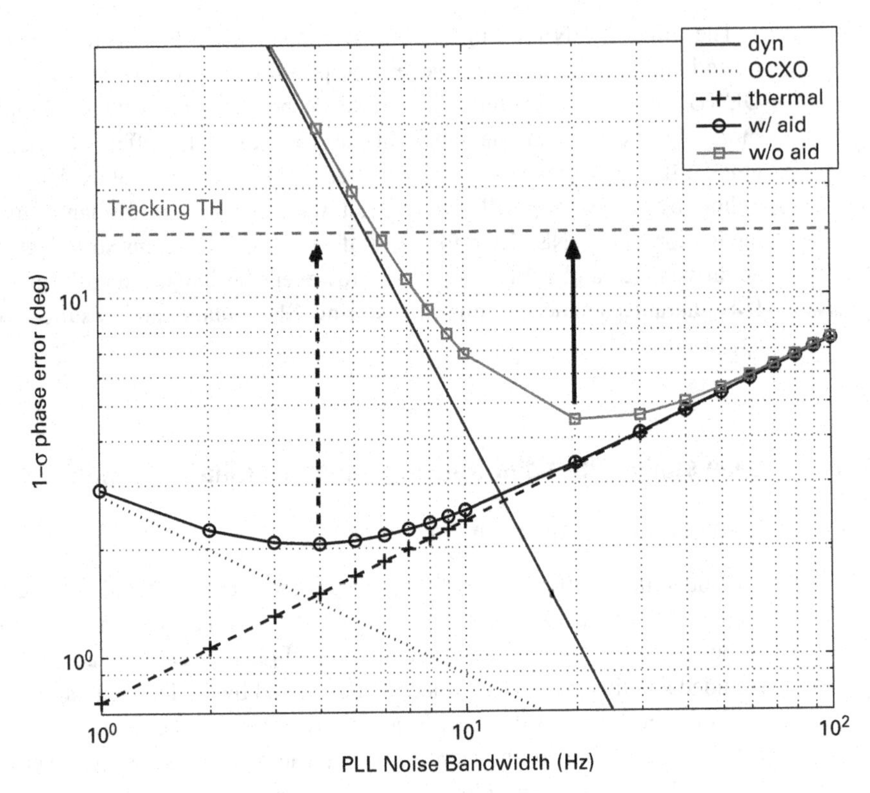

Figure 12.17 Phase error caused by several error sources (without scintillation).

would be disturbed by the irregularities in the ionosphere. This is the difference from signal tracking under interference. In the case of interference, there exists the GPS signal with correct range information. If the signal were tracked in the interference condition, the range would be measured precisely.

12.3.2 Test with simulated signal

In order to evaluate the performance of Doppler aiding implemented in the software receiver, DIF data affected by scintillation are necessary. For signal simulation purposes a device called a Digitized Intermediate Frequency Signal Generator ReGen from iP-Solutions is used. The DIF Signal Generator can create a DIF signal virtually indistinguishable from those which are recorded from live satellites. It also allows us to modify signal environment, to introduce special errors such as scintillation, and to modify error models. The phase and amplitude variation due to scintillation can be simulated by using appropriate shaping filters and probability distribution fitting (see Chapter 10). Also, the real phase/amplitude variation profiles associated with a plasma bubble can be embedded in the DIF data, and this approach is adopted in these analyses. Figure 12.19 depicts the flow of this simulation.

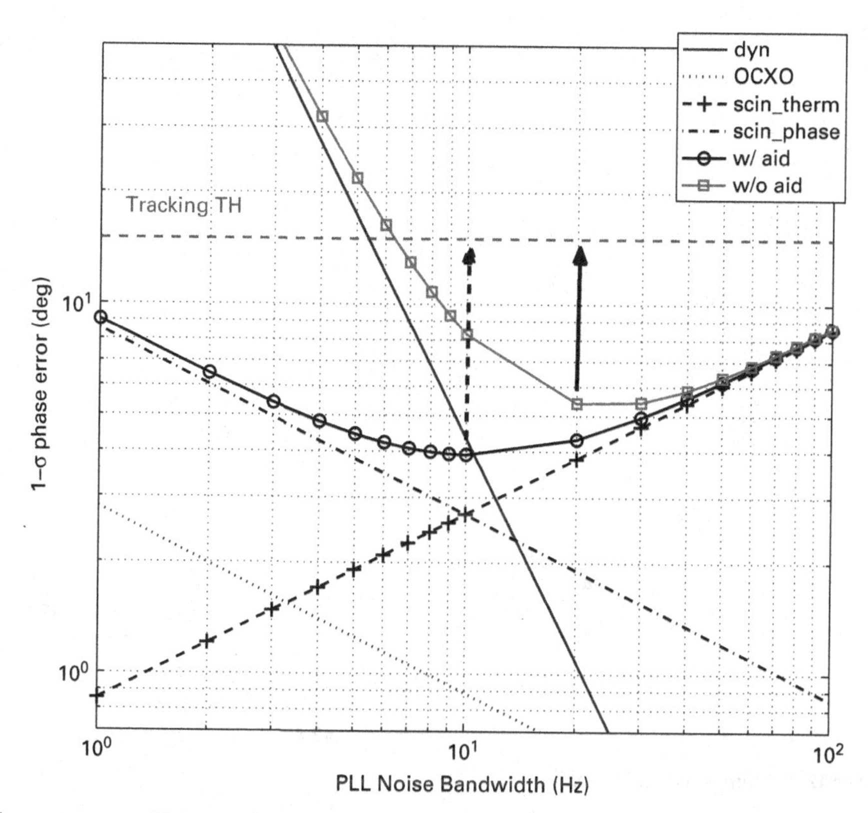

Figure 12.18 Phase error caused by several error sources (with scintillation).

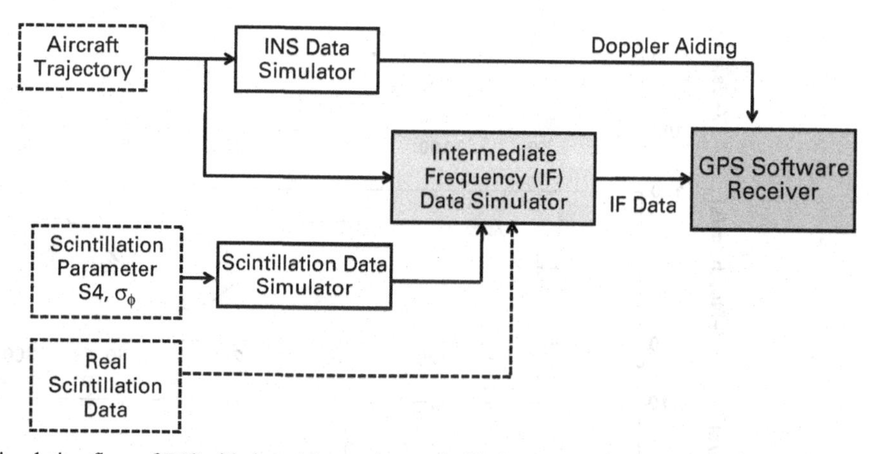

Figure 12.19 Simulation flow of INS-aided tracking under a scintillation environment.

As the first step of the simulation, a trajectory of a light aircraft and INS data are generated, and then corresponding GPS DIF data including scintillation effects are generated. Figures 12.20 and 12.21 show the simulated flight profile and velocity of the aircraft (ENU) used for computing Doppler frequency.

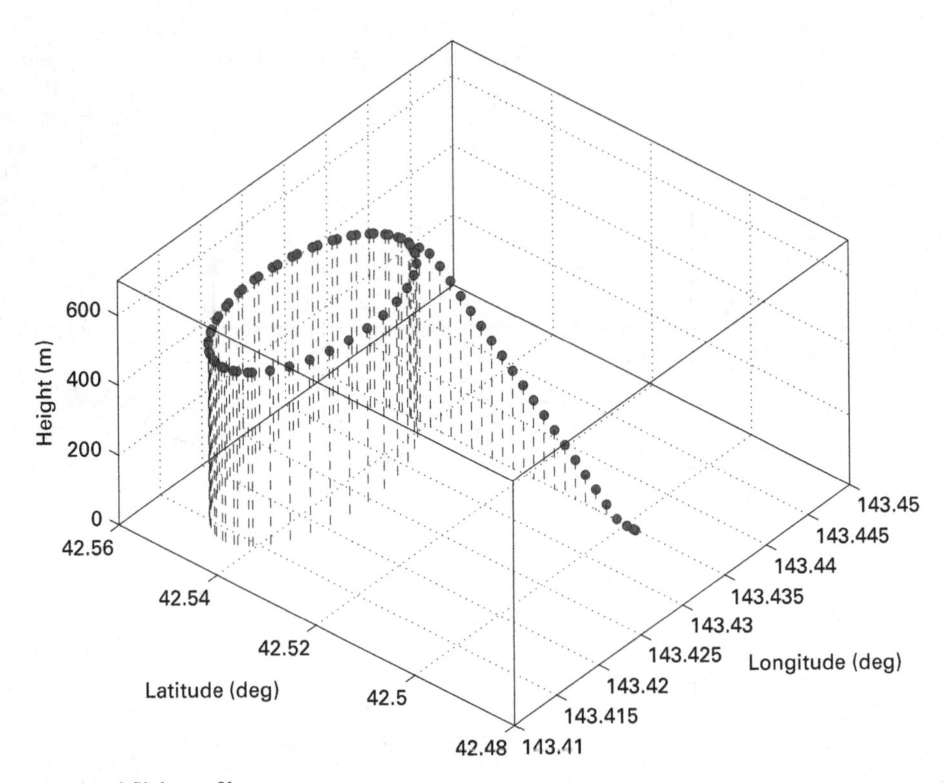

Figure 12.20 Simulated flight profile.

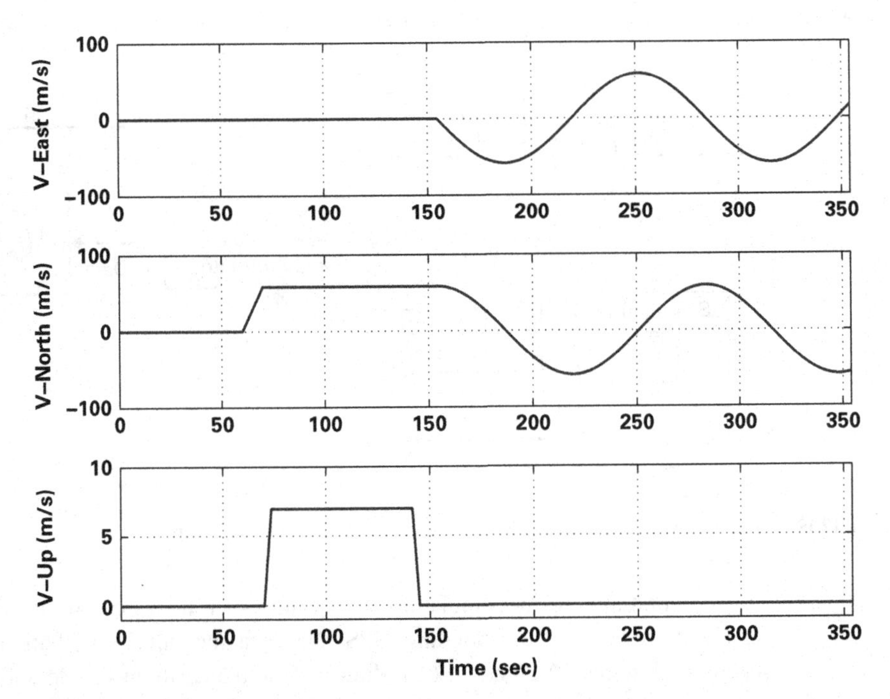

Figure 12.21 Velocity of aircraft (ENU).

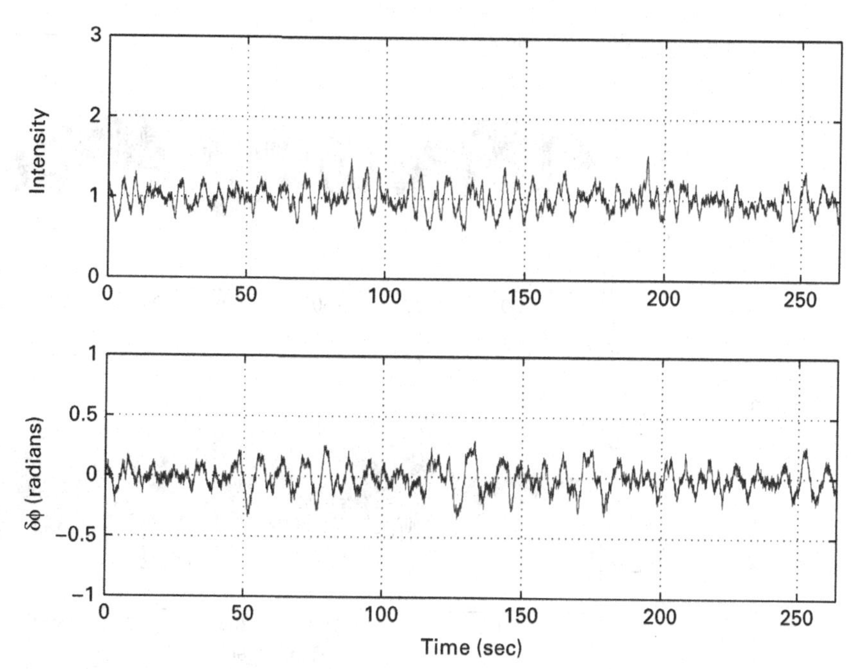

Figure 12.22 Intensity/phase variation used for simulation.

The aircraft took off and made one and a half counter-clockwise rotations. Figure 12.22 shows the variation of intensity and phase embedded in the DIF data, which corresponds to the beginning of the second scintillation (from 14 h 24 m to 14 h 28 m 24 s, see Section 10.4 and Figure 10.10).

The generated DIF including scintillation is processed by iPRx and the resultant carrier phase errors of two satellites are shown in Figure 12.23 where Doppler aid is not applied. The noise bandwidth was set at 8 Hz. Since there is no scintillation for the first satellite, the effect of aircraft dynamics, especially during circling, is clearly seen. The scintillation effect is added to the signal of the second channel after 90 seconds from the time origin. It is clear that the carrier error increases due to scintillation. (Note that the vertical scales are different between first and second channels.)

On the other hand, Figure 12.24 shows the carrier error when Doppler aiding is applied. Compared with Figure 12.23, carrier error trends due to aircraft dynamics are removed. However, cycle slips occurred frequently on the second channel even with aiding. Although the scintillation applied to this test is fairly weak, the observed standard deviation of carrier error is about six degrees as shown in Figure 10.17 and cycle slips are likely to happen. Figure 12.25 represents the probability of carrier error exceeding 45 degrees with/without aiding. If the threshold of the cycle slip is set at 45 degrees (0.125 cycles), the occurrence of cycle slip is reduced by 30 % by applying Doppler aiding. Figure 12.26 shows an example of I-channel amplitude for the second channel (time from 200 to 205 seconds), and the amplitude variation is clearly seen.

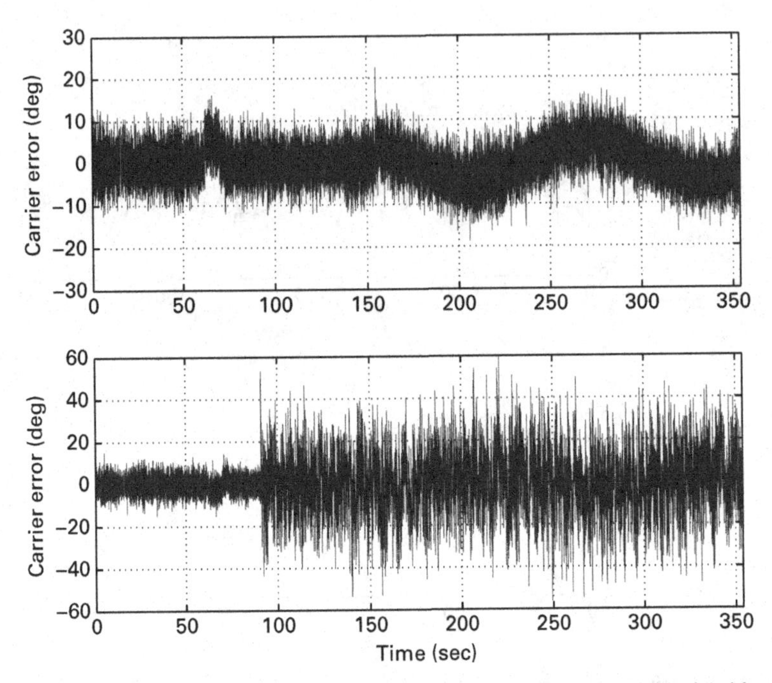

Figure 12.23 Carrier errors (in cycles) for two satellites with scintillation on the second channel (without aid).

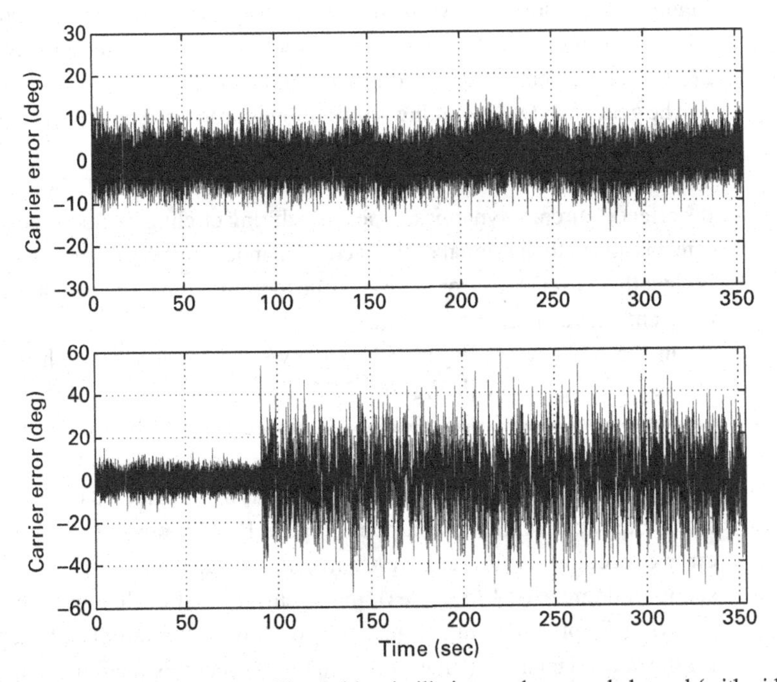

Figure 12.24 Carrier error (in cycles) for two satellites with scintillation on the second channel (with aid).

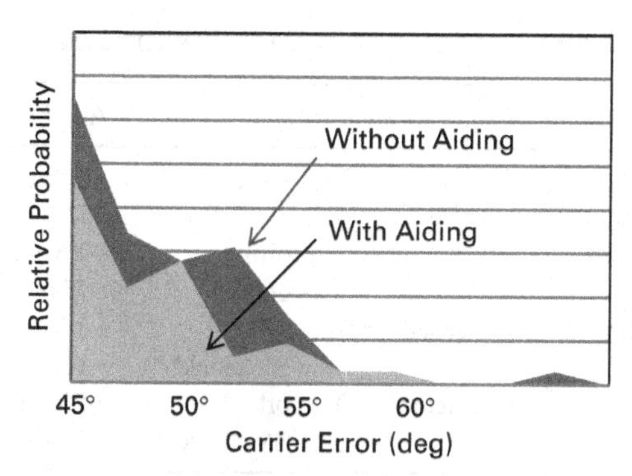

Figure 12.25 Probability of carrier error exceeding 45 degrees with/without aid.

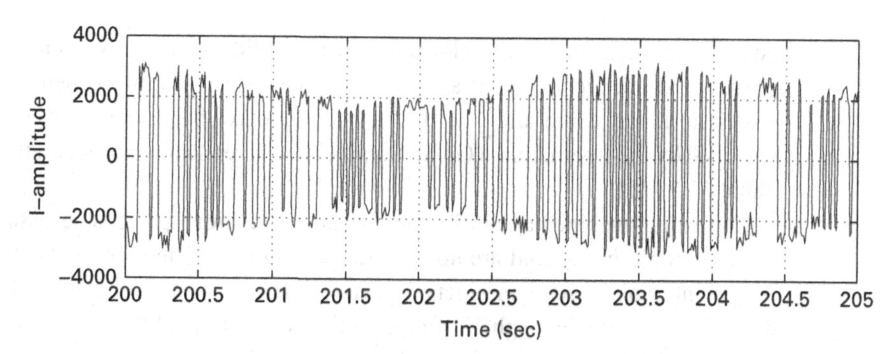

Figure 12.26 I-channel amplitude of the second channel (time from 200 to 205 seconds).

12.4 Precise point positioning for dynamic applications

We use an LSE method described in the Chapter 1 to get a position from code phase (or pseudorange) observables. In this section, we introduce a more accurate method to estimate a stand-alone receiver's position, known as ***precise point positioning (PPP)***. The basic idea behind this method is as follows. If, as in this chapter, we have properly constructed the signal and our models precisely reflect what happens with the signal, then by applying these models during signal processing we can remove all errors from the signal and get coordinates very accurately. The accuracy is limited only by residual errors.

In order to achieve centimeter-level accuracy for GNSS positioning, relative positioning is commonly carried out in which the single/double differences of range measurements are used to cancel common errors. On the other hand, the PPP method has recently been investigated and developed [11],[12]. The PPP method uses un-differenced measurements, so the reference receivers and communication links are not necessary. They offer good benefits for the users on the sea, in the air and in space. However, precise modeling of error sources which are not considered in relative

Table 12.1 Summary of IGS ephemeris/clock products.

		Accuracy	Interval	Latency
Final	orbit	~2.5 cm	15 min	12–18 days
	clock*	~75 ps	30 sec	
Rapid	orbit	~2.5 cm	15 min	17–41 hours
	clock*	~75 ps	5 min	
Ultra-rapid (observed half)	orbit	~3.0 cm	15 min	3–9 hours
	SV clock**	~150 ps		
Ultra-rapid (predicted half)	orbit	~5.0 cm	15 min	Real time
	SV clock**	~3 ns		

* satellite and reference station clocks.
** satellite clock only.

positioning such as satellite clock/ephemeris, solid earth tide, oceanic loading tide, phase wind-up, etc. is necessary. The broadcast GPS satellite ephemeris and clock errors are about 1 min and 5 ns (RMS). On the other hand, the ephemeris and clock provided by the International GNSS Service (IGS) are much more accurate, as shown in Table 12.1 [13].

Regarding the kinematic PPP, where frequent position update is necessary, the clock errors in rapid/ultra-rapid are not accurate enough, and therefore a clock error with a 30 sec interval in final products is very useful. Note that the IGS clock error is determined for the ionospheric-free combination of P1 and P2, and therefore the C/A code user has to correct the pseudorange by using the differential calibration bias (DCB) provided by IGS [14].

The satellite orbit is determined for the center of inertia; however, the range is measured for the distance between phase centers of antennas. Therefore, the offset between the center of inertia and the antenna phase center of the satellite has to be corrected. The standard offset parameters are provided by IGS in the so-called ANTEX (antenna exchange format) file [15]. The phase center of a GPS signal transmitter/receiver antenna varies depending on the direction of radiowave transmission/reception, and the variation is known as PCV (phase center variation). The PCVs for both satellites and IGS stations are included in the same ANTEX file. If the PCV correction is applied for kinematic users, special attention has to be paid to the vehicle attitude and heading, since the PCV parameters are determined for the azimuth/elevation of the antenna. The position of the receiver changes due to the tidal force of the sun and moon by up to a few decimeters. The displacements due to the solid earth tide, ocean loading tide, and polar tide are a few decimeters, several centimeters, and a few centimeters, respectively. These displacements cancel in differential positioning over short baselines (< 100 km). However, for kinematic PPP applications or long baseline relative positioning (> 1000 km), these displacement effects have to be modeled as recommended in IERS Conventions [16]. Since the GPS signal is right hand circularly polarized (RHCP), the relative orientation of the satellite and receiver antennas affects the carrier

Table 12.2 Types of PPP positioning method.

	Ionospheric model	Ephemeris	Tropospheric estimate
PPP-1	broadcast	broadcast	off
PPP-2	GIM	precise	on (wet)

phase measurement. If one of the antennas rotates 360 degrees around its boresight (axis of maximum gain), the phase measurement changes a full wavelength. This is the so-called "phase wind-up" effect, and can be corrected by using the orientation information from the antennas [17]. Regarding rotation of the receiver antenna, the phase wind-up can be included in the receiver clock error estimate.

The ionospheric-free combination of pseudorange and carrier phase is usually used by dual frequency PPP users to achieve high accuracy. We correct the IGS satellite clock by the following:

$$-\frac{f_{L2}^2}{f_{L1}^2 - f_{L2}^2}\mathrm{DCB}, \tag{12.11}$$

where (L1-L2) DCBs are also included in an IONEX file [18].

A result of single-frequency PPP for the flight test, which was described in more detail in Section 8.4.1.4, is introduced here. The PPP algorithm is developed by using the GNSS regression (GR) model [19]. In this analysis, two types of PPP method are implemented, as shown in Table 12.2.

In PPP-1, the broadcast ephemeris and ionospheric model are applied, while the tropospheric delay is not estimated. In PPP-2, the precise ephemeris and GIM (Global ionosphere map) provided by IGS are applied, and the zenith wet delay (ZWD) is estimated. The PPP position errors (east, north, and up) are shown in Figure 12.27 as differences between the results of single-frequency PPP positions and dual-frequency kinematic positions. It can be seen that the PPP algorithms achieve sub-meter level horizontal accuracy and a few meters vertical accuracy. By applying the GIM data and precise ephemeris, the results of PPP-2 achieve better performance than PPP-1.

This single-frequency PPP can be integrated with INS in order to calibrate the accelerometer/gyro errors [20]. It would also be possible to apply carrier smoothing, but as we saw in the previous chapters, the opposite sign of ionospheric error makes it difficult to implement for precise positioning. The position solution of single frequency PPP/INS Kalman filtering is compared with a dual-frequency kinematic solution, and a sub-meter level of accuracy is seen (Figure 12.28). For comparison, the positioning error of GPS/SBAS/INS for the same flight test is shown in Figure 12.29. Since the positioning performance is mainly determined by the GPS range correction methodology, the superior performance of PPP to GPS/SBAS is clearly seen.

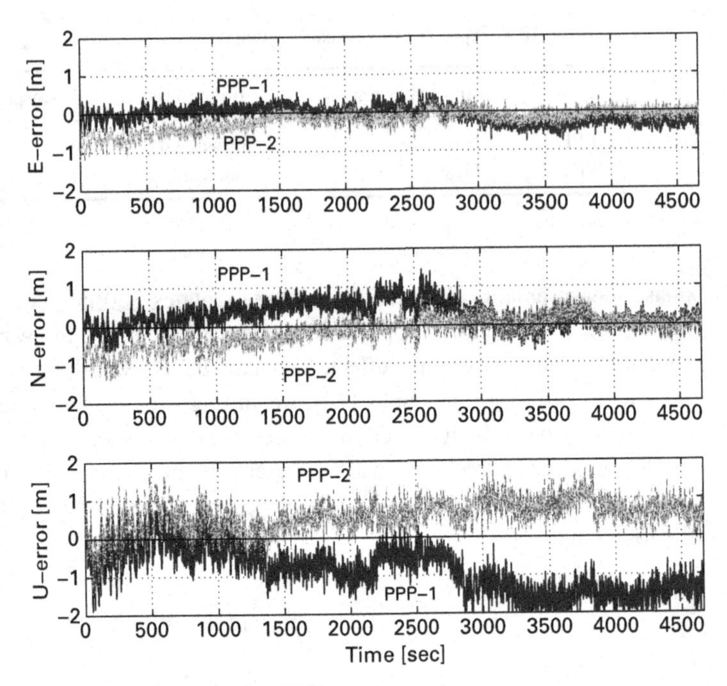

Figure 12.27 Results of PPP positioning error (after [20]).

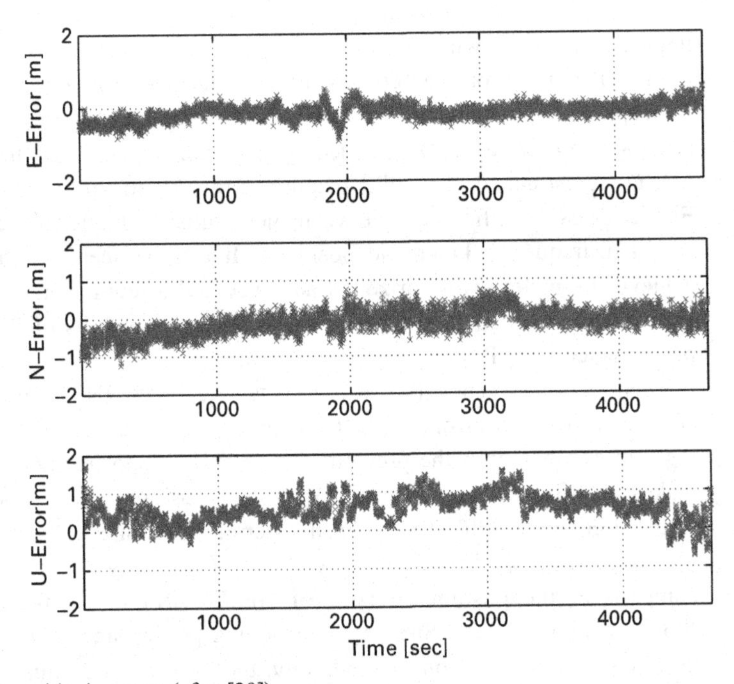

Figure 12.28 PPP/INS positioning error (after [20]).

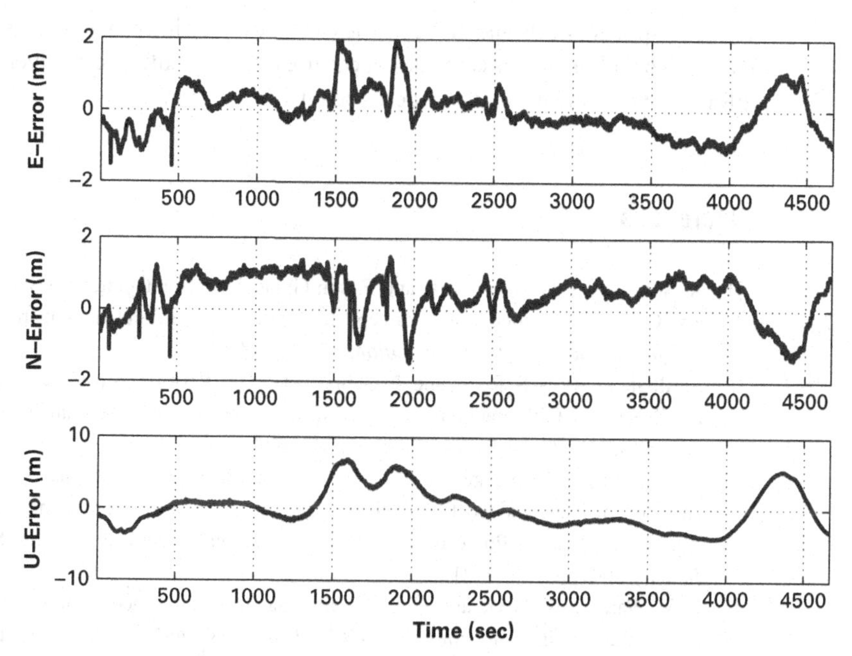

Figure 12.29 GPS/SBAS/INS positioning error (after [20]).

12.5 Project: Processing flight data with software receiver

Use data for a static user simulated in the previous project and bundled flight data for a dynamic user. The book website contains the following data files:

D_FLT_ocxo.bin is a GPS digitized signal record created during a test flight by a recorder with an OCXO clock.

D_FLT_tcxo.bin is a GPS digitized signal record created during a test flight by a recorder with TCXO.

D_INS_mems.txt is an ASCII file of the data from a MEMS grade INS, which includes velocity, coordinates, and output from accelerometers and gyros during the same flight. The data correspond to the signal records.

Format_INS.doc is a readme file explaining the format of the previous file:

1. Process the DIF signal recorded during the test flight with an iPRx software receiver. There are two sets of data for TCXO and OCXO front ends. Compare the signal. Use a predicted ephemeris and Doppler assistance option to show the difference between OCXO and TCXO records.
2. Use various DLL and PLL settings and various options for discriminators to see the features of the particular loop settings for the cases of a static user and an aircraft.
3. Simulate a GPS signal for an aircraft flight with higher dynamics. Observe the behavior of DLL and PLL with various settings.

The INS measurements and raw data file can be used by interested readers for ultra-tight INS-GPS coupling. In this case readers can use their own software receiver or customized iPRx receiver, similar to the one described in this chapter.

References

[1] A. Soloviev, S. Gunawardena, and F. van Graas, Deeply integrated GPS/low-cost IMU for low CNR signal processing: Concept description and in-flight demonstration, navigation, *Journal of The Institute of Navigation*, **55**, (1), 2008, 1–13.

[2] S. Alban, D. Akos, S. Rock, and D. Gebre-Egziobher, Performance analysis and architectures for INS-aided GPS tracking loops, Institute of Navigation National Technical Meeting, Anaheim, CA, 22–24 January 2003 pp. 611–622.

[3] T. Y Chiou, D. Gebre-Egziabher, T. Walter, T., and P. Enge, Model analysis on the performance for an inertial aided FLL-assisted-PLL carrier-tracking loop in the presence of ionospheric scintillation, Proceedings of the ION National Technical Meeting, San Diego, CA, January 2007, pp. 2895–2190.

[4] P. W. Ward, J. W. Betz, and C. J. Hegarty, Satellite signal acquisition, tracking, and data demodulation, in *Understanding GPS, Principles and Applications*, E. Kaplan and C. Hegarty (editors), second edition, Boston/London, Artech House, 2006.

[5] G. Gao and G. Lachapelle, INS-sssisted high sensitivity GPS receivers for degraded signal navigation, Proceedings of GNSS2006, Forth Worth, 26–29 Sep., pp. 2977–2989.

[6] T. Tsujii, T. Fujiwara, Y. Suganuma, H. Tomita, and I. Petrovski, Development of INS-aided GPS tracking loop and flight test evaluation, *SICE Journal of Control, Measurement, and System Integration*, **4**, (1), 2011, 1–7.

[7] J. J. Spilker Jr., Fundamentals of signal tracking theory, in *Global Positioning System: Theory, and Applications*. Vol. I, B. W. Parkinson and J. J. Spilker Jr. (editors), Washington, DC, American Institute of Aeronautics and Astronautics, Inc., 1994, pp. 245–327.

[8] M. Harigae, T. Nishizawa, and H. Tomita, Development of GPS aided inertial navigation avionics for high speed flight demonstrator, Proceedings of the 14th International Technical Meeting of the Satellite Division of the Institute of Navigation, Salt Lake City, UT, 2001, pp. 2665–2675.

[9] T. Fujiwara, H. Tomita, T. Tsujii, and M. Harigae, Performance improvement of MEMS INS/GPS during GPS outage using magnetometer, International Symposium on GPS/GNSS, 11–14 November, 2008, Odaiba, Tokyo, Japan.

[10] R. S. Conker, M. B. El-Arini, C. J. Hegarty, and T. Hsiao, Modeling the effects of ionospheric scintillation on GPS/SBAS availability, *Radio Science*, **38**, (1), 2003, doi: 10.1029/2000RS002604.

[11] Y. Gao and X. Shen, A new method for carrier-phase-based precise point positioning, Navigation, *Journal of The Institute of Navigation*, **49**, (2), Summer, 2002, 109–116.

[12] T. Beran, D. Kim, and R. B. Langley, High-precision single-frequency GPS point positioning, Proc. of the ION GPS/GNSS 2003, Portland, OR, Sep., 2003, pp. 1192–1200.

[13] International GNSS Service, *http://igscb.jpl.nasa.gov/components/prods.html*.

[14] J. Ray, *New pseudorange bias convention*, IGS Mail #2744, (http://igscb.jpl.nasa.gov/mail/igsmail/2000/msg00084.html) 15 March 2000.

[15] IGS absolute antenna file, *ftp://www.igs.org/pub/station/general/igs05.atx*.

[16] *IERS Conventions (2003)*, IERS Technical Note 32, D D. McCarthy and G. Petit (editors), 2003, Frankfurt am Main, Verlag des Bundesamts für Kartographie und Geodäsie, 2004.

[17] J. T. Wu, S, C. Wu, G. A. Hajj, W. I. Bertiger, and S. M. Lichten, Effects of antenna orientation on GPS carrier phase, *Man. Geodetica*, **18**, 1993, 91–98.

[18] J. Kouba, A guide to using international GNSS service (IGS) products, May 2009, *http://igscb.jpl.nasa.gov/components/usage.html.*

[19] S. Sugimoto and Y. Kubo, Carrier-phase-based precise point positioning – a novel approach based on GNSS regression models, Proceedings of The International Symposium on *GPS/GNSS (GNSS2004)*, P94, Sydney, Dec. 2004.

[20] Y. Kubo, N. Munetomo, S. Takehara, *et al.*, Integration algorithms of single frequency precise point positioning and INS – flight test, 2010 International Symposium on *GPS/GNSS*, Taipei, Taiwan, 26–28 October, 2010.

Next step – RF lab

We thought that it would be out of keeping with the book's style if the last word in it was a date in the last reference in the last chapter. Instead, we try to give our view on what the next step might be after the last page has been turned over.

The purpose of this book is to supply readers with a good text book for reference and lab use. What is possible to bundle with the book is limited of course to software and data. We have tried to demonstrate that most of the work can be done and often is done in a digital domain, and therefore in many cases such a digital lab will suffice.

The bundled software packages are free academic versions of the programs which are used in professional high-end applications. They are sufficient in many cases for academic purposes and for self-education. If a reader is engaged in R&D or testing, or is otherwise interested in more capabilities, these programs can be upgraded to professional versions. For those who are interested in development of their own models or algorithms, the development versions are the best solution. Development versions with an application programming interface (API) allow access to source code and use an already developed, optimized, and tested receiver and simulator framework to incorporate models or algorithms developed by the reader. That would eliminate a need to spend time and effort on developing the main receiver and simulator components and instead allow concentration on the subject of interest.

The receiver can also be upgraded to support GLONASS, and the simulator can be upgraded to support various GNSS signals.

The bundled receiver can be used to work with a front end as it is. It doesn't require any upgrades, just a front end. Front ends are available from www.ip-solutions.jp. There are other front ends on the market. One can record data with these front ends and process them with a bundled receiver. However, we do not support other receivers and it would be the responsibility of a user to ensure that all parameters are set correctly.

The same is true for the simulator. The matched hardware can be provided to play a generated DIF signal. There are other playback devices on the market, but the signal format may differ.

The total set of instruments with matching hardware provides one with a complete GNSS lab. Software upgrades to professional or developer versions would provide readers with a high-end multi-purpose GNSS R&D lab.

In conclusion, please check the book website (*www.cambridge.org/petrovski*) regularly for software updates, solutions, and data.

Index

Printed in the United States
By Bookmasters